Chemistry in Quantitative Language

CHEMISTRY IN QUANTITATIVE LANGUAGE

Fundamentals of General Chemistry Calculations

Christopher O. Oriakhi

2009

OXFORD
UNIVERSITY PRESS

Oxford University Press, Inc., publishes works that further
Oxford University's objective of excellence
in research, scholarship, and education.

Oxford New York
Auckland Cape Town Dar es Salaam Hong Kong Karachi
Kuala Lumpur Madrid Melbourne Mexico City Nairobi
New Delhi Shanghai Taipei Toronto

With offices in
Argentina Austria Brazil Chile Czech Republic France Greece
Guatemala Hungary Italy Japan Poland Portugal Singapore
South Korea Switzerland Thailand Turkey Ukraine Vietnam

Published by Oxford University Press, Inc.
198 Madison Avenue, New York, New York 10016

www.oup.com

Library of Congress Cataloging-in-Publication Data

Oriakhi, Christopher O.
Chemistry in quantitative language: fundamentals of general chemistry calculations /
Christopher O. Oriakhi.
p.cm
Includes index.
ISBN 978-0-19-536799-7
1. Chemistry—Problems, exercises, etc. 2. Chemical equations.
3. Chemical reactions. 1. Title.
QD42.O75 2009
540—dc22 2008040392

1 3 5 7 9 8 6 4 2

Printed in the United States of America
on acid-free paper

This book is dedicated to the loving memory of Gabriel, Clement, and Margaret.

Preface

There is a widespread dislike or fear of chemistry among high school and college students. Unfortunately, many science, engineering, and medical programs have chemistry as a prerequisite, thereby making it unavoidable. The subject matter usually covered in general chemistry is not difficult enough to justify the number of students who continue to avoid or fail the course each year. Among the various reasons given for hating the subject, the fear of problem solving ranks the highest. For some students, this correlates with inadequate training in the fundamentals of mathematics. To successfully solve problems, one needs knowledge of the concepts or principles on which chemistry is based as well as a good preparation in basic mathematics. Furthermore, many students lack an appreciation of the excitement of chemistry and are unwilling to exercise the patience and invest the time needed to master chemical principles. Consequently, many are unable to identify the knowledge needed or to extract the relevant information needed to solve problems. They resort to memorization of definitions, equations, and formulas, which is not adequate to allow them to pass their tests with flying colors. The teachers and the available textbooks are partly to blame for this situation. Innovative teaching methods and books with systematically presented subject matter are needed.

The principal objective of this supplementary text is to make chemistry problem-solving more of a pleasure than a pain and to help the student pass exams. In each chapter I have:

- Presented simple, direct and methodical approaches for solving numerical problems.
- Summarized the essential principles necessary for solving problems.
- Provided several worked examples of problems with detailed solutions of the type students will encounter in the chapter covered.
- Furnished a comprehensive list of end-of-chapter problems that allow students to gauge their understanding or mastery of the principles covered.
- Varied the end-of-chapter questions with respect to the degree of difficulty so that they are suitable for exams, homework, recitation, tutorials, or in-class revision.

This book attempts to cover the different types of calculations usually encountered in general chemistry. Apart from chapter 1, which deals with a review of essential mathematics, I have attempted to present the chapters in a logical order as is found in many standard textbooks or syllabus. The contents are organized for easy access to information on any topic to enable the student to select only the necessary materials.

To the Student

Problem solving is one of the most enjoyable aspects of general chemistry. Regrettably, it also accounts for why most students love to hate the subject and why they fail to succeed in it. In my more than 25 years of involvement with chemistry, I have helplessly watched many students who had long-term career objectives of becoming biochemists, chemical engineers, chemists, physicians, or pharmacists, to mention a few, opting for careers in nonscience disciplines because of their failure to pass chemistry! Chemistry problem solving can and should be fun. I wrote this book to give you hope that you can succeed in chemistry. The success strategy requires that you be disciplined and patient and that you persevere and invest time.

Chemistry in Quantitative Language should not replace a standard textbook. Instead, I have written it as a study guide in problem solving. The book covers chemical equations (a.k.a. the language of chemistry) and calculations based on chemical reactions extensively. To do well in problem solving, you need a strong knowledge of high school or at least GCE ordinary level mathematics with emphasis on arithmetic and algebra. The essential mathematics skills you will find most helpful are reviewed in chapters 1 and 2. In each chapter, I have provided some basic concepts followed by several solved examples. There is no substitute for studying these examples until you master them. Several problems are given at the end of each chapter. You should have very little trouble if you have studied and understood the worked examples.

Here are some strategies that will put the odds in your favor to pass general chemistry with flying colors:

- Attitude is everything! Fall in love with chemistry and believe that you can do it. Chemistry is meant for people and not people for chemistry. It is true that some people find the subject hard. But if you spend time working at it, you will do well.
- Go to class on time and prepared. Do not miss class unless it is an emergency. Read your textbook before coming to class. Make use of your professor's and teaching assistant's office hours. Do not be ashamed to ask for help with problems you do not understand.
- Solve lots of problems. The secret of success in problem solving is practice, practice, and more practice. There is no substitute for solving the problems yourself. Do not rely on the problems your professor solved in class or those your classmate solved for you. You have to solve the problems yourself so you can gain the confidence you need.
- Become a "chemistry evangelist." Solve some problems for other students and explain step by step how you arrived at your answer. This will boost your confidence.

- Prepare extensively for tests and examinations. Review concepts. Solve more problems. Attend professor- or teaching assistant-led exam review sessions so you can anticipate what will be on the exam. Be sure to have all your questions answered.
- Finally, review all the relevant parts of this book and solve all the end-of-chapter problems, and you'll be on your way to success in chemistry.

Who Can Use This Book?

Chemistry in Quantitative Language is written for a broad audience including high school (senior secondary schools, GCE O- & A-levels) and first-year college students. In addition, students preparing for various entrance examinations such as AP chemistry, GRE subject test, JAMB, and MCAT will benefit greatly from this book.

Acknowledgments

Although this text bears the name of a single author, it incorporates the constructive contributions of many people. I wish to thank the students and colleagues over the years who inspired me and contributed in various ways to writing of this text. I am deeply indebted to Ted LaPage, my sharpest-thinking and sharpest-seeing friend, for providing a comprehensive critical analysis, working through all the end-of-chapter problems, offering suggestions for improvement, and for doing the initial "dirty" proofreading and editorial work. Many of my colleagues, friends, and mentors contributed their expertise to the book by reviewing the manuscript and providing valuable comments and suggestions to make the book more useful for students. In particular, I want to thank the following people: Michael M. Lerner (Oregon State University), Patrick Woodward (Ohio State University), John S. O. Evans (University of Durham), James Krueger (Oregon State University), Margaret Haak (Oregon State University), and Duane Maycock (Albany, Oregon).

Staff at Oxford University Press have been very supportive throughout the process of writing and producing *Chemistry in Quantitative Language*. I am thankful for their guidance.

Finally, I want to thank my family for their patience and understanding. Although this project removed me from them for many hours, they encouraged and gave me unalloyed support.

Contents

Chemistry in Quantitative Language

1

Essential Mathematics

1.1. Significant Figures

Significant figures are the number of digits in a measured or calculated value that are statistically significant or offer reasonable and reliable information. Significant figures in any measurement usually contain digits that are known with certainty and one digit that is uncertain. To determine the number of significant figures, follow these rules.

1. All nonzero digits (1–9) are significant. For example, 125 has 3 significant figures and 14.44 has 4 significant figures.
2. Leading zeros to the left of the first nonzero digit in the number are not significant. They are only used to fix the position of the decimal. For example:
   - 0.007 has one significant figure
   - 0.000105 has 3 significant figures
   - 0.000000000015 has 2 significant figures
3. Zeros between nonzero digits are significant. For example:
   - 5.005 has 4 significant figures
   - 50.05 has 4 significant figures
   - 500.0000075 has 10 significant figures
4. All zeros to the right of the decimal point in a number greater than 1 are significant. For example:
   - 25.00 has 4 significant figures
   - 0.1250 has 4 significant figures
   - 0.2000 has 4 significant figures
5. Zeros at the end of a number may or may not be significant. They are significant only if there is a decimal point in the number. For example:
   - 1500 has 2 significant figures
   - 1500. has 4 significant figures

- 602,000,000,000,000,000,000,000 has 3 significant figures
- 404,570,000 has 5 significant figures
- 404,590,000. has 9 significant figures

Rounding off

Rounding off is a process of eliminating nonsignificant digits from a calculated number. The rules governing rounding off are summarized below:

1. If the nonsignificant digit is less than 5, round it and all digits to its right off. For example, 100.5129 is equal to 100.5 if rounded off to 4 significant figures.
2. If the nonsignificant digit to the right of the last digit to be retained is a 5 followed by zeros, the last digit is increased by one if it is odd, and is left unchanged if it is even. For example, 64.750 is 64.8 to 3 significant digits, but 25.850 is 25.8 to 3 significant figures.
3. If the nonsignificant digit is more than 5, round it and all digits to its right off, and increase the last reported digit by 1. For example, 15.58729 is 15.59 to 4 significant figures.
4. Do not round off intermediate answers. Wait until you get the final answer before rounding off.

1.2. Significant Figures and Mathematical Operations

Two rules guide the operations of multiplication, division, addition, and subtraction such that the precision of the measurement may be maintained throughout a calculation.

1. Multiplication and division: For operations involving multiplication and division, the number of significant figures in the product or quotient is the same as the original number with the least significant digits. For example: In the operation 5.0085×3.55, the number 3.55 limits our answer to 3 significant figures. Although the calculator gives the product as 17.7802, the answer should be expressed as 17.8 (i.e., $5.0085 \times 3.55 = 17.7802 = 17.8$).

Also, in the operation 2.1259/5.5, the number 5.5 limits the final answer to 2 significant figures (s.f.):

$$\frac{2.1259}{5.5} = 0.3865 = 0.39 \quad (2\text{ s.f.})$$

Example 1.1

Multiply 20.25 by 125.125 and give the answer with the proper significant digits.

Solution

$$20.25 \times 125.125 = 253378.125$$

Final answer is 253400 (4 s.f.)

Example 1.2

Divide 1960.2008 by 2005.252 and give the answer with the proper significant digits.

Solution

$$\frac{1960.2008}{2005.252} = 0.9775333972 = 0.9775334 \ \ (7 \text{ s.f.})$$

2. Addition and subtraction: For operations involving addition and subtraction, the last digit retained in the sum or difference is determined by the position of the first doubtful digit. This means that the answer cannot have more digits to the right of the decimal point than either of the original numbers.

Example 1.3

Add the following and give your answer in the proper number of significant digits.

$$12.08 + 1.2575$$

Solution

$$\begin{array}{r} 12.08 \\ +1.2575 \\ \hline 13.3375 \end{array} \Rightarrow 13.34 \text{ (final answer)}$$

Example 1.4

Subtract 34.2346 from 39.251 and express your answer in the proper significant digits.

Solution

$$\begin{array}{r} 39.251 \\ 34.2346 \\ \hline 5.0164 \end{array} \Rightarrow 5.016 \text{ (final answer)}$$

1.3. Scientific Notation and Exponents

In chemistry one frequently encounters numbers that are either very large or very small. For example, Avogadro's number is 602,200,000,000,000,000,000,000 per mole, and the mass of an electron is 0.00000000000000000000000000000091091 kg. These types of numbers are difficult to handle in routine calculations, and the chances of errors are high. To reduce the chance of error and to convey explicit information about the precision of measurements, scientists usually write such large and small numbers in a more convenient way known as scientific notation. A number is expressed in *scientific notation* or *standard form* if it is written as a power of 10, i.e., $A \times 10^n$, where A is a number between 1 and 10, and the exponent n is a positive or negative integer. To write a number in scientific notation, simply move the decimal in the original number so that the result, A, is greater than or equal to 1 but less than 10. Then multiply A by 10^n, where n is the number of decimal places that the decimal point has been moved. If the decimal point was moved to the left, the exponent n will have a positive value. If the decimal point was moved to the right, n will have a negative value and A will be multiplied by 10^{-n}.

Example 1.5

Express 1,250,000 in scientific notation.

Solution

Place the decimal between 1 and 2 to get the number $A = 1.25$. Since the decimal point was moved 6 places to the left, the power n will be equal to $+6$. Therefore, we can write the number in scientific notation as: $1{,}250{,}000 = 1.25 \times 10^6$.

Example 1.6

Write 0.0000000575 in scientific notation.

Solution

Place the decimal point between the first 5 and 7 to get the number $A = 5.75$. Since the decimal was moved 8 places to the right, the exponent n will be -8. The number in exponential form is $0.0000000575 = 5.75 \times 10^{-8}$.

Addition and subtraction

To add or subtract numbers in scientific notation, it is necessary to first express each quantity to the same power of ten. The digit terms are added or subtracted in the usual manner.

Example 1.7

Carry out the following arithmetic operations and give your answer in scientific notation.

(a) $(2.35 \times 10^{-5}) + (1.8 \times 10^{-6})$
(b) $(3.33 \times 10^{4}) - (1.21 \times 10^{3})$

Solution

(a) $(2.35 \times 10^{-5}) + (1.8 \times 10^{-6}) = (2.35 \times 10^{-5}) + (0.18 \times 10^{-5}) = 2.53 \times 10^{-5}$
(b) $(3.33 \times 10^{4}) - (1.21 \times 10^{3}) = (3.33 \times 10^{4}) - (0.121 \times 10^{4}) = 3.21 \times 10^{4}$

Multiplication and division

To multiply numbers in scientific notation, first multiply the decimals in the usual way, and then add their exponents. In division, the digit term in the numerator is divided by the digit term in the denominator in the usual manner. Then the denominator's exponent is subtracted from the numerator's exponent to obtain the power of 10. In either case, the decimal point may have to be adjusted so that the result has one nonzero digit to the left of the decimal.

Example 1.8

Perform the following arithmetic operations:

(a) $(5.5 \times 10^{3}) \times (2.8 \times 10^{5})$

(b) $\left(\dfrac{2.1 \times 10^{4}}{9.7 \times 10^{7}}\right)$

(c) $\left(\dfrac{4.3 \times 10^{-4}}{9.7 \times 10^{7}}\right) \times \left(\dfrac{5.1 \times 10^{-4}}{3.7 \times 10^{-7}}\right)$

Solution

(a) $(5.5 \times 10^{3}) \times (2.8 \times 10^{5}) = (5.5 \times 2.8) \times 10^{(5+3)}$

$$(15.4 \times 10^{8}) = 1.54 \times 10^{9}$$

(b) $\left(\dfrac{2.1 \times 10^{4}}{9.7 \times 10^{7}}\right) = \left(\dfrac{2.1}{9.7}\right) \times 10^{(4-7)} = 0.217 \times 10^{-3}$

$$= 2.17 \times 10^{-4}$$

(c) $\left(\dfrac{4.3 \times 10^{-4}}{9.7 \times 10^{7}}\right) \times \left(\dfrac{5.1 \times 10^{-4}}{3.7 \times 10^{-7}}\right)$

$= \left(\dfrac{4.3}{9.7}\right) \times 10^{-11} \times \left(\dfrac{5.1}{3.7}\right) \times 10^{3}$

$= (0.443 \times 1.378) \times 10^{(-11+3)}$

$= 0.6106 \times 10^{-8} = 6.11 \times 10^{-9}$

Powers and roots

Generally, a number in scientific notation, $A \times 10^n$, which has been raised to a power b, can be evaluated as follows:

$$(A \times 10^n)^b = A^b \times 10^{n \times b}$$

For example, $(3.05 \times 10^3)^2 = (3.05)^2 \times 10^{3\times 2} = 9.30 \times 10^6$.

The root of an exponential number of the general form $A \times 10^n$ can be extracted by raising the number to a fractional power such as $\frac{1}{2}$, for square root, or $\frac{1}{3}$ for cube root. To evaluate the root of an exponential number, move the decimal point in A such that the exponent in the power of 10 is exactly divisible by the root. Generally:

$$\sqrt[p]{(A \times 10^n)} = \left(\sqrt[p]{A} \times 10^{n/p}\right)$$

Recall the following laws of exponents: $\sqrt[n]{A} = A^{1/n}$ and $\sqrt[n]{A^m} = A^{m/n}$.

Example 1.9

Evaluate the following expressions:

(a) $\sqrt[3]{0.64 \times 10^{-4}}$

(b) $\sqrt{8.1 \times 10^{8}}$

Solution

(a) $\sqrt[3]{0.64 \times 10^{-4}} = \sqrt[3]{64 \times 10^{-6}} = \sqrt[3]{64} \times 10^{-6/3} = 4.0 \times 10^{-2}$

(b) $\sqrt{8.1 \times 10^{8}} = \sqrt{8.1} \times 10^{8/2} = 2.8 \times 10^{4}$

1.4. Logarithms

Many problems, such as those concerning pH, chemical kinetics, or nuclear chemistry, involve the use of logarithms and antilogarithms. There are two types of logarithm, common logarithms and natural logarithms.

Common logarithms

The common logarithm of a number N (abbreviated as $\log N$) is the exponent to which 10 must be raised to obtain that number. Mathematically, this is expressed as:

$$N = 10^x \text{ (10 is the base; } x \text{ is the exponent), and}$$

$$\log_{10} N = x \text{ (10 is the base; } x \text{ is the logarithm)}$$

For example:

$$\begin{aligned} 10{,}000 &= 10^4 & \log 10{,}000 &= 4 \\ 1{,}000 &= 10^3 & \log 1{,}000 &= 3 \\ 10 &= 10^1 & \log 10 &= 1 \\ 1 &= 10^0 & \log 1 &= 0 \\ 0.1 &= 10^{-1} & \log 0.1 &= -1 \\ 0.001 &= 10^{-3} & \log 0.001 &= -3 \\ 0.0001 &= 10^{-4} & \log 0.0001 &= -4 \end{aligned}$$

The common logarithm of other numbers (not powers of 10) may be obtained from a logarithm table or from a calculator. For example, the logarithm of 45 is 1.6532. Hence 45 may be expressed as follows: $45 = 10^{1.6532}$. The logarithm of a number greater than 1 is positive, and that of a number less than 1 is negative.

Antilogarithms

The common antilogarithm (or inverse logarithm) of a number (abbreviated as antilog) is simply 10 raised to that power. Thus finding the antilogarithm of a number is the reverse of evaluating the logarithm of the number. For example, if $\log_{10} N = 3$, the antilog of $N = 10^3$, or 1000.

Natural logarithms

The natural logarithm of a number N (abbreviated as $\ln N$) is the power to which the number e (where e = 2.718) must be raised to give N. The base e is a mathematical constant encountered in many science and engineering problems.

The natural antilogarithm of x is also referred to as the power of e, represented as e^x.

$$N = e^x = 2.718^x, \text{ and}$$
$$\ln e^x = x$$
$$\Rightarrow \ln e^5 = 5$$

There is a relationship between the common and natural logarithm. It is derived as follows:

$$\ln x = \ln 10.\log x$$
$$= 2.303 \log x$$

Thus it is possible to express the natural logarithm of a number in terms of the common logarithm (to the base 10).

1.4.1. Calculations involving logarithms

The following relationships hold true for all logarithms:

1. $\log_a(AB) = \log_a A + \log_a B,$ and $\ln(AB) = \ln A + \ln B$
2. $\log_a\left(\frac{A}{B}\right) = \log_a A - \log_b B,$ and $\ln\left(\frac{A}{B}\right) = \ln A - \ln B$
3. $\log_a A^x = x \log_a A$ and $\ln A^x = x \ln A$
4. $\log_a \sqrt[x]{A} = \log_a A^{1/x} = \frac{1}{x}\log_a A$ and $\ln \sqrt[x]{A} = \ln A^{1/x} = \frac{1}{x}\ln A$

Example 1.10

Calculate:

1. $\log(2.51 \times 10^5)$
2. $\log(8.6 \times 10^{-4})$

Solution

1. $\log(2.51 \times 10^5) = \log 2.51 + \log 10^5$

$$= 0.3997 + 5.0000$$
$$= 5.3997$$

2. $\log(8.6 \times 10^{-4}) = \log 8.6 + \log 10^{-4}$
$$= 0.9345 + (-4.0000)$$
$$= 3.0655$$

Example 1.11

Find the antilog of the following:

(a) 23.7796
(b) −15.5515

Solution

(a) antilog of $23.7796 = 10^{23.7796} = 6.02 \times 10^{23}$
The antilog of 23.7796 is 6.02×10^{23}
(b) antilog of −15.5515

The antilog of a negative log such as −15.5515 cannot be found directly. The following steps are helpful in finding the antilogarithm of a negative number:

1. Rewrite the log as the sum of a negative whole number and a positive decimal number $-15.5515 = (-16 + 0.4485)$
2. Then find the antilog of the decimal portion antilog of $0.4485 = 10^{0.4485} = 2.8087$
3. Multiply the antilog by 10 raised to the power of the negative whole number; i.e., 2.8087×10^{-16}
4. Thus the antilog of −15.7796 is equal to 2.8×10^{-16}.

1.5. Algebraic Equations

Chemistry students will invariably encounter mathematical problems that require a basic knowledge of algebraic operations. Two common types are simple linear equations and quadratic equations.

1.5.1. Linear equations

A linear equation is an equation that contains one unknown whose exponent is exactly 1. It can be expressed in the following general form:

$$bx^1 = c \quad (b \neq 0, c = \text{ real number}) \quad \text{or} \quad a + bx = c \quad (a \neq 0)$$

Basic rules of algebraic transformation are needed to solve for the unknown in a linear equation. The necessary steps are:

1. Clear any parentheses.
2. Isolate the unknown or collect like terms.
3. Solve for the unknown.

Recall the following:

- Multiplication and division must be completed before addition and subtraction.
- Adding or subtracting the same number on both sides of the equation does not alter the equation.
- Multiplying or dividing both sides of the equation by the same number does not alter the equation.

Example 1.12

Solve for x in each of the following equations:

1. $4x = 12$
2. $\frac{3x}{5} = \frac{2}{7}$
3. $106 = \frac{9}{5}x + 32$
4. $\frac{5}{3x} = \frac{2}{7}$

Solution

1. $4x = 12$

   Divide both sides by 4

   $$\frac{4x}{4} = \frac{12}{4} = 3$$

   $x = 3$

2. $\frac{3x}{5} = \frac{2}{7}$

   Multiply both sides by $\frac{5}{3}$:

   $$\frac{\cancel{5} \times \cancel{3}x}{\cancel{3} \times \cancel{5}} = \frac{5 \times 2}{3 \times 7}$$

   $$x = \frac{10}{21}$$

3. $106 = \frac{9}{5}x + 32$

First subtract 32 from both sides of the equation and rearrange:

$$106 - 32 = \frac{9}{5}x + 32 - 32$$

$$\frac{9}{5}x = 74$$

Multiply both sides by $\frac{5}{9}$:

$$\frac{\not{5}}{\not{9}} \times \frac{\not{9}}{\not{5}}x = \frac{5}{9} \times 74$$

$$x = \frac{370}{9} \quad \text{or} \quad 41.1$$

4. $\frac{5}{3x} = \frac{2}{7}$

First cross-multiply the expression as follows:

$$3x \times 2 = 5 \times 7 \quad \Rightarrow \quad 6x = 35$$

Now divide both sides of the equation by 6:

$$\frac{\not{6}x}{\not{6}} = \frac{35}{6}$$

$$x = 5.83$$

1.5.2. Straight-line graphs and linear equations

The equation of a straight-line graph can be written as $y = mx + b$. In graphing this linear equation, y is treated as the dependent variable, while x is the independent variable. The constant m is called the slope (or gradient) of the line, while b is the point at which the line intercepts the y-axis.

The slope of a line is obtained as the ratio of the rate of change in y to that in x.

$$m = \frac{\Delta y}{\Delta x} = \frac{y_2 - y_1}{x_2 - x_1}$$

Let us consider the linear equation $y = 4x + 25$. The slope is 4, and the intercept is 25, occurring when $x = 0$. The plot is illustrated by figure 1-1.

The slope can be verified directly from the graph by considering any two points, say, $(x_1, y_1 = 5, 45)$ and $(x_2, y_2 = 15, 85)$.

$$m = \frac{\Delta y}{\Delta x} = \frac{y_2 - y_1}{x_2 - x_1} = \frac{85 - 45}{15 - 5} = \frac{40}{10} = 4$$

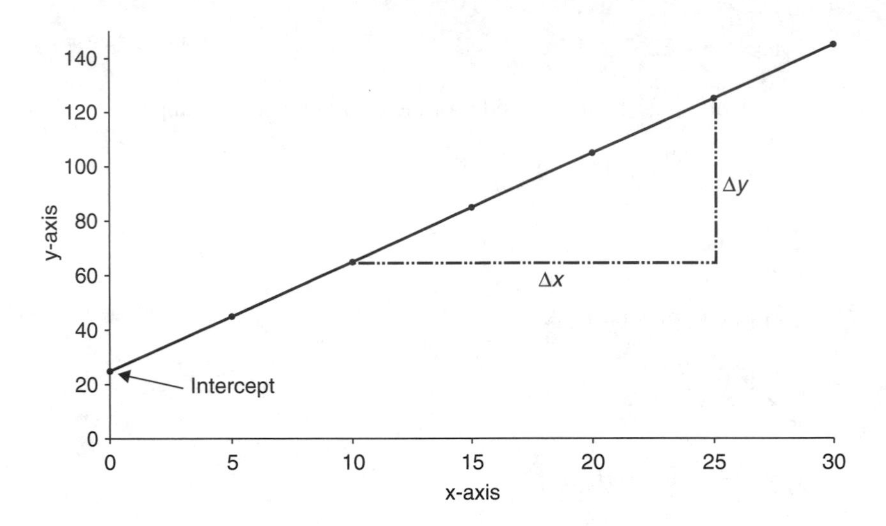

Figure 1-1 The graph of $y = 4x + 25$.

Examples of some linear equations you will encounter in chemistry include:

1. First-order rate law

$$\underbrace{\ln [A]_t}_{y} = \underbrace{-kt}_{mx} + \underbrace{\ln [A]_0}_{b}$$

2. Second-order rate law

$$\underbrace{\frac{1}{[A]_t}}_{y} = \underbrace{kt}_{mx} + \underbrace{\frac{1}{[A]_0}}_{b}$$

3. The Arrhenius equation

$$\underbrace{\ln k}_{y} = -\underbrace{\left(\frac{E_a}{R}\right)}_{m}\underbrace{\left(\frac{1}{T}\right)}_{x} + \underbrace{\ln A}_{b}$$

4. The van't Hoff equation

$$\underbrace{\ln K_{eq}}_{y} = \underbrace{\left(-\frac{\Delta H_{vap}}{R}\right)}_{m}\underbrace{\left(\frac{1}{T}\right)}_{x} + \underbrace{C}_{b}$$

1.5.3. Quadratic equations

A quadratic equation is a polynomial equation in which the largest exponent of the unknown is 2. It can be expressed in the following general form:

$$ax^2 + bx + c = 0 \quad (\text{where } a, b, \text{ and } c \text{ are real, and } a \neq 0)$$

The second-degree term, ax^2, is called the quadratic term. The first-degree term, bx, is called the linear term, while the numerical term, c, is the constant term.

Many general chemistry problems involving chemical equilibrium require quadratic equations. Here we will summarize some of the methods that can be used to solve these equations.

Solution of quadratic equations

We can solve quadratic equations by three methods:

1. Extraction of the square root
2. Factorization
3. The quadratic formula

A quadratic equation will in general have two solutions, though only one of these may be physically meaningful. It is also possible to have two identical solutions, or none at all (which usually means you solved it wrong, or set it up wrong).

Extraction of the square root

If $b = 0$, the general equation above becomes $ax^2 + c = 0$. The roots or solutions of this type of quadratic equation are numerically equal but opposite in sign; the equation is easily solved by rearranging and taking the square root of both sides. Note, however, that if a and c have the same sign, the equation has no solution in real numbers.

Example 1.13

Solve the following quadratic equations

(a) $x^2 - 25 = 0$
(b) $3x^2 - 6.02 \times 10^{23} = 0$

Solution

(a) $x^2 - 25 = 0$

$$\sqrt{x^2} = \sqrt{25}$$

$$x = \pm 5$$

(b) $3x^2 - 6.02 \times 10^{23} = 0$

$$3x^2 = 6.02 \times 10^{23}$$

$$x^2 = \frac{6.02 \times 10^{23}}{3} = 2.01 \times 10^{23}$$

$$x = \pm\sqrt{(2.01 \times 10^{23})} = \pm 4.48 \times 10^{11}$$

The factor method

Every reducible quadratic trinomial of the form $ax^2 + bx + c = 0$ can be solved by the factor method. It involves binomial factors represented as $Ax + B$ and $Cx + D$. The product of the two binomial terms gives:

$$(Ax+B)(Cx+D) = ACx^2 + (AD + BC)x + BD$$

Therefore solving a quadratic equation by the factor method involves finding the values of the coefficients A, B, C, and D that will satisfy the above condition; i.e., make $AC = a$, $AD + BC = b$, and $BD = c$.

Example 1.14

Solve $x^2 + 7x + 12$ by the factor method.

Solution

1. Identify the possible factors—integers A and C which will multiply together to give A, and B and D whose product is C. Try to find a combination such that $AD + BC = b$, or 7.

   $AC = 1, AD + BC = 7, BD = 12$

   $(x + 2)(x + 6) = x^2 + 8x + 12$

   $(x+3)(x+4) = x^2 + 7x + 12$

   $(x+1)(x+12) = x^2 + 13x + 12$

   The factors of 12 which make $AD + BC = 7$ and $BD = 12$ are 3 and 4.
2. Set each factor equal to zero and solve the resulting linear equations.

   $(x + 3)(x + 4) = 0$

   $x + 3 = 0$ or $x + 4 = 0$

   $x = -3$ or $x = -4$

Quadratic formula

Many quadratic equations cannot be solved by the factor method. In such cases the quadratic formula may be used. The solution of a quadratic equation of the general form $ax^2 + bx + c = 0$ is given by:

$$x = \frac{-b \pm \sqrt{b^2 - 4ac}}{2a}$$

Example 1.15

Solve $2x^2 + x - 16 = 0$ using the formula method.

Solution

1. Comparing this to the general equation, $a = 2$, $b = 1$, and $c = -16$.
2. Solve for x by substituting the values of a, b, and c in the quadratic formula.

$$x = \frac{-b \pm \sqrt{b^2 - 4ac}}{2a}$$

$$= \frac{-1 \pm \sqrt{(-1)^2 - 4(2)(-16)}}{2(2)}$$

$$= \frac{-1 \pm \sqrt{129}}{4}$$

$$= \frac{-1 \pm 11.36}{4}$$

$$x = 2.56 \text{ or } -3.09$$

Thus the solution set is $\{2.56, -3.09\}$.

1.6. Problems

1. Determine the number of significant figures in each of the following:
   (a) 0.0038 (b) 2.0038 (c) 125 (d) 50,008 (e) 0.00022500 (f) 40.000
2. Round off the final answer in each of the following:

   (a) $(2.254)(5.55) = 12.5097$

   (b) $\frac{18.322}{1.91750} = 9.55515$

   (c) $\frac{(4.3)(8.51)(20.5360)}{6.750} = 11.32946$

3. Carry out the following mathematical operations and express the answer to the proper number of significant figures.

   (a) $55.58\ \text{g} + 20.0015\ \text{g} + 33.2\ \text{g}$
   (b) $18.25\ \text{mL} - 1.55\ \text{mL}$
   (c) $\dfrac{(98{,}365)(0.02365)}{(44.0556)} - 2.3$
   (d) $(3.1258 - 2.303)(0.00022)$

4. Express the following numbers in standard scientific notation
   (a) 525000 (b) 0.00000911 (c) 602,000,000,000,000,000,000,000
   (d) – 0.00010

5. Express the following in decimal form:
   (a) 2.345×10^0 (b) 7.85×10^5 (c) -2.345×10^{-7} (d) 0.0345×10^4
   (e) 25.0×10^{-3}

6. Carry out the indicated mathematical operations and leave your final answer in exponential form:

   (a) $(1.25 \times 10^2) + (7.25 \times 10^3) + (2.5 \times 10^4)$
   (b) $(3.15 \times 10^{-2}) - (7.25 \times 10^{-3})$
   (c) $(5.1 \times 10^{-2}) + (6.23 \times 10^3) - (1.15 \times 10^2)$
   (d) $(0.81 \times 10^6) - (2.25 \times 10^4) - (1.15 \times 10^0)$

7. Perform the indicated mathematical operations, expressing the result in exponential form:

   (a) $(2.55 \times 10^4)(1.85 \times 10^{-6})$
   (b) $(4.45 \times 10^{-12})(9.55 \times 10^{-4})(2.55 \times 10^9)$
   (c) $\sqrt{(8.2 \times 10^7)}$
   (d) $(3.2 \times 10^{-5})^2$
   (e) $\left(\sqrt[3]{6.4 \times 10^{10}}\right)^2$

8. Carry out the following calculations, expressing your result in exponential form:

   (a) $\dfrac{(2.3 \times 10^{-3})}{(5.3 \times 10^{-6})}$
   (b) $\dfrac{(9.3 \times 10^{-3})}{(0.35 \times 10^6)}$
   (c) $\dfrac{(8.3 \times 10^3)(7.4 \times 10^{-9})}{(0.35 \times 10^6)(1.3 \times 10^{-13})}$
   (d) $\dfrac{(5.3 \times 10^{-5})}{(7.5 \times 10^6)}(6.3 \times 10^{12}) - (9.3 \times 10^{-3})$

9. Find the approximate value for

$$\frac{(987{,}000{,}234)(0.000{,}000{,}000{,}2156)}{(675.87)(0.000{,}000{,}321456)}$$

10. If $a = 3,000$, $b = 0.0009825$, $c = 3.142$, determine the approximate value for $\frac{a^2b}{c^3}$.

11. Solve for x in the following:

    (a) $\log_{10} x = 6$
    (b) $\log_3 81 = x$
    (c) $\log_x 25 = 2$
    (d) $\log_{1/3} 9 = x$

12. Evaluate X in the following expressions:

    (a) $\log_{10} X = \log_{10} 5 + \log_{10} 3$
    (b) $\log_{10} X = \log_{10} 8 - \log_{10} 12$
    (c) $\log_{10} X = 1 + \log_{10} 6$
    (d) $X = -\log_{10} 10^{-12.5}$
    (e) $\log_{10}(X+1) - \log_{10}(X-1) = 1$

13. Solve for x in the following:

    (a) $12x - 12 = 36$
    (b) $\frac{x}{5} = \frac{5}{75}$
    (c) $6x + 12 = 3x - 6$
    (d) $5(2x+1) - 3(x-2) = 4$

14. Solve for E_a (in kJ) in the following expression:

    $$\ln \frac{k_2}{k_1} = \frac{-E_a}{R}\left(\frac{1}{T_2} - \frac{1}{T_1}\right)$$

    given that $k_1 = 7.5 \times 10^{-8}\ \mathrm{s}^{-1}$ at $T_1 = 298\,\mathrm{K}$, $k_2 = 4.7 \times 10^{-6}\ \mathrm{s}^{-1}$ at $T_2 = 318\,\mathrm{K}$, and $R = 8.314\,\mathrm{J/K\text{-}mol}$.

15. Determine k at $T = 1500$ K from the following expression:

    $$\ln k = A + \frac{\ln T}{2} - \frac{22000}{T}$$

    where $A = 21.3$ L/mol-s.

16. Solve for x in the following equations:

    (a) $x^2 + 7x + 12 = 0$
    (b) $(x+3)^2 = 16$
    (c) $10x^2 - 9x + 2 = 0$
    (d) $\frac{2}{x} - \frac{3}{x+1} - 1 = 0$
    (e) $\log_{10}(x+2)(x-1) = 1$

2

Systems of Measurement

2.1. Measurements in Chemistry

Chemistry is an experimental science that involves measurement of the magnitude of various properties of substances. Measurements generate numbers, and we need units attached to these numbers so we can tell what exactly is being measured. Some examples of quantities measured include amount, mass, pressure, size, temperature, time, volume, etc. There are several systems of units, for example, the English system and the metric system. The metric system is the most commonly used system of measurement in chemistry. In 1960, a modernized version of the metric system known as the Systeme Internationale (or SI system) was recommended for worldwide adoption. The SI, based on the metric system, consists of seven fundamental units and several units derived from them. These units serve all scientific measurements. The seven fundamental units are listed in table 2-1.

The fundamental units do have some shortcomings. For example, in some cases the base unit is inconveniently large or small. To overcome this, the SI units can be modified through the use of prefixes. The prefixes define multiples or fractions of the base or fundamental units. Some examples are listed in table 2-2.

2.2. Measurement of Mass, Length, and Time

Mass, length, and time are important quantities commonly measured in science, which are assigned fundamental units.

Measurement of mass

The standard unit of mass in the SI is the kilogram (kg), which is defined as the mass of a certain block of platinum–iridium alloy, also known as the prototype kilogram, kept in a vault at the International Bureau of Weights and Measures, in Sevres, France.

Table 2-1 The Seven Fundamental Units of Measurement (SI)

Physical quantity	*Name of unit*	*Symbol*
Length	meter	m
Mass	kilogram	kg
Time	second	s
Amount of substance	mole	mol
Electric current	ampere	A
Temperature	kelvin	K
Luminous intensity	candela	cd

Table 2-2 SI Prefixes

Prefix	*Symbol*	*Numerical value*	*Factor*
exa	E	1,000,000,000,000,000,000	10^{18}
peta	P	1,000,000,000,000,000	10^{15}
tera	T	1,000,000,000,000	10^{12}
giga	G	1,000,000,000	10^{9}
mega	M	1,000,000	10^{6}
kilo	k	1,000	10^{3}
hecto	h	100	10^{2}
deka	da	10	10^{1}
deci	d	0.1	10^{-1}
centi	c	0.01	10^{-2}
milli	m	0.001	10^{-3}
micro	μ	0.000001	10^{-6}
nano	n	0.000000001	10^{-9}
pico	P	0.00000000000 1	10^{-12}
femto	f	0.00000000000000 1	10^{-15}
atto	a	0.00000000000000000 1	10^{-18}

Commonly used prefixes are highlighted in bold.

Other mass units that are related to the kilogram include gram (g), milligram (mg), and microgram (μg). Here is how they are related:

$$1000 \text{ g} = 1 \text{ kg}$$

$$1000 \text{ mg} = 1 \text{ g}$$

$$1{,}000{,}000\ \mu\text{g} = 1 \text{ g}$$

Note that the symbol for grams is g, not gm, gms, gr, or anything else.

Measuring length

The SI unit of length is the meter (m), which is defined as the distance light travels in approximately $\frac{1}{299{,}792{,}458}$ second. Another convenient unit of length is the centimeter (cm), which is related to the meter as follows:

$$100 \text{ cm} = 1 \text{ m} \quad \text{or} \quad 1 \text{ cm} = \frac{1}{100} \text{ (or 0.01) m}$$

Measuring time

The SI unit of time is the second (s), which is defined as 9,192,631,770 oscillations of the light emitted by cesium atoms.

2.3. Temperature

Temperature is a measure of the intensity of heat, or hotness of an object. Three temperature scales are commonly used in chemistry. They are:

1. The Fahrenheit scale (°F)
2. The Celsius (centigrade) scale (°C)
3. The Kelvin (absolute) scale (K).

Note that no degree symbol is used with Kelvin temperatures.

On the Fahrenheit scale, water freezes at 32°F and boils at 212°F. On the Celsius scale, water freezes at 0°C and boils at 100°C. On the Kelvin scale, water freezes at 273 K and boils at 373 K. There are 100 degrees between the freezing and boiling points of water on both the Celsius and Kelvin scales and 180 degrees difference on the Fahrenheit scale. Thus,

$$1 \text{ Fahrenheit degree} = \frac{100}{180}\left(\text{or } \frac{5}{9}\right) \times \text{Celsius degree or}$$

$$1 \text{ Celsius degree} = \frac{180}{100}\left(\text{or } \frac{9}{5}\right) \times \text{Fahrenheit degree}$$

We can convert a temperature from one scale to another by using the following equations:

1. Converting from Celsius to Fahrenheit

$$°\text{F} = \left(\frac{9}{5} \times °\text{C}\right) + 32$$

2. Converting from Fahrenheit to Celsius

$$°C = \frac{5}{9} \times (°F - 32)$$

3. Converting from Celsius to Kelvin

$$K = °C + 273$$

4. Converting from Kelvin to Celsius

$$°C = K - 273$$

Example 2.1

Express 37°C on the Fahrenheit scale

Solution

$$°F = \left(\frac{9}{5} \times °C\right) + 32$$

$$°F = \left(\frac{9}{5} \times 37\right) + 32 = 98.6°F$$

Example 2.2

Helium boils at −452°F. Convert this temperature to °C and K.

Solution

1. Converting −452°F to °C.

$$°C = \frac{5}{9} \times (°F - 32)$$

$$°C = \frac{5}{9} \times (-452 - 32)$$

$$= -269°C$$

2. Converting −452°F to K.

From part 1 above, −452°F = −269°C

$$K = °C + 273$$

$$= -269 + 273$$

$$= 4\ K$$

Thus −452°F corresponds to −269°C and 4 K.

Table 2-3 Examples of Some Derived Units

Quantity	*Derived unit*	*Symbol for unit*
Area	square meter	m^2
Volume	cubic meter	m^3
Density	kilogram per cubic meter	$kg\,m^{-3}$
Force	Newton	N (i.e., $kg\,m\,s^{-1}$)
Energy	joule	J (i.e., $kg\,m^2\,s^{-2}$)
Heat capacity	joule per kelvin	J/K
Electric charge	coulomb (ampere.second)	Q (A.s)
Gas constant	joule per kelvin per mole	J/K-mol

2.4. Derived Units

Derived units are units obtained by applying algebraic operations (multiplication or division) to the fundamental units. For example, the unit for volume is the cubic meter, which is derived as follows:

$$\text{Volume of a solid} = \text{Length} \times \text{Width} \times \text{Height} = \text{m} \times \text{m} \times \text{m} = \text{m}^3$$

Table 2-3 shows several such derived units.

2.5. Density and Specific Gravity

2.5.1. Density

Density is defined as the mass of a substance per unit volume. In mathematical terms:

$$\text{Density } (\rho) = \frac{\text{Mass}}{\text{Volume}}$$

Any units of mass and volume may be used, but g/mL (or cm^3) and kg/m^3 are the most common. Like the volume of a substance, density varies with temperature. For example, water has a density of 1.0000 g/mL at 4°C and 0.9718 g/mL at 80°C.

Example 2.3

The density of carbon tetrachloride is 1.595 g/mL at 20°C. Calculate the volume occupied by 770 g of carbon tetrachloride at this temperature.

Solution

The unknown in this case is volume. Use the density formula and solve for the unknown.

$$\text{Density} = \frac{\text{Mass}}{\text{Volume}} \quad \text{or} \quad \text{Volume} = \frac{\text{Mass}}{\text{Density}}$$

$$\text{Volume} = \frac{770\text{ g}}{1.595\text{ g/mL}}$$

$$= 484.8\text{ mL} = 485\text{ mL}$$

2.5.2. Specific gravity

The specific gravity of a substance is the ratio of the density of that substance to the density of some substance taken as a standard. The standards commonly used are water in the case of liquids or solids, and air in the case of gases.

$$\text{Specific gravity} = \frac{\text{Density of a liquid or solid}}{\text{Density of water}} \quad \text{or} \quad \frac{\text{Density of a gas}}{\text{Density of air}}$$

In terms of masses, specific gravity may also be expressed as:

$$\text{Specific gravity} = \frac{\text{Mass of a substance}}{\text{Mass of an equal volume of water at } 4°\text{C}}$$

A careful look at the information given will allow one to decide on the appropriate formula to use. For calculation purpose densities must be expressed in the same units, hence specific gravity does not carry any unit.

Example 2.4

The specific gravity of concentrated acetic acid, CH_3COOH, is 1.05. Calculate the mass of the acid that would have a volume of 500 mL. (The density of water is 1.00 g/mL.)

Solution

First calculate the density of acetic acid. Then use the result to calculate the mass of the acid which corresponds to 500 mL.

$$\text{Specific gravity of acid} = \frac{\text{Density of acetic acid}}{\text{Density of water}}, \text{ and}$$

$$\text{Density of acetic acid} = \text{Specific gravity} \times \text{Density of water}$$
$$= 1.05 \times 1.00\,\text{g/mL}$$
$$= 1.05\,\text{g/mL}$$

$$\text{Density} = \frac{\text{Mass}}{\text{Volume}}$$

$$\text{Mass} = \text{Density} \times \text{Volume} = 1.05\ \frac{\text{g}}{\cancel{\text{mL}}} \times 500\,\cancel{\text{mL}} = 525\,\text{g}$$

Example 2.5

A block of aluminum (Al) measuring 5.0 cm × 10.0 cm × 15.0 cm weighs 2027 g. Calculate the specific gravity of aluminum. The density of water is 1.00 g/mL.

Solution

First calculate the volume of aluminum. Use the result to calculate the density of Al. Then calculate the specific gravity.

$$\text{Volume of Al} = 5.0\,\text{cm} \times 10.0\,\text{cm} \times 15.0\,\text{cm} = 750\,\text{cm}^3$$

$$\text{Density} = \frac{\text{Mass}}{\text{Volume}} = \frac{2027\,\text{g}}{750\,\text{cm}^3} = 2.70\,\text{g/cm}^3$$

$$\text{Specific gravity} = \frac{\text{Density of Al}}{\text{Density of H}_2\text{O}} = \frac{2.70\,\text{g/cm}^3}{1.00\,\text{g/cm}^3} = 2.70$$

2.6. Dimensional Analysis and Conversion Factors

Most numerical calculations in science require conversion from one system of units to another. The unit conversion may be from metric to metric (e.g., grams to kilograms) or English-to-metric (e.g., inches to centimeters) equivalents. Therefore chemistry students and scientists should be able to correctly carry out the needed transformations. The best and simplest method is called the *factor-unit method or dimensional analysis*.

The dimensional analysis method of solving problems makes use of *conversion factors* to convert one unit to another unit. Table 2.4 shows representative examples of conversion factors.

The general setup is as follows:

$$\text{Unit A} \times \text{conversion factor} = \text{Unit B}$$

For example, if you want to find out how many pounds (lb) are in 10 kg, you will need to determine the conversion factor. Once you figure out the conversion factor,

Table 2-4 Common Conversion Factors

Physical quantity	*Symbol*	*Conversion factor*
Length		
Meter	m	1 m = 1.0936 yd = 39.37 in = 100 cm
Centimeter	cm	1 cm = 0.3937 in
Inch	in	1 in = 2.54 cm
Feet	ft	1 ft = 12 in
Yard	yd	1 yd = 0.9144 m
Mile	mi	1 mi = 1.609 km = 5280 ft
Kilometer	km	1 km = 1000 m = 0.6215 mi
Angstrom	Å	1 Å = 10^{-10} m
Mass		
Kilogram	kg	1 kg = 1000 g = 2.205 lb
Gram	g	1 g = 1000 mg = 0.03527 oz
Pound	lb	1 lb = 453.59 g
Ton	ton	1 ton = 907.2 kg = 2000 lb
Ounce	oz	1 oz = 28.35 g
Atomic mass unit	amu	1 amu = 1.66056×10^{27} kg
Volume		
Liter	L	1 L = 1000 cm^3 = 0.001 m^3 = 1.056 qt
Cubic decimeter	dm^3	1 dm^3 = 1 L
Cubic centimeter	cm^3	1 cm^3 = 10^{-6} m^3
Milliliter	mL	1 cm^3 = 0.001 L = 10^{-6} m^3
Gallon	gal	1 gal = 4 qt = 3.7854 L
Quart	qt	1 qt = 0.9463 L
Cubic foot	ft^3	1 ft^3 = 7.474 gal = 28.316 L
Time		
Minute	min	1 min = 60 s
Hour	h	1 h = 3,600 s
Day		1 day = 86,400 s
Energy		
Calorie	cal	1 cal = 4.184 J
Joule	J	1 J = 1 kg-m^2/s^2 = 0.2390 cal
Electron volt	eV	1 eV = 1.602×10^{-19} J

use it to multiply the known (10 kg) so as to obtain the unknown:

$$\text{quantity in kg} \times \text{conversion factor} = \text{quantity in lb}$$

A conversion factor is a fraction in which the numerator and the denominator are the same type of measurement, but expressed in different units. For example, we can convert kilograms to pounds by using the following equivalence statement:

$$1\,\text{kg} = 2.205\,\text{lb}$$

If we divide both sides of this expression by 1 kg or 2.205 lb, we get the conversion factors:

$$\frac{1\,\text{kg}}{2.205\,\text{lb}} = 1 \quad \text{or} \quad \frac{2.205\,\text{lb}}{1\,\text{kg}} = 1$$

A conversion factor is always equal to 1, hence it is sometimes called a factor-unit. If it is not equal to 1, it is not a valid conversion factor. To convert from pounds to kilograms, we simply multiply the given quantity in pounds by the first conversion factor. This ensures that the pound (lb) in the numerator and denominator cancel out, yielding the desired unit (kg). For example, 155.0 lb may be converted to kilograms:

$$155.0\,\cancel{\text{lb}} \times \frac{1\,\text{kg}}{2.205\,\cancel{\text{lb}}} = 70.3\,\text{kg}$$

Example 2.6

How many kilometers are there in a marathon (26 miles)? (1 mile = 1.6093 km).

Solution

1. Find the relationship between the units involved in the problem and give the equivalence statement.

   The relationship between mile and kilometer is: 1 mile = 1.6093 km
2. Next generate the conversion factors.

   The conversion factors are:

   $$\frac{1\,\text{mile}}{1.6093\,\text{km}} \quad \text{and} \quad \frac{1.6093\,\text{km}}{1\,\text{mile}}$$
3. Multiply the quantity to be converted (26 miles) by the conversion factor that gives the desired unit.

   $$26.0\,\text{miles} \times \frac{1.6093\,\text{km}}{1.0\,\text{mile}} = 41.8\,\text{km}$$

   In many cases, problems may require a combination of two or more conversion factors in other to obtain the desired units. For example, we may want to express 1 week in seconds. This would require the following sequence:

$$\text{Weeks} \longrightarrow \text{days} \longrightarrow \text{hours} \longrightarrow \text{minutes} \longrightarrow \text{seconds}$$

There are 7 days in 1 week; 24 hours in 1 day; 60 minutes in 1 hour; and 60 seconds in 1 minute. From these relationships, we can generate conversion factors, which we can multiply to get the desired result.

$$1\,\cancel{\text{wk}} \times \frac{7\,\cancel{\text{day}}}{1\,\cancel{\text{wk}}} \times \frac{24\,\cancel{\text{h}}}{1\,\cancel{\text{day}}} \times \frac{60\,\cancel{\text{min}}}{1\,\cancel{\text{h}}} \times \frac{60\,\text{s}}{1\,\cancel{\text{min}}} = 604{,}800\,\text{s}$$

Example 2.7

Express 4.5 gallons of vegetable oil in milliliters.

Solution

1. The quantity given is 4.5 gallons, and the result is to be expressed in milliliters. Use the sequence:

$$\text{gallons} \longrightarrow \text{quarts} \longrightarrow \text{liters} \longrightarrow \text{milliliters}$$

2. State the relationship and the conversion factors:

$$1\,\text{gal} = 4\,\text{qt}; \quad \frac{1\,\text{gal}}{4\,\text{qt}} \quad \text{or} \quad \frac{4\,\text{qt}}{1\,\text{gal}}$$

$$1\,\text{qt} = 0.9463\,\text{L}; \quad \frac{0.9463\,\text{L}}{1\,\text{qt}} \quad \text{or} \quad \frac{1\,\text{qt}}{0.9463\,\text{L}}$$

$$1\,\text{L} = 1000\,\text{mL}; \quad \frac{1000\,\text{mL}}{1\,\text{L}} \quad \text{or} \quad \frac{1\,\text{L}}{1000\,\text{mL}}$$

3. Set up and multiply the appropriate conversion factors:

$$4.5\,\cancel{\text{gal}} \times \frac{4\,\cancel{\text{qt}}}{1\,\cancel{\text{gal}}} \times \frac{0.9463\,\cancel{\text{L}}}{1\,\cancel{\text{qt}}} \times \frac{1000\,\text{mL}}{1\,\cancel{\text{L}}} = 17{,}033\,\text{mL} \approx 20{,}000\,\text{mL}$$

2.7. Problems

1. Convert the following temperatures:

   (a) 30°C to °F
   (b) −196°C to °F
   (c) 0.0500°C to °F
   (d) 1200°F to °C
   (e) −68°F to °C
   (f) 173°F to °C

2. Convert the following temperatures:

   (a) 300°C to K
   (b) −196°C to K

(c) 105°F to K
(d) 1200 K to °C
(e) 350 K to °F
(f) 173°F to K

3. Propane gas (a) boils at −42°C and (b) melts at −188°C. Express these temperatures in °F and K.
4. What is the temperature at which the Celsius and Fahrenheit thermometers have the same numerical value?
5. A steel tank has the dimensions 200 cm × 110 cm × 45 cm. Calculate the volume of the tank in liters.
6. The volume of a cylindrical container of radius r and height h is given by $\pi r^2 h$.
   (a) What is the volume of a cylinder with a radius of 8.15 cm and a height of 106 cm?
   (b) Calculate the mass in kg of a volume of carbon tetrachloride equal to the volume of the cylinder obtained in part (a) given that the density of carbon tetrachloride is 1.60 g/cm^3.
7. The density of gold is 19.3 g/cm^3. Calculate the mass of 500 cm^3 of gold.
8. A block of a certain precious stone 8.00 cm long, 5.50 cm wide, and 3.8 cm thick has a mass of 0.80 kg. Calculate its density in g/cm^3.
9. A steel ball weighing 3.250 g has a density of 8.00×10^3 kg/m^3. Calculate the volume (in mm^3) and the radius (in mm) of the steel ball (volume of a sphere with radius r is $V = \frac{4}{3}\pi r^3$).
10. The density of a sphere of edible clay is 1.4 g/cm^3. Calculate the mass of the clay if the radius of the sphere is 10 cm. (Hint: the volume of a sphere with radius r is $V = \frac{4}{3}\pi r^3$)
11. Concentrated sulfuric acid assays 98% by weight of sulfuric acid (the remainder is water) and has a density of 1.84 g/cm^3. Calculate the mass of pure sulfuric acid in 80.00 cm^3 of concentrated acid.
12. Express 0.55 cm in nanometers (nm).
13. The density of blood plasma is 1.027 g/cm^3. Express the density in kg/m^3.
14. Which has the greater mass, 125 cm^3 of olive oil or 130 cm^3 of ethyl alcohol? The density of olive oil is 0.92 g/cm^3, and the density of ethyl alcohol is 0.79 g/cm^3.
15. The specific gravity of concentrated hydrochloric acid, HCl, is 1.18. Calculate the volume of the acid that would have a mass of 750 g. (The density of water is 1.00 g/cm^3.)
16. Calculate the length, in millimeters, of a 250-ft steel rod given that 12 in = 1 ft, 1 m = 39.37 in, and 1000 mm = 1 m.
17. How many seconds are there in 365 days?

18. What is the mass in pounds of 25.5 kg of lead?
19. Express 350 lb in kg.
20. A common unit of energy is the electron volt (eV) (see table 2-4). Convert 12 eV to the following units of energy: (a) joule, (b) kilojoule, (c) calorie, (d) ergs ($1\ J = 1 \times 10^7$ ergs).

3

Atomic Structure and Isotopes

3.1. Atomic Theory

John Dalton proposed his theory of the atom in 1808 based on experimental data and chemical laws known in his day. The theory states that:

1. All chemical elements are made up of tiny indivisible particles called atoms.
2. Atoms cannot be created or destroyed. Chemical reactions only rearrange the manner in which atoms are combined.
3. Atoms of the same element are identical in all respects and have the same masses and physical and chemical properties. Atoms of different elements have different masses as well as different physical and chemical properties.
4. Combination of elements to form a compound occurs between small, whole-number ratios of atoms.

Dalton's theory resulted in the formulation of the *law of conservation of mass* and the *law of multiple proportions.*

3.1.1. The law of conservation of mass

In a chemical reaction matter is neither created nor destroyed. The mass of products is equal to the mass of the reactants.

3.1.2. The law of multiple proportions

If two elements form more than one compound between them, the masses of one element that combine with a fixed mass of the second element are in a ratio of whole numbers. Nitrogen and oxygen combine to form different compounds such as NO, NO_2, and N_2O. According to this law the number of nitrogen to oxygen atoms in these compounds should be a simple ratio of two small whole numbers. This is one of the basic laws of stoichiometry, as we shall see in chapter 9.

Table 3-1 Properties of Subatomic Particles

Fundamental particle	*Symbol*	*Location*	*Mass (g)*	*Mass (amu)*	*Charge*
Electron	e	Outside the nucleus	9.110×10^{-28}	0.00549	−1
Proton	p	Inside the nucleus	1.673×10^{-24}	1.0073	1
Neutron	n	Inside the nucleus	1.675×10^{-24}	1.0087	0

3.2. The Structure of the Atom

An atom consists of a central nucleus, which contains roughly 99.9% of the total mass of the atom, and a surrounding cloud of electrons. The nucleus is composed of two kinds of particles, the protons and the neutrons, which are collectively known as the nucleons. The proton is positively charged while the neutron is electrically neutral. The electrons have a negative charge and surround the nucleus in "shells" of definite energy levels. (Note: energy level will be discussed in chapter 10.) In a neutral (unreacted) atom, the number of electrons is equal to the number of protons, so the atom has a charge of zero. It must be mentioned that the chemistry of a given atom comes from its electrons; all chemical changes take place entirely with regard to the electrons—the nucleus is never affected by chemical reactions. The properties of the three sub-atomic particles are summarized in table 3-1.

3.2.1. Atomic number

The atomic number (Z) of an element is defined as the number of protons in the nucleus of an atom of the element. It is also equal to the number of electrons in a neutral atom. Atomic number is a characteristic of a given element, and determines its chemical properties.

3.2.2. Mass number

The mass number (A) of an element is the sum of protons and neutrons in the nucleus of the atom.

$$\text{Mass number } (A) = \text{No. of protons } (Z) + \text{No. of neutrons } (N)$$

$$A = Z + N$$

The general symbol for an element, showing its mass number and atomic number, is:

$${}^{A}_{Z}E$$

$A =$ mass number in atomic mass units (amu)

$Z =$ atomic number, and

$E =$ symbol of the element as shown on the periodic table

For example, the symbols for carbon and magnesium, showing their masses and atomic numbers, are:

$$^{12}_{6}C \quad \text{and} \quad ^{24}_{12}Mg$$

3.2.3. Ions

An ion is an atom or a group of atoms that has gained or lost electron(s). A positively charged ion results when an atom loses one or more electrons. Conversely, a negatively charged ion is formed when an atom gains one or more electrons. For example, a sodium atom, Na, loses one electron to form a sodium ion, Na^+; a nitrogen atom, N, gains three electrons to form a nitrogen (or nitride) ion, N^{3-}. (In an ion, the number of protons is not equal to the number of electrons. A positively charged ion has more protons than electrons. Conversely, a negatively charged ion has more electrons than protons). Let's try to determine the number of protons and electrons in the $^{40}_{20}Ca^{2+}$ ion. In a neutral Ca atom, the number of electrons is the same as the number of protons (= atomic number), that is, 20. However, the calcium ion carries 2 positive charges. This implies that Ca^{2+} has 2 less electrons than the atom. Therefore, there are 20 protons and 18 electrons (for a net charge of +2) in Ca^{2-}.

Example 3.1

Determine the number of protons, neutrons, and electrons in:
(a) $^{23}_{11}Na$, (b) $^{16}_{8}O^{2-}$, (c) $^{56}_{26}Fe^{3+}$, (d) $^{63}_{29}Cu$

Solution

The solutions are summarized in the table below:

	Symbol	*Protons*	*Neutrons*	*Electrons*
(a)	$^{23}_{11}Na$	11	12	11
(b)	$^{16}_{8}O^{2-}$	8	8	10
(c)	$^{56}_{26}Fe^{3+}$	26	30	23
(d)	$^{63}_{29}Cu$	29	34	29

Example 3.2

Complete the following table.

			Number of				
Particle	*Atomic no.*	*Mass no.*	*Protons*	*Neutrons*	*Electrons*	*Net charge*	*Symbol*
A			25	30	25	0	Mn
B	13			14			Al^{3+}

Continued

Continued

Particle	Atomic no.	Mass no.	Number of Protons	Number of Neutrons	Number of Electrons	Net charge	Symbol
C	35	80			36		
D	15	31			18		
E			82	125	80	2	
F		39	19			0	K
G	9			10		−1	
H		56	26		23		

Solution

To complete this table, you need to recall the basic definitions and the general equation, mass number (A) = no. of protons (Z) + no. of neutrons (N). Use the periodic table to identify the corresponding symbols. Also remember that the atomic number is equal to the number of protons in the nucleus of an atom. In a neutral atom (zero charge), the number of electrons is always equal to the number of protons. The difference between the number of protons and the number of electrons is the charge on the atomic particle. If the number of protons is greater than the number of electrons, the particle takes on a net positive charge, and vice versa. The complete table is:

Particle	Atomic no.	Mass no.	Number of Protons	Number of Neutrons	Number of Electrons	Net charge	Symbol ${}^{A}_{Z}E^{n\pm}$
A	25	55	25	30	25	0	Mn
B	13	27	13	14	10	3	Al^{3+}
C	35	80	35	45	36	−1	Br^{-}
D	15	31	15	16	18	−3	P^{3-}
E	82	207	82	125	80	2	Pb^{2+}
F	19	39	19	20	19	0	K
G	9	19	9	10	11	−1	F^{-}
H	26	56	26	30	23	3	Fe^{3+}

3.3. Isotopes

Isotopes are atoms which have the same atomic number, and hence are the same element, but which have different mass numbers (that is, different numbers of neutrons). The difference in mass is due to the variation in the neutron number. Isotopes have practically the same chemical properties but differ slightly in physical properties. Most elements exhibit isotopy. For example, hydrogen has three isotopes: hydrogen, deuterium, and tritium. The atomic numbers and mass numbers are shown table 3-2.

Table 3-2 The Hydrogen Isotopes

	Hydrogen (H)	*Deutrium (D)*	*Tritium (D)*
Atomic number	1	1	1
Mass number	1	2	3

Example 3.3

Which of the following are isotopes?

(a) $^{18}_{9}F$ and $^{18}_{10}Ne$
(b) $^{2}_{1}H$ and $^{3}_{1}H$
(c) $^{40}_{19}K$ and $^{40}_{20}Ca$
(d) $^{24}_{12}Mg$ and $^{25}_{12}Mg$

Solution

Isotopes are atoms with the same atomic number but different mass number. Here only (b) and (d) are isotopes.

3.4. Relative Atomic Mass

The relative mass of an atom is called atomic mass (or atomic weight) and is characteristic of each element. The mass of an individual atom is difficult to measure, but the relative masses of the atoms of different elements can be measured, and from this the atomic mass is determined.

The masses of individual atoms and molecules are very small and cannot be conveniently expressed in grams or kilograms. For example, one atom of Ca has a mass of 6.64×10^{-23} g. For this reason, scientists use a relative atomic mass scale instead. Scientists have chosen the carbon-12 isotope as a reference, and its mass is defined as exactly 12 atomic mass units (amu). Therefore one atomic mass unit is equal to $\frac{1}{12}$ of the mass of the carbon-12 atom. One amu has been determined experimentally to be 1.66×10^{-24} g.

The relative atomic mass of an element is defined as the mass of an atom of that element compared with $\frac{1}{12}$ of the mass of the carbon-12 isotope.

The atomic masses of elements given in the periodic table are not the same as their mass numbers. In nature, most elements occur as a mixture of isotopes. Therefore, the atomic mass of an element is an average based on the abundance of the isotopes.

3.4.1. Calculating atomic masses

The information needed to calculate the atomic mass of an element is: the number of isotopes for the element; the relative mass of each isotope on the carbon-12 scale; and the percent abundance of each isotope.

The atomic mass of a given element is then obtained as the sum of the product of the exact mass of each isotope and its percent abundance. For example, naturally occurring carbon consists of a mixture of 98.89% carbon-12 ($^{12}_{6}C$) with a mass of 12.00000 amu, and 1.11% carbon-13 ($^{13}_{6}C$) with a mass of 13.00335 amu. The atomic mass is obtained as follows:

$$\text{Atomic mass} = (12.00000\ \text{amu})\left(\frac{98.89}{100}\right) + (13.00335\ \text{amu})\left(\frac{1.11}{100}\right) = 12.011\ \text{amu}$$

Example 3.4

Calculate the average atomic mass of lithium, which is a mixture of 7.5% ^{6}Li and 92.5%^{7}Li. The exact masses of these isotopes are 6.01 amu and 7.02 amu.

Solution

Multiplying the abundance of each isotope by its atomic mass and then adding these products give the average atomic mass.

$$\text{Average atomic mass of Li} = (6.01\ \text{amu})\left(\frac{7.5}{100}\right) + (7.02\ \text{amu})\left(\frac{92.5}{100}\right) = 6.94\ \text{amu}$$

Example 3.5

Boron, B, with an atomic mass of 10.81, is composed of two isotopes, ^{10}B and ^{11}B, weighing 10.01294 and 11.00931 amu, respectively. What is the fraction and percentage of each isotope in the mixture?

Solution

Let the abundance of ^{10}B be X and that of ^{11}B be $1 - X$ since there are only two isotopes of boron, which together total 100% (or 1.00). Therefore:

$$\text{Atomic mass} = 10.81 = (X)(10.01294) + (1 - X)(11.00931)$$

Now solve for X:

$$10.81 = 10.01294X + 11.0093 - 11.0093X$$

$$10.0129X - 11.0093X = 10.81 - 11.0093$$

$$-0.9964X = -0.1993$$

$$X = 0.20 \text{ or } 20\% \quad \text{and} \quad 1 - X = 0.80 \text{ or } 80\%$$

The fraction and % of ^{10}B is 0.20 or 20%. Also, the fraction of ^{11}B is 0.80 or 80%.

3.5. Problems

1. How many protons, neutrons, and electrons are in each of the following?
   (a) $^{197}_{79}Au$ (b) $^{10}_{5}B$ (c) $^{40}_{20}Ca$ (d) $^{163}_{66}Dy$
2. How many protons, neutrons, and electrons are in each of the following?
   (a) $^{84}_{36}Kr$ (b) $^{24}_{12}Mg$ (c) $^{69}_{31}Ga$ (d) $^{75}_{33}As$
3. Find the number of protons, neutrons and electrons in the following ions:
   (a) $^{59}_{27}Co^{2+}$ (b) $^{24}_{12}Mg^{2-}$ (c) $^{69}_{31}Ga^{3-}$ (d) $^{118}_{50}Sn^{2-}$
4. Identify the following atoms or ions:

   (a) A halogen (an element in group VII of the periodic table) anion with 36 electrons and 35 protons.
   (b) An alkali metal (an element in group I of the periodic table) cation with 55 protons and 54 electrons.
   (c) A transition metal cation with 25 protons and 23 electrons. Note: transition elements are a group of elements embedded between group II and III in the periodic table.
5. Fill in the blanks in the following table:

Symbol	$^{40}_{20}Ca^{2+}$		$^{79}_{34}Se^{2-}$			$^{137}_{56}Ba^{2+}$
Mass number				238	15	
Protons		15				
Neutrons		16		146	8	
Electrons				88		
Net charge		−3			−3	

6. With the aid of a periodic table, identify the following elements:
   (a) $^{56}_{26}X$ (b) $^{131}_{53}X$ (c) $^{202}_{80}X$ (d) $^{19}_{9}X$
7. Fill in the blanks in the following table:

			Number of				
Particle	*Atomic no.*	*Mass no.*	*Protons*	*Neutrons*	*Electrons*	*Net charge*	*Symbol*
Br	35	80				−1	
B			13	14		3	
C		31	15				P^{3-}
D	24			28	24		
E		35	17			0	
F	54	131				0	
G		204	81		80		
H			40	51		4	Zr^{4+}

8. Find the number of protons, neutrons and electrons in the following ions:
   (a) $^{32}_{16}S^{2-}$ (b) $^{80}_{35}Br^{-}$ (c) $^{128}_{52}Te^{2-}$ (d) $^{14}_{7}N^{3-}$
9. Naturally occurring chlorine has two isotopes, which occur in the following abundance:

   75.76% $^{35}_{17}Cl$, with an isotopic mass of 34.9689 amu, and 24.24% $^{37}_{17}Cl$, with an isotopic mass 36.9659 amu. Calculate the atomic mass of chlorine.
10. Boron has two naturally occurring isotopes: 80% of $^{11}_{5}B$, and 20% of $^{10}_{5}B$ which has an isotopic mass of 10.02 amu. If the atomic mass of boron is 10.81, what is the isotopic mass of $^{11}_{5}B$?
11. Naturally occurring oxygen gas consists of three isotopes, as shown below:

Isotope	*Exact mass*	*Abundance*
$^{16}_{8}O$	15.995	99.96%
$^{17}_{8}O$	16.999	0.037%
$^{18}_{8}O$	17.999	0.024%

   Use this information to calculate the atomic weight of oxygen gas.
12. Naturally occurring gallium (Ga) exists in two isotopic forms, ^{69}Ga and ^{71}Ga. The atomic mass of Ga is 69.72. What is the percentage abundance of each isotope? (The exact isotopic masses of ^{69}Ga and ^{71}Ga are 68.9259 and 70.9249, respectively.)

4

Formula and Molecular Mass

Many chemists use the terms *formula mass* and *molecular mass* interchangeably when dealing with chemical compounds of known formula. But there is a slight difference between the two terms, as explained below.

4.1. Formula Mass

The formula mass of a compound is the sum of the atomic masses of all the atoms in a formula unit of the compound, whether it is ionic or molecular (covalent). The formula mass is based on the ratio of different elements in a formula, as opposed to the molecular mass, which depends on the actual number of each kind of atom (compare section 6.2, "Empirical Formula"). Formula masses are relative since they are derived from relative atomic masses. For example, the formula mass of phosphoric acid, H_3PO_4, is 97.98 atomic mass units (amu), which is obtained by adding the atomic masses (taken from the periodic table) of the elements in one formula unit (i.e., 3 H + 1 P + 4 O).

$$(3 \times \text{At. wt. of H}) + (1 \times \text{At. wt. of P}) + (4 \times \text{At. Wt. of O})$$

$$= (3 \times 1.00) + (1 \times 30.97) + (4 \times 16.00) = 97.97$$

4.2. Molecular Mass

Once the actual formula of a chemical substance is known, the molecular mass can be determined in a manner similar to calculating the formula mass. *The molecular mass of a compound is the sum of the atomic masses of all the atoms in one molecule of the compound.* The term applies only to compounds that exist as molecules, such as H_2O, SO_2, and glucose, $C_6H_{12}O_6$. For example, the molecular mass of ethanol, C_2H_5OH, is:

$$(2 \times \text{C}) + (6 \times \text{H}) + (1 \times \text{O})$$

$$(2 \times 12.0) + (6 \times 1.0) + (1 \times 16.0) = 46$$

When ionic compounds such as NaCl, $Zn(NO_3)_2$, and NH_4Cl, are in the crystalline state or in solution form, they do not contain physically distinct uncharged molecular entities. Therefore chemists often use the term formula mass to represent the total composition of such substances.

Example 4.1

Calculate the formula mass (FM) of NaOH using a table of atomic masses (AM).

Solution

$$\begin{aligned}
1 \times \text{AM of Na} &= 1 \times 23 \text{ amu} = 23 \text{ amu} \\
1 \times \text{AM of O} &= 1 \times 16 \text{ amu} = 16 \text{ amu} \\
1 \times \text{AM of H} &= 1 \times \ \ 1 \text{ amu} = \underline{\ \ 1 \text{ amu}} \\
\text{FM} &= 40 \text{ amu}
\end{aligned}$$

The formula mass of NaOH is 40 g.

Example 4.2

Calculate the formula mass of $MgSO_4 \cdot 7H_2O$ using the table of atomic masses.

Solution

$$\begin{aligned}
1 \times \text{AM of Mg} &= \ \ 24.3 \text{ amu} \\
1 \times \text{AM of S} &= \ \ 32.1 \text{ amu} \\
4 \times \text{AM of O} &= \ \ 64.0 \text{ amu} \\
7 \times 2 \times \text{AM of H} &= \ \ 14.0 \text{ amu} \\
7 \times 1 \times \text{AM of O} &= \underline{112.0 \text{ amu}} \\
\text{FM} &= 246.4 \text{ amu}
\end{aligned}$$

The formula mass of $MgSO_4 \cdot 7H_2O$ is 246.4 amu.

4.3. Molar Mass

The term *molar mass* is now commonly used as a general term for both formula mass and molecular mass. The molar mass of any substance is the mass in grams of one mole of the substance, and is numerically equal to its formula mass (expressed in amu). (The mole will be defined and explained in chapter 5.) For example, the formula mass of glucose, $C_6H_{12}O_6$, is 180.0 amu. So the molar mass or the mass in grams of 1 mol of glucose is 180.0 g.

Let me mention that for most purposes you can round atomic and molecular masses to approximately 1 decimal place when solving problems.

Example 4.3

Calculate the formula mass and molar mass of $MgSO_4$.

Solution

1. To calculate the formula mass of $MgSO_4$, multiply the atomic mass of each component element by the number of times it appears in the formula and then add the results.

$$
\begin{array}{lll}
1\ \text{Mg atom} & = 1 \times 24.30\ \text{amu} = & 24.30\ \text{amu} \\
1\ \text{S atom} & = 1 \times 23.07\ \text{amu} = & 23.07\ \text{amu} \\
4\ \text{O atoms} & = 4 \times 16.00\ \text{amu} = & \underline{64.00\ \text{amu}} \\
 & & 111.37\ \text{amu}
\end{array}
$$

The formula mass of $MgSO_4$ is 111.4 amu.

2. Molar mass of a substance is defined as the mass of 1 mol or the formula mass of the substance expressed in grams. Thus the molar mass of $MgSO_4$ is 111.4 g/mol.

Example 4.4

Mycomycin is an antibiotic produced naturally by the fungus *Nocardoa acidophilus.* The molecular formula is $C_{13}H_{10}O_2$. What is the molar mass?

Solution

Multiply the atomic mass of each element by the number of times it appears in the molecular formula and then add the results.

$$
\begin{array}{lll}
13\ \text{C atoms} & = 13 \times 12.0\ \text{g} = & 156.0\ \text{g} \\
10\ \text{H atoms} & = 10 \times 1.0\ \text{g} = & 10.0\ \text{g} \\
2\ \text{O atoms} & = 2 \times 16.0\ \text{g} = & \underline{32.0\ \text{g}} \\
 & & 198.0\ \text{g}
\end{array}
$$

The molar mass of $C_{13}H_{10}O_2$ is 198.0 g/mol.

4.4. Problems

1. Calculate the formula masses of the following:
   (a) $NaHCO_3$ (b) $CaCO_3$ (c) NaH_2PO_4 (d) $Ag_2S_2O_3$ (e) $Al_2(SO_4)_3$
2. Determine the molecular masses of the following substances:
   (a) $C_{12}H_{22}O_{11}$ (b) H_2S (c) Mg_3N_2 (d) $Mg_3(BO_3)_2$ (e) Na_2SO_4
3. When hydrogen sulfide gas is bubbled into an aqueous solution of $SbCl_3$, the compound Sb_2S_3 is precipitated according to the following equation:

   $$2\,SbCl_3(aq) + 3\,H_2S(g) \longrightarrow Sb_2S_3(s) + 6\,HCl(aq)$$

   Calculate the formula mass of the precipitated Sb_2S_3.
4. Hydroxyapatite, a chemical compound present in tooth enamel, has the molecular formula $Ca_{10}(PO_4)_6(OH)_2$. Calculate the molar mass in grams.
5. Calculate the formula mass of aspirin, a mild pain reliever with the formula $C_9H_8O_4$.
6. Penicillin V is an important antibiotic. Its chemical formula is $C_{10}H_{14}N_2$. What is the formula mass of penicillin V?
7. The chemical formula of TNT (trinitrotoluene), an industrial explosive, is $C_7H_5O_6N_3$. Calculate its molecular mass.
8. The major chemical component of eucalyptus oil, extracted from eucalyptus leaves, is eucalyptol. Its chemical formula is $C_{10}H_{18}O$. Calculate its formula mass.
9. The formula mass of the compound K_2MCl_6 is 483.14 amu. What is the atomic mass of M? Use the periodic table to identify M.
10. The formula mass of the inorganic acid $H_4X_2O_7$ is 178. What is the atomic mass of X? Use the periodic table to identify X.

5

Measuring Chemical Quantities: the Mole

5.1. The Mole and Avogadro's Number

A *mole* is defined as the amount of a given substance that contains the same number of atoms, molecules, or formula units as there are atoms in 12 g of carbon-12. For example, one mole of glucose contains the same number of glucose molecules as there are carbon atoms in 12 g of carbon-12. The number of atoms in exactly 12 g of carbon-12 has been determined to be 6.02×10^{23}. This number, 6.02×10^{23}, is called Avogadro's number (N_A). Therefore, a mole is the amount of a substance that contains Avogadro's number of atoms, ions, molecules, or particles. For example:

$$1 \text{ mol He atoms} = 6.02 \times 10^{23} \text{ atoms}$$

$$1 \text{ mol } CH_3 OH \text{ molcculcs} = 6.02 \times 10^{23} \text{ molecules}$$

$$1 \text{ mol } SO_4^{2-} \text{ ions} = 6.02 \times 10^{23} \text{ ions}$$

5.2. The Mole and Molar Mass

The term *molar mass* is now commonly used as a general term for both formula mass and molecular mass. The molar mass of any substance is the mass in grams of one mole of the substance, and it is numerically equal to its formula mass (expressed in amu). For example, the formula mass of glucose, $C_6H_{12}O_6$, is 180.0 amu. So the molar mass or the mass in grams of 1 mol of glucose is 180.0 g.

5.3. Calculating the Number of Moles

In terms of chemical arithmetic, the mole is the most important number in chemistry. It provides useful stoichiometric information about reactants and products in any given chemical reaction. The quantities commonly encountered in chemical problems include the number of moles of a substance; the number of atoms, molecules, or formula units of a substance; and the mass in grams. These quantities are related and can be readily interconverted with the aid of the molar mass and Avogadro's number. Calculations based on the mole can be carried out by using conversion factors, or with simple equations based on the conversion factor.

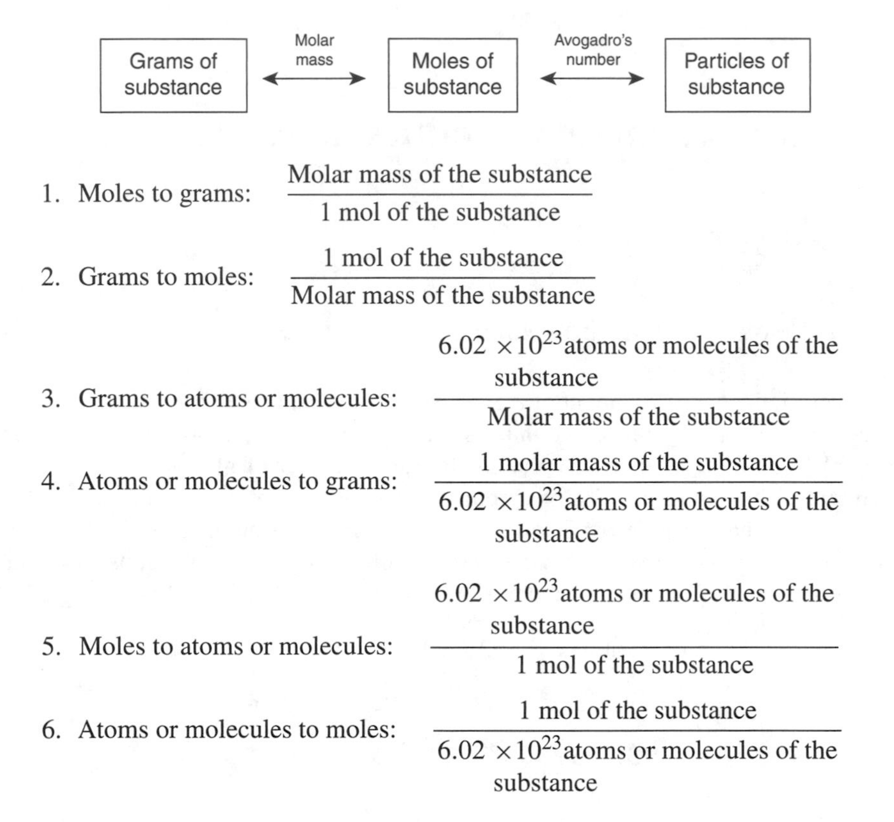

1. Moles to grams: $\dfrac{\text{Molar mass of the substance}}{\text{1 mol of the substance}}$
2. Grams to moles: $\dfrac{\text{1 mol of the substance}}{\text{Molar mass of the substance}}$
3. Grams to atoms or molecules: $\dfrac{6.02 \times 10^{23}\text{ atoms or molecules of the substance}}{\text{Molar mass of the substance}}$
4. Atoms or molecules to grams: $\dfrac{\text{1 molar mass of the substance}}{6.02 \times 10^{23}\text{ atoms or molecules of the substance}}$
5. Moles to atoms or molecules: $\dfrac{6.02 \times 10^{23}\text{ atoms or molecules of the substance}}{\text{1 mol of the substance}}$
6. Atoms or molecules to moles: $\dfrac{\text{1 mol of the substance}}{6.02 \times 10^{23}\text{ atoms or molecules of the substance}}$

Depending on the information provided, the number of moles of a chemical substance can be calculated from the following equations:

1. $\text{Number of moles} = \dfrac{\text{Mass in grams}}{\text{Molar mass}}$
2. $\text{Number of moles} = \dfrac{\text{Number of atoms or molecules}}{6.02 \times 10^{23}\text{ atoms or molecules/mol}}$

Example 5.1

Calculate the number of moles contained in 5.4 g of each of the following substances:

(a) Ag
(b) H_2SO_4
(c) $Na_2CO_3 \cdot 10\,H_2O$

Solution

First determine the molar mass of each substance. From the information given, decide on the appropriate conversion factor or formula to use. This is a grams to moles conversion. Therefore the following conversion factor can be used:

$$\text{Grams to moles:} \quad \frac{\text{1 mol of the substance}}{\text{1 molar mass of the substance}}$$

(a) $5.4\ \text{g Ag} \times \dfrac{1\text{mol Ag}}{108\ \text{g Ag}} = 0.050\ \text{mol Ag}$

(b) $5.4\ \text{g } H_2SO_4 \times \dfrac{1\ \text{mol } H_2SO_4}{98\ \text{g } H_2SO_4} = 0.055\ \text{mol } H_2SO_4$

(c) $5.4\ \text{g } Na_2CO_3 \cdot 10\ H_2O \times \dfrac{1\ \text{mol } Na_2CO_3 \cdot 10\ H_2O}{286\ \text{g } Na_2CO_3 \cdot 10\ H_2O}$
$= 0.0189\ \text{mol } Na_2CO_3 \cdot 10\ H_2O$

Example 5.2

How many atoms are present in 0.0125 mol of elemental sodium?

Solution

This is a moles to atoms conversion, moles Na $\longrightarrow$ atoms Na, so the following conversion factor is applicable:

$$\text{Moles to atoms or molecules:} \quad \frac{6.02 \times 10^{23}\ \text{atoms or molecules of the substance}}{\text{1 mol of the substance}}$$

$$?\ \text{Na atoms} = 0.0125\ \text{mol Na} \times \frac{6.02 \times 10^{23}\ \text{atoms Na}}{1\ \text{mol Na}} = 7.5 \times 10^{21}\ \text{Na atoms}$$

Example 5.3

Find the number of moles in each of the following samples:

(a) 0.027 g Al
(b) 1 atom nitrogen
(c) 1.0×10^{15} HPO_4^{2-} ions

Solution

For this problem, I am going to use the formula method. As demonstrated in the previous examples, we can use the conversion factor method to solve the problem as well.

(a) $\text{No. of moles Al} = \dfrac{\text{Mass of Al}}{\text{Molar mass of Al}} = \dfrac{0.027\ \text{g}}{27\ \text{g/mol}} = 0.0010\ \text{mol Al}$

(b) $\text{No. of moles } N_2 = \dfrac{\text{Number } N_2 \text{ atoms}}{6.02 \times 10^{23}\ \text{atoms/mol}} = \dfrac{1.0\ \text{atom}}{6.02 \times 10^{23} \text{atoms/mol}}$
$= 1.66 \times 10^{-23} \text{mol } N_2$

(c) $\text{No. of moles } HPO_4^{2-} = \dfrac{\text{Number } HPO_4^{2-} \text{ ions}}{6.02 \times 10^{23}\ \text{ions/mol}} = \dfrac{1.0 \times 10^{15}\ \text{ions}}{6.02 \times 10^{23}\ \text{ions/mol}}$
$= 1.66 \times 10^{-9}\ \text{mol } HPO_4^{2-}$

Example 5.4

Phosphoric acid (H_3PO_4) is used widely in the formulation of detergents, carbonated beverages, fertilizers, and toothpastes. Calculate the number of moles of each type of atom present in 3.5 mol of H_3PO_4.

Solution

The formula given represents 1 mol of H_3PO_4, and it contains 3 mol of H atoms, 1 mole of P atoms, and 4 mol of O atoms. Use the following conversion factors to calculate the number of moles of each atom present in 3.5 mol of the acid:

$$\frac{3\ \text{mol H atoms}}{1\ \text{mol } H_3PO_4}; \quad \frac{1\ \text{mol P atom}}{1\ \text{mol } H_3PO_4}; \quad \frac{4\ \text{mol O atoms}}{1\ \text{mol } H_3PO_4}$$

$$\text{Moles of H atoms} = 3.5\ \cancel{\text{mol } H_3PO_4} \times \frac{3\ \text{mol H atoms}}{1\ \cancel{\text{mol } H_3PO_4}} = 10.5\ \text{mol H atoms}$$

$$\text{Moles of P atoms} = 3.5\ \cancel{\text{mol } H_3PO_4} \times \frac{1\ \text{mol P atom}}{1\ \cancel{\text{mol } H_3PO_4}} = 3.5\ \text{mol P atoms}$$

$$\text{Moles of O atoms} = 3.5\ \cancel{\text{mol } H_3PO_4} \times \frac{4\ \text{mol O atoms}}{1\ \cancel{\text{mol } H_3PO_4}} = 14\ \text{mol O atoms}$$

Example 5.5

Hemoglobin, which is the oxygen carrier in the blood, has the molecular formula $C_{2952}H_{4664}N_{812}S_8Fe_4$. (a) Calculate its molar mass. (b) What is the number of moles contained in 25,000 g of this sample? (c) Calculate the gram atoms of N, S, and Fe present in 25,000 g of hemoglobin.

Solution

(a)

$$\begin{aligned}
&\text{2952 atoms of C weigh } 2952 \times 12 = 35424 \\
&\text{4664 atoms of H weigh } 4664 \times 1 = 4664 \\
&\text{812 atoms of N weigh } 812 \times 14 = 11368 \\
&\text{8 atoms of S weigh } 8 \times 32 = 256 \\
&\text{4 atoms of Fe weigh } 4 \times 56 = \underline{224} \\
&\text{Formula mass} = 51{,}936
\end{aligned}$$

(b) Number of moles

$$C_{2952}H_{4664}N_{812}S_8Fe_4 = 25{,}000\text{ g} \times \frac{1\text{ mol } C_{2952}H_{4664}N_{812}S_8Fe_4}{51{,}936\text{ g } C_{2952}H_{4664}N_{812}S_8Fe_4}$$

$$= 0.4814\text{ mol}$$

(c)

$$\text{Number of g atom of N} = 0.4814\text{ mol Hemo} \times \frac{11{,}368\text{ g N}}{1\text{ mol Hemo}}$$

$$= 5472\text{ g N}$$

$$\text{Number of g atom of S} = 0.4814\text{ mol Hemo} \times \frac{256\text{ g S}}{1\text{ mol Hemo}}$$

$$= 123.2\text{ g S}$$

$$\text{Number of g atom of Fe} = 0.4814\text{ mol Hemo} \times \frac{224\text{ g Fe}}{1\text{ mol Hemo}}$$

$$= 107.8\text{ g Fe}$$

Example 5.6

White phosphorus burns in excess oxygen to give tetraphosphorus decaoxide, P_4O_{10}. The oxide P_4O_{10} is highly deliquescent, and absorbs water from the atmosphere to form phosphoric acid (H_3PO_4) according to the following equation:

$$P_4O_{10}(s) + 6H_2O(l) \longrightarrow 4H_3PO_4(aq)$$

How many kg of phosphoric acid (H_3PO_4) could be obtained from 950 g of white phosphorus (P)?

Solution

In solving this problem, it is not necessary to find the mass of P_4O_{10}. We need to focus only on H_3PO_4. The formula H_3PO_4 indicates that 1 mol of P yields 1 mol of H_3PO_4. Using the appropriate conversion factors, we have:

$$\text{Mass of } H_3PO_4 = (950 \text{ g P})\left(\frac{1 \text{ mol P}}{30.97 \text{ g P}}\right)\left(\frac{1 \text{ mol } H_3PO_4}{1 \text{ mol P}}\right)\left(\frac{97.97 \text{ g } H_3PO_4}{1 \text{ mol } H_3PO_4}\right)$$

$$= 3005.2 \text{ g} = 3.0 \text{ kg}$$

5.4. Problems

1. Calculate the number of moles present in 100 g of each of the following compounds:
(a) $CaCO_3$ (b) H_2O (c) NaOH (d) H_2SO_4 (e) N_2
2. Calculate the number of moles present in each of the following compounds:
(a) 12.5 g $CaSO_4$ (b) 0.25 g H_2O_2 (c) 0.04 g $NaHCO_3$ (d) 9.8 g H_2SO_4
3. Calculate the mass in grams present in 0.025 mol of each of the following compounds:
(a) MgO (b) Na_2O (c) NH_3 (d) C_2H_4 (e) $H_4P_2O_7$
4. What mass in grams is present in each of the following?
(a) 0.50 mol CaO (b) 0.005 mol Li_2O (c) 2.5 mol P_4O_{10}
(d) 1.5 mol C_2H_4 (e) 0.01 mol $B_3N_3H_6$
5. What mass in grams is present in each of the following?
(a) 5.0×10^{20} atoms of Ca (b) 2.5×10^{26} atoms of Fe
(c) 3.0×10^{23} atoms of Ca (d) 10×10^{0} atoms of K
(e) 6.0×10^{3} atoms of H
6. Calculate the number of atoms of each element contained in 1 mol of
(a) Al_2O_3 (b) H_2S (c) CO_2 (d) HCl
7. Calculate the number of moles present in each of the following:
(a) 5.0×10^{20} atoms of Mg (b) 2.5×10^{26} ions of Fe^{3+}
(c) 1.15×10^{23} molecules of H_2O (d) 6.0×10^{0} ions of SO_4^{2-}
(e) 3.0×10^{3} molecules of $C_6H_{12}O_6$
8. Calculate the mass of H_3PO_4 that could be produced from 2000 kg of phosphorus.
9. Calculate the number of oxygen atoms present in 500 g of the following:
(a) H_2O_2 (b) $C_{12}H_{22}O_{11}$ (c) $CaCO_3$ (d) $Ca_3(PO_4)_2$
10. Iron (III) sulfate ($Fe_2(SO_4)_3$), ionizes according to the following equation:

$$Fe_2(SO_4)_3 \longrightarrow 2\,Fe^{3+} + 3\,SO_4^{2-}$$

Calculate the number of each ions produced from 2.00 mol of $Fe_2(SO_4)_3$.

11. A superconductor is a material that loses all resistance to the flow of electrical current below a characteristic temperature known as the superconducting transition temperature (T_c). The copper-containing oxide $YBa_2Cu_3O_7$, the so-called 1–2–3 compound (1 yttrium, 2 bariums, and 3 coppers), is an example of a superconductor with a high T_c. Determine (a) the number of moles of Y, Ba, and Cu in 500 g of the oxide, (b) the number of atoms of Y, Ba, and Cu in 500 g of the oxide.
12. Silicon nitride (Si_3N_4) is a high-temperature ceramic that is used in the fabrication of some engine components. How many atoms of Si and N are present in 0.50 mol of Si_3N_4?

6

Formulas of Compounds and Percent Composition

6.1. Percent Composition

Percent means "parts per 100" or "part divided by whole, multiplied by 100." If the formula of a compound is known, one can easily calculate the mass percent of each element in the compound. By definition, the percent composition of a compound is the mass percent of each element in the compound. The molar mass represents 100% of the compound.

$$\%\ \text{Element} = \frac{\text{Mass of element}}{\text{Formula mass}} \times 100$$

The following steps are helpful in calculating the percent composition if the formula of the compound is known.

1. Calculate the formula mass or molar mass of the compound
2. Calculate the percent composition by dividing the total mass of each element in the formula unit by the molar mass and multiplying by 100.

Example 6.1

Aspirin, which is commonly used as a pain reliever, has the chemical composition $C_9H_8O_4$. Calculate the percent composition of aspirin.

Solution

First, calculate the molar mass of aspirin.

$$9\ \text{C} = 9 \times 12.0 = 108.0$$

$$8\ \text{H} = 8 \times 1.0 = 8.0$$

$$4\ \text{O} = 4 \times 16 = \underline{64.0}$$

$$\text{Molar mass} = 180.0\ \text{g}$$

Next, calculate the percent composition.

$$\% \text{ C} = \frac{9 \text{ C}}{C_9H_8O_4} \times 100 = \frac{9 \times 12}{180} \times 100 = \frac{108}{180} \times 100 = 60.0\%$$

$$\% \text{ H} = \frac{8 \text{ H}}{C_9H_8O_4} \times 100 = \frac{8 \times 1}{180} \times 100 = \frac{8}{180} \times 100 = 4.44\%$$

$$\% \text{ O} = \frac{4 \text{ O}}{C_9H_8O_4} \times 100 = \frac{64}{180} \times 100 = 35.5\%$$

Now, check to ensure they all add to 100%.

$$60.00 + 4.44 + 35.56 = 100.0\%$$

Example 6.2

Arsenic occurs in the following native compounds (or ores): arsenopyrite, FeAsS, realgar, As_2S_2, and orpiment, As_2S_3. Which of these ores has the highest sulfur content?

Solution

First calculate the molar mass of each ore:

$$\text{FeAsS} = 1 \times \text{Fe} + 1 \times \text{As} + 1 \times \text{S} = 55.85 + 74.92 + 32.06 = 162.83 \text{ g}$$

$$As_2S_2 = 2 \times \text{As} + 2 \times \text{S} = 149.84 + 64.12 = 213.96 \text{ g}$$

$$As_2S_3 = 2 \times \text{As} + 3 \times \text{S} = 149.84 + 96.16 = 246.02 \text{ g}$$

Now calculate the percent of S in each ore:

$$\% \text{ S in FeAsS} = \frac{\text{S}}{\text{FeAsS}} \times 100 = \frac{32.06}{162.83} \times 100 = 19.69\%$$

$$\% \text{ S in } As_2S_2 = \frac{2 \text{ S}}{As_2S_2} \times 100 = \frac{64.12}{213.96} \times 100 = 13.72\%$$

$$\% \text{ S in } As_2S_3 = \frac{3 \text{ S}}{As_2S_3} \times 100 = \frac{96.16}{246.02} \times 100 = 39.09\%$$

Orpiment (As_2S_3) has the highest sulfur content.

Example 6.3

Calculate the percentage of water of hydration in the following hydrates:

(a) $CuSO_4 \cdot 5\,H_2O$
(b) $Na_2CO_3 \cdot 10\,H_2O$
(c) $Zn(NO_3)_2 \cdot 6\,H_2O$

Solution

First calculate the molar mass. Then express water as a mass percent of the compound:

1) $CuSO_4 \cdot 5\,H_2\,O = 63.55 + 32.06 + 4 \times 16 + 5\,(2 \times 1 + 16) = 249.61$

$$\%\,H_2O = \frac{5\,H_2O}{CuSO_4 \cdot 5H_2O} \times 100 = \frac{90}{249.61} \times 100 = 36.06\%$$

2) $Na_2CO_3 \cdot 10\,H_2O = 2 \times 23 + 12 + 3 \times 16 + 10\,(2 \times 1 + 16) = 286$

$$\%\,H_2O = \frac{10\,H_2O}{Na_2CO_3 \cdot 10\,H_2O} \times 100 = \frac{180}{286} \times 100 = 62.94\%$$

3) $Zn(NO_3)_2 \cdot 6\,H_2O = 65.38 + 2\,(14 + 3 \times 16) + 6\,(2 \times 1 + 16) = 297.38$

$$\%\,H_2O = \frac{6H_2O}{Zn(NO_3)_2 \cdot 6\,H_2O} \times 100 = \frac{108}{297.38} \times 100 = 36.32\%$$

Example 6.4

Titanium disulfide is a semiconductor with a layered structure. The lithiated form, Li_xTiS_2, is useful as an electrode in rechargeable batteries. Calculate the mass percent of lithium and sulfur present in 25.25 g of Li_xTiS_2 when $x = 0.25$.

Solution

First, calculate the molar mass of $Li_{0.25}TiS_2$.

$$\begin{aligned} Li &= 6.94 \times 0.25 = 1.735 \\ Ti &= 47.9 \times 1 \quad = 47.900 \\ S &= 32.1 \times 2 \quad = \underline{64.200} \\ \text{Molar mass} &\quad\quad = 113.835\ \text{g} \end{aligned}$$

Now calculate the % of Li and S in $Li_{0.25}TiS_2$.

$$\%\,Li = \frac{1.735}{113.835} \times 100\% = 1.52\%$$

$$\%\,S = \frac{64.2}{113.835} \times 100\% = 56.4\%$$

Example 6.5

The boron content of a Pyrex glass was reported as 17% B_2O_3. Determine the percentage of boron in this glass.

Solution

1. Determine the molar mass of B_2O_3.
   $B_2O_3 = 10.81 \times 2 + 16.0 \times 3 = 69.62$
2. Determine the percentage of B in B_2O_3.
   $\% \text{ B} = \dfrac{21.62}{69.62} \times 100\% = 31.05\%$, or 31.05 g of B in every 100 g of B_2O_3
3. Determine the percentage of B in the glass.
   $(17\% \text{ B}_2\text{O}_3)\left(\dfrac{31.05 \text{ g B}}{100 \text{ g B}_2\text{O}_3}\right) = 5.38\% \text{ B in the glass}$

Example 6.6

To determine the amount of silver in an alloy sample, a chemist dissolved 0.75 g of the sample in dilute nitric acid. He precipitated the silver content of the solution as AgCl by adding excess NaCl solution. The precipitated AgCl weighed 0.85 g. Calculate the mass of silver in the dissolved sample.

Solution

1. First, calculate the percent of Ag in AgCl:
   Molar mass of AgCl = 143.4
   $$\% \text{ Ag in AgCl} = \frac{\text{Mass of Ag}}{\text{Mass of AgCl}} \times 100\%$$
   $$= \frac{107.9}{143.4} \times 100 = 75.2\%$$
2. Calculate the mass of Ag in the dissolved sample.
   $$\text{Mass of Ag in the sample} = 75.2\% \text{ of the precipitated AgCl}$$
   $$= (0.752)(0.85 \text{ g})$$
   $$= 0.64 \text{ g}$$
3. The percentage of Ag in the original sample is simply the actual amount of silver divided by the total weight:
   $$\% \text{ of Ag in the sample} = \frac{0.64}{0.75} \times 100\%$$
   $$= 85.3\%$$

6.2. Empirical Formula

Chemical formulas are often determined from experimental data. Generally, two types of formula can be obtained, depending on the amount of experimental information available. These are the *empirical formula* and the *molecular formula*. Usually the empirical formula is determined first, and the molecular formula is calculated from that.

Empirical formula

The empirical formula of a compound represents the smallest whole-number ratio of atoms present in a formula unit of the compound. An empirical formula is found by measuring the mass percent of each element in the compound. The mass percent of each element is converted to the number of moles of atoms of each element per 100 g of the compound. Then the mole ratios between the elements are expressed in terms of small whole numbers.

Steps for determining empirical formula

1. Find the number of grams of each element in a given mass of the compound. Often we use 100 g for simpler calcuation. Find the relative number of moles by dividing the mass of each element by its molar mass.
2. Express the mole ratio between the elements in terms of small whole numbers by dividing through by the smallest number. If this does not result in simple whole numbers for all the elements in the compound, multiply through by the smallest suitable integer to convert all to whole numbers.
3. Obtain the empirical formula by inserting the whole numbers after the symbol of each element.

Example 6.7

Analysis of a nicotine product yielded 74% C, 8.7% H, and 17.3% N. What is the empirical formula?

Solution

First assume you have 100 g of sample, so that the given percentage compositions will correspond to masses in grams, i.e., 74 g C, 8.7 g H, and 17.3 g N. Now, convert the relative masses of each element to the relative number of moles of each element. Since this is a mass to moles conversion, use the following conversion factor:

$$\text{Grams to moles:} \quad \frac{\text{1 mol of the substance}}{\text{Molar mass of the substance}}$$

$$\text{C: } 74 \text{ g C} \times \frac{1 \text{ mol C}}{12.0 \text{ g C}} = 6.17 \text{ mol C}$$

$$\text{H: } 8.7 \text{ g H} \times \frac{1 \text{ mol H}}{1.0 \text{ g H}} = 8.70 \text{ mol H}$$

$$\text{N: } 17.3 \text{ g N} \times \frac{1 \text{ mol N}}{14.0 \text{ g N}} = 1.24 \text{ mol N}$$

So, 100 g of the compound contains 6.17 mol of C, 8.70 mol H, and 1.24 mol N. Convert these to whole number ratios by dividing by the smallest of the numbers (i.e., 1.24).

$$\text{C: } \frac{6.17 \text{ mol}}{1.24 \text{ mol}} = 4.99 \approx 5.0$$

$$\text{H: } \frac{8.7 \text{ mol}}{1.24 \text{ mol}} = 7.02 \approx 7.0$$

$$\text{N: } \frac{1.24 \text{ mol}}{1.24 \text{ mol}} = 1.0$$

The empirical formula of the compound is C_5H_7N.

Example 6.8

A chemist extracted a medicinal compound believed to be effective in treating high blood pressure. Elemental analysis of this compound gave the composition 60.56% C, 11.81% H, and 28.26% N. What is the empirical formula of this compound?

Solution

To solve the problem, we assume that we are starting with 100 g of the material, which will contain 60.56 g C, 11.8 g H, and 28.26% N.

Next, convert the gram amount to moles for each element.

$$\text{C}: 60.56 \text{ g C} \times \frac{1 \text{ mol C}}{12.0 \text{ g C}} = 5.05 \text{ mol C}$$

$$\text{H: } 11.18 \text{ g H} \times \frac{1 \text{ mol H}}{1.0 \text{ g H}} = 11.18 \text{ mol H}$$

$$\text{N: } 28.26 \text{ g N} \times \frac{1 \text{ mol N}}{14.0 \text{ g N}} = 2.02 \text{ mol N}$$

Determine the smallest whole-number ratio of moles by dividing each number of moles by the smallest number of moles, 2.02:

$$\text{C: } \frac{5.05 \text{ mol}}{2.02 \text{ mol}} = 2.5$$

$$\text{H: } \frac{8.7 \text{ mol}}{2.02 \text{ mol}} = 4.3$$

$$\text{N: } \frac{2.02 \text{ mol}}{2.02 \text{ mol}} = 1.0$$

We still do not have whole-number ratios. Therefore, multiply by the lowest integer that will convert all to whole numbers. The number 10 seems to do the trick.

$$C: 2.5 \times 10 = 25$$

$$H: 4.3 \times 10 = 43$$

$$N: 1.0 \times 10 = 10$$

The empirical formula of the compound is $C_{25}H_{43}N_{10}$

Example 6.9

A chemical engineer heated 1.8846 g of pure iron powder in an atmosphere of pure oxygen in a quartz tube until all the iron was converted to the oxide, which weighed 2.6946 g. Calculate the empirical formula of the oxide formed.

Solution

First, determine the mass of oxygen gas that has reacted. Recall that, in a reaction, mass is conserved. Hence:

$$\text{Mass of Fe} + \text{Mass of O} = \text{Mass of } Fe_x\,O_y$$

$$\text{Mass of O} = \text{Mass of } Fe_x\,O_y - \text{Mass of Fe}$$

$$= 2.6946 \text{ g} - 1.8846 \text{ g} = 0.81 \text{ g}$$

Now, convert relative masses to relative number of moles.

$$\text{Fe: } 1.8846 \text{ g Fe} \times \frac{1 \text{ mol Fe}}{55.85 \text{ g Fe}} = 0.0337 \text{ mol Fe}$$

$$\text{O: } 0.8100 \text{ g O} \times \frac{1 \text{ mol O}}{16.00 \text{ g O}} = 0.0506 \text{ mol O}$$

Determine the smallest whole-number ratio of atoms:

$$\text{Fe: } \frac{0.0337 \text{ mol}}{0.0337 \text{ mol}} = 1.0$$

$$\text{O: } \frac{0.0506 \text{ mol}}{0.0337 \text{ mol}} = 1.5$$

Multiply the ratios by 2 to clear the remaining decimals. The gives Fe : O = 2 : 3. The empirical formula of the oxide is thus Fe_2O_3.

6.2.1. Empirical formula from combustion analysis

Combustion analysis is one of the most common methods used to determine the empirical formula of an unknown compound containing carbon and hydrogen.

When a known amount of a compound containing carbon and hydrogen is burned in a combustion apparatus in the presence of oxygen, all the carbon is converted to CO_2, and the hydrogen to H_2O. The CO_2 formed is trapped by NaOH, and the water is trapped by magnesium perchlorate, $Mg(ClO_4)_2$. The mass of CO_2 is measured as the increase in the mass of the CO_2 trap, and the mass of H_2O produced is measured as the increase in mass of the water trap. From the masses of CO_2 and H_2O, the masses and moles of carbon and hydrogen present in the compound can be determined. (If nitrogen is present in the compound, it must be determined by the Kjeldahl method in a separate analysis, as NH_3). Oxygen in the compound can then be determined by difference — that is, the percentage remaining when C, H, and N (if present) have been found.

Example 6.10

A 39.0-mg sample of a compound containing C, H, O, and N is burned. The C is recovered as 97.7 mg CO_2, and the H is recovered as 20.81 mg H_2O. A separate Kjeldahl nitrogen analysis gives a nitrogen content of 3.8%. Calculate the empirical formula of the compound.

Solution

First convert grams of CO_2 and H_2O to grams of C and H.

$$97.7\ \text{mg CO}_2 \times \frac{1\ \text{g CO}_2}{1000\ \text{mg CO}_2} \times \frac{1\ \text{mol CO}_2}{44.0\ \text{g CO}_2} \times \frac{1\ \text{mol C}}{1\ \text{mol CO}_2} \times \frac{12\ \text{g C}}{1\ \text{mol C}}$$

$$= 0.0266\ \text{g} \quad \text{or} \quad 26.6\ \text{mg C}$$

$$20.81\ \text{mg H}_2\text{O} \times \frac{1\ \text{g H}_2\text{O}}{1000\ \text{mg H}_2\text{O}} \times \frac{1\ \text{mol H}_2\text{O}}{18.0\ \text{g H}_2\text{O}} \times \frac{2\ \text{mol H}}{1\ \text{mol H}_2\text{O}} \times \frac{1\ \text{g H}}{1\ \text{mol H}}$$

$$= 0.0023\ \text{g} \quad \text{or} \quad 2.3\ \text{mg H}$$

For nitrogen, find 3.8% of the starting material.

$$\left(\frac{3.8}{100} \times 39.3\ \text{mg}\right)\left(\frac{1\ \text{g N}}{1000\ \text{mg N}}\right) = 0.001493\ \text{g or } 1.5\ \text{mg N}$$

Next, calculate the mass of O as the difference between the mass of the sample and the masses of C, H, and N.

$$\text{Mass of O} = 39.3\ \text{mg} - 26.7\ \text{mg} - 2.3\ \text{mg} - 1.5\ \text{mg} = 8.8\ \text{mg O}$$

Now convert the mass of each element to the corresponding moles.

$$\text{C: } 0.0267\ \text{g C} \times \frac{1\ \text{mol C}}{12.0\ \text{g C}} = 0.0022\ \text{mol C}$$

$$\text{H: } 0.0023\text{ g H} \times \frac{1\text{ mol H}}{1.01\text{ g H}} = 0.0023\text{ mol H}$$

$$\text{N: } 0.0015\text{ g N} \times \frac{1\text{ mol N}}{14.0\text{ g N}} = 0.00011\text{ mol N}$$

$$\text{O: } 0.0088\text{ g O} \times \frac{1\text{ mol O}}{16.0\text{ g O}} = 0.00055\text{ mol O}$$

Divide each molar value by the smallest of them.

$$C = \frac{0.0022\text{ mol}}{0.000107\text{ mol}} = 20$$

$$H = \frac{0.0023\text{ mol}}{0.00011\text{ mol}} = 21$$

$$N = \frac{0.00011\text{ mol}}{0.00011\text{ mol}} = 1$$

$$O = \frac{0.00055\text{ mol}}{0.00011\text{ mol}} = 5$$

The empirical formula of the compound is $C_{20}H_{21}NO_5$.

6.3. Molecular Formula

The molecular formula of a compound is the true formula, which shows the actual numbers of atoms in a molecule of the compound. The molecular formula may be the same as the empirical formula or a multiple of it. For example, a compound with the empirical formula CH_2O can have any of the following molecular formulas: $C_2H_4O_2$, $C_3H_6O_3$, or $C_5H_{10}O_5$.

6.3.1. Determination of molecular formula

To determine the molecular formula of a molecular compound, the empirical formula (obtained from percent composition), and the molecular mass (obtained experimentally) must be known. Since the molecular formula for a compound is either the same as, or a multiple of, the empirical formula, the following relationship is helpful in determining molecular formula:

$$\text{Molecular weight} = n\,(\text{Empirical formula weight}) \quad \text{or} \quad n = \frac{\text{Molecular weight}}{\text{Empirical formula weight}}$$

Then, multiplying the coefficients in the empirical formula by n gives the actual molecular formula.

Example 6.11

The empirical formula of resorcinol, a common chemical used in the manufacture of plastics, drugs, and paper products, is C_3H_3O. The molecular mass is 110.0. What is the molecular formula?

Solution

We are given the empirical and molecular mass of the compound. The unknown is the molecular formula.

Begin by calculating the formula mass of C_3H_3O.

$$3 \times \mathrm{C} + 3 \times \mathrm{H} + 1 \times \mathrm{O} = 3 \times 12 + 1 \times 3 + 1 \times 16 = 55 \text{ amu}$$

Next, calculate the multiplication factor, n, from the expression:

$$\text{Molecular weight} = n(\text{Empirical formula weight}) \quad \text{or}$$

$$n = \frac{\text{Molecular weight}}{\text{Empirical formula weight}}$$

$$n = \frac{110}{55} = 2$$

Now, multiply the subscripts in the empirical formula by 2. This yields $C_6H_6O_2$ as the molecular formula.

Example 6.12

The empirical formula of ibuprofen, an active component in the pain remedy Advil, is $C_{13}H_{18}O_2$. Its molecular mass is 206.0. What is its molecular formula?

Solution

The empirical formula is $C_{13}H_{18}O_2$ and has formula weight of 206.0 amu. That is:

$$13 \times 12 + 18 \times 1 + 2 \times 16 = 206 \text{ amu}$$

Since the empirical formula mass and the molecular mass are equal (206.0), n is equal to 1. Hence the molecular formula is the same as the empirical formula, i.e., $C_{13}H_{18}O_2$.

Example 6.13

Epinephrine is an adrenaline hormone secreted into the bloodstream in times of danger or stress. Analysis indicates it has the following composition by mass:

59% C, 7.1% H, 26.2% O, and 7.7% N. The molecular mass is 183.0. Determine the empirical and molecular formula of epinephrine.

Solution

Begin by assuming you have 100g of sample which contains 59 g C, 7.1 g H, 26.2 g O, and 7.7 g N.

Next convert each mass of an element into the number of moles of that element:

$$\text{C: } 59\text{ g C} \times \frac{1\text{ mol C}}{12.0\text{ g C}} = 4.92\text{ mol C}$$

$$\text{H: } 7.1\text{ g H} \times \frac{1\text{ mol H}}{1.0\text{ g H}} = 7.1\text{ mol H}$$

$$\text{O: } 26.2\text{ g O} \times \frac{1\text{ mol O}}{16.0\text{ g O}} = 1.64\text{ mol O}$$

$$\text{N: } 7.7\text{ g N} \times \frac{1\text{ mol N}}{14.0\text{ g N}} = 0.55\text{ mol N}$$

Next, determine the relative number of moles of each element by dividing the above numbers by the smallest of the four elements in the compound:

$$\text{C: } \frac{4.92\text{ mol}}{0.55\text{ mol}} = 8.95 \approx 9$$

$$\text{H: } \frac{7.10\text{ mol}}{0.55\text{ mol}} = 13.0$$

$$\text{O: } \frac{1.64\text{ mol}}{0.55\text{ mol}} = 3.0$$

$$\text{N: } \frac{0.55\text{ mol}}{0.55\text{ mol}} = 1.00$$

The empirical formula is $C_9H_{13}O_3N$ and has formula mass of 183 amu. The molecular mass is 183.0. Since the empirical formula mass and the molecular mass are equal, the molecular formula is the same as the empirical formula, i.e., $C_9H_{13}O_3N$.

6.4. Problems

1. Calculate the percentage by mass of sulfur in the following:
   (a) H_2S (b) SO_3 (c) H_2SO_4 (d) $Na_2S_2O_3$
2. Calculate the percent composition of each of the following compounds:
   (a) K_2CO_3 (b) $Ca_3(PO_4)_2$ (c) $Al_2(SO_4)_3$ (d) C_6H_5OH
3. Calculate the percent composition of each of the following organic compounds:
   (a) $C_{20}H_{25}N_3O$ (b) $C_{10}H_{16}N_5P_3O_{13}$ (c) $C_6H_4N_2O_4$ (d) $C_6H_5CONH_2$

4. Calculate the percent by mass of magnesium, chlorine, hydrogen, and oxygen in hydrated magnesium chloride, $MgCl_2 \cdot 6H_2O$.
5. Find the mass of water of crystallization present in 10.85 g of hydrated sodium carbonate,$Na_2CO_3 \cdot 10H_2O$.
6. Caffeine, a stimulant in coffee and tea, has the composition 49.5% C, 5.19% H, 28.9% N, and 16.5% O. It has the molecular formula $C_8H_{10}N_4O_2$. How many grams of carbon and nitrogen are present in 77.25 g of caffeine?
7. Large quantities of hematite, an important ore of iron, are found in Australia, Ukraine, and USA. A 100 g sample of a crude ore contains 46.75 g of Fe_2O_3. What is the percentage of iron (Fe) in the ore? (Assume no other Fe compounds are present.)
8. The element lithium is commonly obtained from the ore spodumene, $LiAlSi_2O_6$, which is found in Brazil, Canada, and the USA. What mass of lithium is present in 17.85 g $LiAlSi_2O_6$?
9. Determine the empirical formulas for compounds having the following compositions:
   (a) 40% S and 60% O
   (b) 14.7% Ca, 67.7% W and 17.6% O
   (c) 85.63% C and 14.37% H
   (d) 9.90% C, 58.6% Cl and 31.5% F
   (e) 26.52% Cr, 24.52% S and 48.96% O
10. Given the following percent compositions, determine the empirical formulas:
    (a) 21.85% Mg, 27.83% P and 50.32% O
    (b) 19.84% C, 2.50% H, 66.08% O and 11.57% N
    (c) 47.3% C, 2.54% H and 50.0% Cl
    (d) 23.3% Co, 25.3% Mo and 51.4% Cl
    (e) 41.87% C, 2.34% H and 55.78% O
11. Write the empirical formula for each of the following compounds:
    (a) C_3H_6 (b) Fe_2S_3 (c) $C_6H_9O_6$ (d) $C_8H_{10}N_4O_2$ (e) $Al_2(SO_4)_3$
12. A compound which has the empirical formula P_2O_5 is found to have a molecular mass of 283.9. What is the molecular formula of the compound?
13. Cholesterol is the compound thought to be responsible for hardening of the arteries. An analysis of the compound gives the following percent composition by mass: 84.0% C, 11.9% H, and 4.1% O. Determine the empirical formula. What is the molecular formula given that the molar mass is 386?
14. The molar mass of estradiol, a female sex hormone, is 272 g/mol. Is the molecular formula $C_9H_{12}O$ or $C_{18}H_{24}O_2$?
15. A hydrate of sodium carbonate, $Na_2CO_3 \cdot nH_2O$, was found to contain 62.9% by mass of water of crystallization. Determine the empirical formula for the hydrate.
16. A colorless organic liquid, *A*, was found to contain 47.37% C, 10.6% H, and 42% O. The molar mass of the compound is 228 g/mol. Calculate the empirical and molecular formula of *A*.

17. On strong heating, 25.07 g of a hydrated salt, $FeCl_3 \cdot xH_2O$, gave 15.06 g of the anhydrous salt. Determine the empirical formula of the salt.
18. Combustion analysis of 4.86 g of a sugar yielded 7.92 g of CO_2 and 2.70 g of H_2O. If the compound contains only carbon, hydrogen, and oxygen, what is its empirical formula? If the molar mass of 324 g/mol, what is its molecular formula?
19. 3.2 mg of an unknown organic compound containing C, H and O gave, on combustion analysis, 3.48 mg CO_2 and 1.42 mg H_2O. What is the empirical formula of the compound? If the compound has a molecular mass of 244 g/mol, what is its molecular formula?
20. A combustion analysis was carried out on 0.4710 g of an organic compound containing only carbon, hydrogen, nitrogen, and oxygen. The products of combustion were 0.9868 g of CO_2 and 0.2594 g of H_2O. In another experiment, combustion of 0.3090 g of the compound produced 0.0357 g NH_3. (a) What is the mass percent composition of the compound? (b) Determine the empirical formula of the compound.

7

Chemical Formulas and Nomenclature

7.1. General Background

7.1.1. Elements

- An element is a pure substance that cannot be split up into simpler substances. There are 106 known different elements. Of these, only 92 occur naturally in the earth's crust and atmosphere. Scientists have made the other 14 elements artificially. All the elements can be arranged in a *periodic table*, which displays the important relationships between the elements and the structures of their atoms.
- Each element has a name and a symbol. The symbol is usually derived from the name of the element by taking the capital form of the first letter of the name. For example, the symbol for nitrogen is N. In most other cases, the capital letter followed by the next letter in lowercase is used. For example, the symbol for aluminum is Al.
- The symbols for some elements are derived from their Latin names. For example, the symbol for sodium is Na, from its Latin name natrium. Potassium has the symbol K, taken from its Latin name, kalium.
- Most of the known elements are metals. Only 22 are nonmetallic.
- All elements are made up of atoms. Atoms of the same element are identical in all respects (except for isotopic differences, see chapter 3) but are different from atoms of other elements.

7.1.2. Some basic definitions

- *Atom*: the smallest particle of an element that can take part in a chemical combination.
- *Molecule*: a collection of two or more atoms of the same or different elements held together by covalent bonds.

- *Atomicity of an element*: the number of atoms present in one molecule of the element.

  1. *Monatomic* elements exist as single atoms. Examples include sodium (Na), iron (Fe), and the noble gases (He, Ne, Ar, Kr, Xe, and Rn).
  2. *Diatomic* elements contain two atoms per molecule of the element. The atomicity is written as a subscript after the symbol of the element. Examples of diatomic molecules include H_2, O_2, Cl_2, N_2, etc.
  3. *Triatomic* elements contain three atoms per molecule. An example is ozone (O_3).
  4. *Tetraatomic* elements contain four atoms of the element per molecule. An example is yellow phosphorus (P_4).

- *Ions*: an electrically charged atom or a group of atoms formed by the gain or loss of electrons.

  1. *Anion*: An anion is a negatively charged ion.
  2. *Cation*: A cation is positively charged ion.

- *Polyatomic ions (or radicals)*: groups of covalently bonded atoms with an overall charge.
- A *compound* is a pure substance formed when two or more elements combine with each other. For example, water is a compound formed by the combination of two atoms of hydrogen and one atom of oxygen.

  1. The process of forming a compound involves a *chemical reaction*.
  2. The composition of a compound is fixed.

- The *formula* of a compound consists of the symbols of its elements and the number of each kind of atom.
- In every compound, the constituent elements are combined in a fixed proportion. This is known as the *law of definite composition*, which states that the atoms in a compound are combined in a fixed proportion by mass.
- There are two types of compounds: *molecular compounds* and *ionic compounds*. Molecular compounds are composed of molecules and are not charged. A good example is water, H_2O. Ionic compounds are composed of metallic (positively charged) and nonmetallic (negatively charged) ions held together by attractive forces. An example of ionic compound is sodium chloride (table salt, NaCl).

7.2. Chemical Formula

- A chemical formula is used to show the number and kinds of atoms in a molecule. For example, ammonia gas is a compound which contains one atom of nitrogen per three atoms of hydrogen. Its formula is NH_3.
- A chemical formula only shows the number and kind of each atom contained in the compound. It usually does not show how the atoms are linked together or the nature of the chemical bonds.

7.3. Oxidation Numbers

The oxidation number of an element is a positive or negative number that expresses the combining capacity of an element in a particular compound or polyatomic ion. The numbers are used to keep track of electron transfer in chemical reactions. The oxidation number finds application in writing chemical formulas and equations, balancing oxidation–reduction reactions involving electron transfer, and predicting the properties of compounds. Some general rules can help you determine the oxidation number.

7.3.1. Rules for assigning oxidation numbers

1. Any atom in an uncombined (or free) element (e.g., N_2, Cl_2, S_8, O_2, O_3, and P_4) has an oxidation number of zero.
2. Hydrogen has an oxidation number of $+1$ except in metal hydrides (e.g., NaH, MgH_2) where it is -1.
3. Oxygen has an oxidation number of -2 in all compounds except in peroxides (e.g., H_2O_2, Na_2O_2) where it is -1.
4. Metals generally have positive oxidation numbers.
5. In simple monoatomic ions such as Na^+, Zn^{2+}, Al^{3+}, Cl^-, and C^{4-}, the oxidation number is equal to the charge on the ion.
6. The algebraic sum of the oxidation numbers in a neutral molecule (e.g., $KMnO_4$, NaClO, H_2SO_4) is zero.
7. The algebraic sum of the oxidation numbers in a polyatomic ion (e.g., SO_4^{2-}, $Cr_2O_7^{2-}$) is equal to the charge on the ion.
8. Generally, in any compound or ion, the more electronegative atom is assigned the negative oxidation number while the less electronegative atom is assigned the positive oxidation number.

7.3.2. Oxidation numbers in formulas

The oxidation number (ON) of an atom in a compound or ion can be determined using the rules above. You will find the following steps helpful in determining the oxidation number of an element within a compound or a polyatomic ion.

1. Write known oxidation numbers below each atom in the formula.
2. Write an algebraic expression summing the product of the number of each atom and its oxidation number and equate it to zero for a compound or to the net charge on the ion in the case of a polyatomic ion.
3. Solve the equation to obtain the unknown oxidation number.

Example 7.1

What is the oxidation number of silicon (Si) in silicon dioxide (SiO_2)?

Solution

Step 1: ON of O = − 2
Required, ON of Si
Step 2: Set up the algebraic expression
$1 \times \text{ON of Si} + 2 \times \text{ON of O} = \text{charge on the compound} = 0$
Step 3: Solve for Si
$Si + 2\,(-2) = 0$
$Si - 4 = 0$
$Si = +4$

The oxidation number of silicon is + 4.

Example 7.2

Determine the oxidation number of K in $K_2Cr_2O_7$.

Solution

Step 1: ON of K = +1
ON of O = −2
ON of Cr = ?
Step 2: $2 \times \text{ON of K} + 2 \times \text{ON of Cr} + 7 \times \text{ON of O} = 0$
Step 3: $2 \times 1 + 2\,Cr + 7 \times (-2) = 0$
$2 + 2\,Cr - 14 = 0$
$2\,Cr - 12 = 0$
$Cr = +6$

The oxidation number of Cr is +6.

Example 7.3

Determine the oxidation number of Mn in MnO_4^-.

Solution

Step 1: ON of O = − 2
Required, ON of Mn
Step 2: Set up the algebraic expression
$1 \times \text{ON of Mn} + 4 \times \text{ON of O} = \text{charge on the ion}$
Step 3: Solve for Mn
$Mn + 4 \times (-2) = -1$

$$Mn - 8 = -1$$
$$Mn = +7$$

7.4. Writing the Formulas of Compounds

In writing the formula of a compound, one must have an accurate knowledge of the ionic charges of anions and cations. Tables 7-1 and 7-2 show the formulas and charges of some simple and polyatomic ions.

The following rules serve as general guides:

- Where the charge on the cation is not equal to the charge on the anion, use appropriate subscripts to balance the charges.
- The cation (positive ion) is usually written before the anion (negative ion).
- The sum of the oxidation numbers of all the atoms in a compound is equal to zero.
- The sum of the oxidation numbers of all the atoms in a polyatomic ion is equal to the charge on the ion.

Table 7-1 Some Simple Cations and Anions

Name	*Common name*	*Formula*	*Charge*	*Common name*	*Name*	*Formula*	*Charge*
Cations							
Cesium		Cs^{+}	1	Tin (II)	Stannous	Sn^{2+}	2
Copper (I)	Cuprous	Cu^{+}	1	Aluminum		Al^{3+}	3
Hydrogen		H^{+}	1	Bismuth (III)		Bi^{3+}	3
Lithium		Li^{+}	1	Chromium (III)	Chromic	Cr^{3+}	3
Mercury (I)	Mercurous	Hg^{+}	1	Cobalt (III)	Cobaltic	Co^{3+}	3
Potassium		K^{+}	1	Iron (III)	Ferric	Fe^{3-}	3
Rubidium		Rb^{-}	1	Manganese (IV)	Manganic	Mn^{4+}	4
Silver		Ag^{+}	1	Lead (IV)	Plumbic	Pb^{4-}	4
Sodium		Na^{-}	1	Tin (IV)	Stannic	Sn^{4-}	4
Barium		Ba^{2+}	2				
Calcium		Ca^{2+}	2	**Anions**			
Chromium (II)		Cr^{2-}	2	Bromide		Br^{-}	−1
Cobalt (II)		Co^{2+}	2	Chloride		Cl^{-}	−1
Copper (II)	Cupric	Cu^{2+}	2	Fluoride		F^{-}	−1
Iron (II)	Ferrous	Fe^{2+}	2	Hydride		H^{-}	−1
Lead (II)	Plumbous	Pb^{2+}	2	Iodide		I^{-}	−1
Magnesium		Mg^{2-}	2	Nitride		N^{3-}	−3
Manganese (II)	Manganous	Mn^{2-}	2	Oxide		O^{2-}	−2
Mercury (II)	Mercuric	Hg^{2-}	2	Phosphide		P^{3-}	−3
Nickel (II)		Ni^{2+}	2	Sulfide		S^{2-}	−2
Strontium		Sr^{2+}	2				

Table 7-2 Some Common Polyatomic Ions

Name of polyatomic ions	*Formula*	*Charge*	*Name of polyatomic ions*	*Formula*	*Charge*
Cations					
Ammonium	NH_4^+	1	Hydrogen sulfate	HSO_4^-	−1
Hydroxonium	H_3O^+	1	Hydrogen sulfite	HSO_3^-	−1
			Hydroxide	OH^-	−1
Anions			Hypochlorite	ClO^-	−1
Acetate	CH_3COO^-	−1	Hydrosulfide	HS^-	−1
Arsenate	AsO_4^{3-}	−3	Nitrate	NO_3^-	−1
Borate	BO_3^{3-}	−3	Nitrite	NO_2^-	−1
Bromate	BrO_3^-	−1	Oxalate	$C_2O_4^{2-}$	−2
Carbonate	CO_3^{2-}	−2	Perchlorate	ClO_4^-	−1
Chlorate	ClO_3^-	−1	Permaganate	MnO_4^-	−1
Chlorite	ClO_2^-	−1	Peroxide	O_2^{2-}	−2
Chromate	CrO_4^{2-}	−2	Phosphate	PO_4^{3-}	−3
Cyanide	CN^-	−1	Pyrophosphate	$P_2O_7^{4-}$	−4
Dichromate	$Cr_2O_7^{2-}$	−2	Silicate	SiO_3^{2-}	−2
Dihydrogen phosphate	$H_2PO_4^-$	−1	Sulfate	SO_4^{2-}	−2
Hydrogen carbonate	HCO_3^-	−1	Sulfite	SO_3^{2-}	−2
Hydrogen phosphate	HPO_4^{2-}	−2	Thiosulfate	$S_2O_3^{2-}$	−2

The following examples will illustrate how to write the formula of a compound:

Example 7.4

Write the correct formula for each of the following compounds. See tables 7-1 and 7-2 for charges and formulas of the various ions.

(a) Magnesium chloride
(b) Aluminum oxide
(c) Iron (III) sulfate
(d) Sodium phosphate
(e) Potassium chromium sulfate

Solution

(a) Magnesium chloride is composed of magnesium and chloride ions. The formulas of these ions are $Mg^{2+}$and Cl^-. Since Mg has a charge of +2, it will need two Cl ions to form a compound with a net charge of zero. So we write $Mg^{2+}Cl^-Cl^-$ or $Mg^{2+}(Cl^-)_2$. Since we do not write individual charges in chemical formulas, the correct formula is $MgCl_2$.

Charge balance:
$[Mg^{2+}] + 2[Cl^{-}] = 0$
$[+2] + 2[-1] = 2 - 2 = 0$
The correct formula is $MgCl_2$.

(b) The formulas for the cation and anion in aluminum oxide are Al^{3+} and O^{2-}.
Charge balance:
$2[Al^{3+}] + 3[O^{2-}] = 0$
$2[+3] + 3[-2] = 6 - 6 = 0$
The correct formula is Al_2O_3.

(c) The formulas for the cation and anion in iron (III) sulfate are Fe^{3+} and SO_4^{2-}.
Charge balance:
$2[Fe^{3+}]+3[SO_4^{2-}]= 0$
$2[+3] + 3[-2] = 6 - 6 = 0$
The correct formula of iron (III) sulfate is $Fe_2(SO_4)_3$.

(d) The formulas for the cation and anion in sodium phosphate are Na^+ and PO_3^{3-}.
Charge balance:
$3[Na^{-}]+1[PO_4^{3-}] = 0$
$3[+1] + 1[-3] = 3 - 3 = 0$
The correct formula of sodium phosphate is Na_3PO_4.

(e) The formulas for the cation and anion in potassium chromium sulfate are K^+, Cr^{3+}, and SO_4^{2-}.
Charge balance:
$[K^{-}]+1[Cr^{3+}]+2[SO_4^{2-}]= 0$
$[+1] + [+3] + 2[-2] = 1 + 3 - 4 = 0$
The correct formula of potassium chromium sulfate is $KCr(SO_4)_2$.

7.5. Nomenclature of Inorganic Compounds

Chemical compounds are named systematically according to the following rules.

(a) All binary compounds (those containing two elements) are named as ionic compounds, even though some of them may be covalent compounds. The name of the more electropositive (metallic) element is given first, followed by the name of the more electronegative (nonmetallic) element.
(b) Monatomic cations retain the name of the parent element while monatomic anions have the ending of the parent element changed to *-ide*. For example, NaCl is named as sodium chlor*ide*.
(c) When elements exhibiting variable valency or oxidation numbers combine to form more than one compound, differentiate the various compounds by using

Table 7-3 Compound Names Using Greek Prefixes

Name	*Formula*
Carbon monoxide	CO
Carbon dioxide	CO_2
Dinitrogen monoxide	N_2O
Nitrogen monoxide	NO
Nitrogen dioxide	NO_2
Dinitrogen trioxide	N_2O_3
Dinitrogen tetroxide	N_2O_4
Dinitrogen pentoxide	N_2O_5
Phosphorus trichloride	PCl_3
Phosphorus pentachloride	PCl_5

Greek prefixes such as mono- (one), di- (two), tri- (three), tetra- (four), penta- (five), hexa- (six), hepta- (seven), etc. Table 7-3 gives some examples.

(d) Binary compounds can also be named by using Roman numerals in parentheses to indicate the oxidation number of the more electropositive element, followed by the name of the more electronegative element with the ending *-ide*. Examples are shown in table 7-4. Exception to the *-ide* ending rule: A few nonbinary compounds are also named according to the *-ide* rule. These exceptions include compounds of ammonium NH_4^+ such as NH_4 I, ammonium iodide; those of hydroxide, OH^-, such as calcium hydroxide ($Ca(OH)_2$); cyanides such as KCN, potassium cyanide; and hydrosulfide, such as NaSH, sodium hydrosulfide.

(e) *Binary acids*: These acids consist of hydrogen and a nonmetal anion. To name a binary acid, add the prefix *hydro-* to the front of the name of the nonmetallic element, and add the suffix *–ic* after the nonmetal name. Then add the word *acid*. For example, HCl and H_2S are named as *hydro*-chlor-*ic acid* and *hydro*-sulfur-*ic acid*. Table 7-5 gives additional examples.

Table 7-4 Compound Names Using Roman Numerals

Name	*Formula*
Iron (II) chloride	$FeCl_2$
Iron (III) chloride	$FeCl_3$
Lead (II) oxide	PbO
Lead (IV) oxide	PbO_2
Nitrogen (II) oxide	NO
Nitrogen (IV) oxide	NO_2
Phosphorus (III) chloride	PCl_3
Phosphorus (V) chloride	PCl_5
Sulfur (IV) oxide	SO_2
Sulfur (VI) oxide	SO_3

Table 7-5 Names and Formulas of Some Binary Acids

Formula	*Acid name*	*Covalent compound name*
HF	Hydrofluoric acid	Hydrogen fluoride
HCl	Hydrochloric acid	Hydrogen chloride
HBr	Hydrobromic acid	Hydrogen bromide
HI	Hydroiodic acid	Hydrogen iodide
H_2S	Hydrosulfuric acid	Hydrogen sulfide
H_2Se	Hydroselenic acid	Hydrogen selenide

(f) *Ternary Compounds*: These contain three different elements and are usually made up of a cation and an anion. The anion is normally a polyatomic ion (a radical). Ternary compounds are named like binary compounds. The cationic group is named first, followed by the name of the anion. For example, $Mg(NO_3)_2$ is magnesium nitrate. Most polyatomic ions contain oxygen and usually have the suffix *–ate* or *–ite*. The *–ate* form normally indicates more oxygen or higher oxidation number than the *–ite* form. For example, the nitrogen atom in sodium nitrate ($NaNO_3$) is in a +5 oxidation state and has more oxygen than sodium nitrite ($NaNO_2$), in which the nitrogen atom is in a +3 oxidation state. It is very helpful to memorize the names of the various polyatomic ions.

(g) *Ternary acids*: These contain hydrogen and an oxygen-containing polyatomic ion known as an oxyanion. The name of a ternary acid is formed by adding the suffix *–ic* or *–ous acid* to the the root name of the anion. For anions ending with *–ate*, the *–ate* is replaced with *–ic acid*. For example, HNO_3 contains the nitrate (NO_3^-) anion and is named nitric acid. For anions ending with *–ite*, the *–ite* is replaced with *–ous acid*. For example, HNO_2 contains the nitrite (NO_2^-) anion and is named nitrous acid. Table 7-6 list examples of some ternary or oxyacids.

(h) Oxy-halo acids are named from the corresponding anion depending on the number of oxygen atoms present. Table 7-7 illustrates the rule for naming the oxy-halogen acids.

(i) *Ternary bases*: These consist of a metallic or polyatomic cation and a hydroxide ion. Bases have the ending *–ide*, like binary compounds. For example, NaOH is sodium hydroxide, $Al(OH)_3$ is aluminum hydroxide, NH_4OH is ammonium hydroxide, and $Ca(OH)_2$ is calcium hydroxide. See table 7-7 for further examples.

Example 7.5

Give systematic names for the following compounds:

(a) BaO
(b) MnO_2

(c) Fe_2O_3
(d) SO_3
(e) N_2O_5
(f) AlN
(g) Na_3P
(h) V_2O_5

Table 7-6 Names of Some Oxy-acids and Bases

Ternary acids		*Bases*	
Formula	*Name*	*Formula*	*Name*
$HClO$	Hypochlorous acid	$LiOH$	Lithium hydroxide
$HClO_2$	Chlorous acid	$NaOH$	Sodium hydroxide
$HClO_3$	Chloric acid	KOH	Potassium hydroxide
$HClO_4$	Perchloric acid	$Mg(OH)_2$	Magnesium hydroxide
HNO_3	Nitric acid	$Ca(OH)_2$	Calcium hydroxide
HNO_2	Nitrous acid	$Fe(OH)_2$	Iron (II) hydroxide
H_2SO_3	Sulfurous acid	$Fe(OH)_3$	Iron (III) hydroxide
H_2SO_4	Sulfuric acid	$Al(OH)_3$	Aluminum hydroxide
H_2CO_3	Carbonic acid	$Ba(OH)_2$	Barium hydroxide
H_3PO_3	Phosphorous acid	$Pb(OH)_2$	Lead hydroxide
H_3PO_4	Phosphoric acid	$Cu(OH)_2$	Copper (II) hydroxide
$HC_2H_3O_2$	Acetic (ethanoic) acid	$Cd(OH)_2$	Cadmium hydroxide
$H_2C_2O_4$	Oxalic acid	$Sr(OH)_2$	Strontium hydroxide
H_3BO_3	Boric acid	$Zn(OH)_2$	Zinc hydroxide

Table 7-7 Examples of Some Oxy-halo Acids

Formula	*No. of oxygen atom*	*Prefix*	*Suffix*	*Name of the oxyhalo-acids*
$HClO$	1	Hypo-	-ous	Hypochlorous acid
$HClO_2$	2		-ous	Chlorous acid
$HClO_3$	3		-ic	Chloric acid
$HClO_4$	4	Per-	-ic	Perchloric acid
$HBrO$	1	Hypo-	-ous	Hypobromous acid
$HBrO_2$	2		-ous	Bromous acid
$HBrO_3$	3		-ic	Bromic acid
$HBrO_4$	4	Per-	-ic	Perbromic acid
HIO	1	Hypo-	-ous	Hypoiodous acid
HIO_2	2		-ous	Iodous acid
HIO_3	3		-ic	Iodic acid
HIO_4	4	Per-	-ic	Periodic acid

Solution

(a) BaO barium oxide
(b) MnO_2 manganese (IV) oxide
(c) Fe_2O_3 iron (III) oxide
(d) SO_3 sulfur (VI) oxide (also known as sulfur trioxide)
(e) N_2O_5 nitrogen (V) oxide (dinitrogen tetroxide)
(f) AlN aluminum nitride
(g) Na_3P sodium phosphide
(h) V_2O_5 vanadium (V) oxide

Example 7.6

Give the systematic names of the following compounds:

(a) $Cu_3(PO_4)_2$
(b) Cs_2SO_3
(c) $KMnO_4$
(d) $Na_2Cr_2O_7$
(e) CaC_2O_4
(f) $Ca_3(BO_3)_2$
(g) $NaClO_4$
(h) $Hg_3(PO_4)_2$

Solution

(a) $Cu_3(PO_4)_2$ copper (II) phosphate
(b) Cs_2SO_3 cesium sulfite
(c) $KMnO_4$ potassium permanganate
(d) $Na_2Cr_2O_7$ sodium dichromate
(e) CaC_2O_4 calcium oxalate
(f) $Ca_3(BO_3)_2$ calcium borate
(g) $NaClO_4$ sodium perchlorate
(h) $Hg_3(PO_4)_2$ mercury (II) phosphate

7.6. Problems

1. Assign oxidation numbers to each element in the following:
   (a) CO_2 (b) SO_3 (c) SF_6 (d) N_2H_4 (e) PbO
2. Assign oxidation numbers to nitrogen (N) in the following:
   (a) NO (b) NO_2 (c) N_2O (d) N_2 (e) NH_3
3. Determine the oxidation numbers of the atoms in the following ions:
   (a) N^{3-} (b) Te^{2-} (c) PO_3^{3-} (d) SO_4^{2-} (e) ClO_3^-

4. Determine the oxidation numbers of the atoms in the following ions:
(a) MnO_4^- (b) $Cr_2O_7^{2-}$ (c) UO_2^{2+} (d) $S_2O_3^{2-}$ (e) $S_4O_6^{2-*}$
(*Hint: The average charge of the sulfur atoms in this ion may not be a whole number. This is uncommon but it does happen.)
5. What is the oxidation state of the metal present in each species?
(a) $Fe_2(SO_4)_3$ (b) Cu_2O (c) $NiSO_4$ (d) $Rh_2(CO_3)_3$ (e) Fe_3O_4
6. Assign oxidation numbers to the underlined elements in each of the following:
(a) $Na\underline{Au}Cl_4$ (b) $H_2\underline{S}O_3$ (c) $Mg(\underline{Cl}O_3)_2$ (d) $K_2\underline{Cr_2}O_7$ (e) $Li_4\underline{P_2}O_7$
7. Assign oxidation numbers to the underlined elements in each of the following ions:
(a)$\underline{Mn}O_4^2-$ (b)$\underline{Fe}(CN)_6^{3-}$ (c)$\underline{Cl}O_4^-$ (d) $H\underline{V}_{10}O_{28}^5-$ (e) $Li\underline{Al}H_4$
8. Write the formulas of the compounds that would be formed between the following pairs of elements:
(a) Mg and N (b) Sn and F (c) H and S (d) In and I (e) Al and Br
9. Write the formulas of the compounds that would be formed between the following pairs of elements:
(a) Li and N (b) B and O (c) Ca and O (d) Rb and Cl (e) Cs and S
10. What compounds would form between the following pairs of anions and cations?
(a) $Fe^{3+}$and CO_3^{2-} (b) $NH_4^+$and PO_4^{3-} (c) $Ca^{2+}$and NO_3^-
(d) Li and ClO_4^- (e) $K^+$and $Cr_2O_7^{2-}$
11. What are the formulas of the compounds formed between the following pairs of anions and cations?
(a) $Mn^{2+}$and CO_3^{2-} (b) Sn^{2+} and AsO_4^{3-} (c) $Ca^{2+}$and $C_2H_3O_2^-$
(d) Li and ClO_4^- (e) $Na^+$and BO_3^{3-}
12. Write the formulas of the following binary compounds:
(a) Sulfur dioxide (b) Carbon dioxide (c) Nitrogen dioxide
(d) Dinitrogen pentoxide (e) Carbon tetrachloride (f) Chlorine dioxide
(g) Lithium iodide (h) Selenium dioxide (i) Iron (II) chloride
(j) Barium phosphide
13. Write the formula of each of the following compounds:
(a) Vanadium (V) oxide (b) Copper (II) sulfide (c) Iron (III) sulfide
(d) Gallium nitride (e) Mercury (II) chloride
14. Write the formula for each compound:
(a) Sodium carbonate (b) Ammonium chloride (c) Potassium phosphate
(d) Calcium hydrogen phosphate (e) Iron (III) chromate
(f) Palladium (II) phosphate (g) Aluminum hydrogen carbonate
(h) Potassium dichromate (i) Iron (III) hydroxide (j) Magnesium borate
15. Write the names of the following binary compounds:
(a) PCl_3 (b) PCl_5 (c) CO (d) CO_2 (e) SO_2 (f) SO_3 (g) SiO_2
(h) P_2S_5 (i) N_2O_5
16. Name the following compounds:
(a) Ni_3N_2 (b) $FeCl_3$ (c) Al_2S_3 (d) GeS_2 (e)TiS_2 (f) CaH_2 (g) $HgCl_2$
(h) Hg_6P_2 (i) Cu_3N

17. Name the following acids:
    (a) HF (b) HBr (c) H_2Se (d) H_2S (e) HCN
18. Name the following oxygen-containing acids:
    (a) HNO_3 (b) $HBrO_4$ (c) HClO (d) $H_2C_2O_4$ (e) H_3PO_3 (f) HIO_3
19. Write the name of each salt and the name of the acid from which the salt may be derived:
    (a) $Ga(NO_3)_3$ (b) $CoSO_4$ (c) $Fe(C_2H_3O_2)_2$ (d) $Ca_3(BO_3)_2$ (e) PbC_2O_4
    (f) $CrPO_4$ (g) $NiCO_3$ (h) $Fe(CN)_3$ (i) AlI_3 (j) $RbBrO_3$
20. Write the name of each of the following inorganic bases:
    (a) KOH (b) NH_4OH (c) $Co(OH)_2$ (d) $Ba(OH)_2$ (e) $Cr(OH)_3$

8

Chemical Equations

8.1. Writing Chemical Equations

A chemical equation is a shorthand way of describing a chemical reaction. It uses symbols of elements and formulas of compounds in place of words to describe a chemical change or reaction.

General rules for writing chemical equations

1. The formulas and symbols of reactants are written on the left side of the equation.
2. The formulas and symbols of products are written on the right side of the equation.
3. A plus sign (+) is placed between different reactants and different products.
4. The reactants are separated from the products by an arrow ($\longrightarrow$) pointing in the direction of the reaction. For a reversible reaction, double arrow ($\rightleftarrows$) is used.
5. The physical states of substances may be indicated by the symbols (s) for solid, (l) for liquid, (aq) for aqueous solution, and (g) for gases.
6. The equation is then balanced by inserting appropriate coefficients for the products and reactants.

For the word equation

$$\text{Sulfur trioxide} + \text{Water} \longrightarrow \text{Sulfuric acid}$$

we can substitute the formulas for reactants and products, following the above rules, and obtain a chemical equation for the reaction as:

$$SO_3(g) + H_2O(l) \longrightarrow H_2SO_4(aq)$$

8.2. Balancing Chemical Equations

A chemical equation is balanced when it has the same number of atoms of each element on either side of the equation. Thus, a balanced equation obeys the law of conservation of mass. That is, atoms are not 'created' or 'destroyed' in writing a chemical equation.

Guidelines for balancing a chemical equation

1. Formulate and write a word equation if necessary from experiment or problem.
2. Write the unbalanced equation using the rules for writing chemical formulas.
3. Balance the equation to make sure the law of conservation of mass is observed.

   (a) Inspect both sides of the equation to identify atoms that need to be balanced.

   (b) Balance one element at a time by placing a suitable coefficient to the left of the formula containing the element. Note that a coefficient placed in front of the formula affects all the atoms in the formula. For example, 2 H_2O implies 2 molecules of water containing 4 atoms of hydrogen and 2 atoms of oxygen.

   (c) Never attempt to balance an equation by changing subscripts because this will change the formulas of the compounds. For example, to balance the equation

$$H_2(g) + Cl_2(g) \longrightarrow HCl(g)$$

   you cannot change the subscripts:

$$H_2(g) + Cl_2(g) \longrightarrow HCl_2(g)$$

   since HCl_2 is not the same as the compound HCl.

   (d) Make sure the elements already balanced aren't changed (if possible) since we are essentially doing this one element at a time.

   (e) Balance polyatomic ions (e.g., NH_4^- and CO_3^{2-}) as single entities if they appear unchanged on both sides of the equation.

   (f) The balanced equation should contain the smallest possible set of whole number coefficients. For example, a balanced equation showing the decomposition of copper (II) oxide should be expressed as:

$$2\ CuO(s) \longrightarrow 2\ Cu(s) + O_2(g)$$

   And not as:

$$4\ CuO(s) \longrightarrow 4\ Cu(s) + 2\ O_2(g)$$

   (g) Do a final check by counting atoms on each side of the equation.

You may find some equations difficult to balance using these rules, particularly redox equations. These will be treated under Oxidation and Reduction Reactions, in chapter 22.

Example 8.1

Balance the equation

$$H_2 + O_2 \longrightarrow H_2O \quad \text{(Unbalanced)}$$

Solution

Hydrogen is balanced, since there are two atoms of hydrogen on each side of the equation. Oxygen is not balanced. There are 2 atoms on the left-hand side and 1 atom on the right-hand side. To balance oxygen, place the coefficient 2 in front of H_2O.

$$H_2 + O_2 \longrightarrow 2\,H_2O \quad \text{(Unbalanced)}$$

Oxygen is now balanced but hydrogen is no longer balanced. To balance hydrogen, place the coefficient 2 in front of H_2.

$$2\,H_2 + O_2 \longrightarrow 2\,H_2O \quad \text{(Balanced)}$$

There are 4 atoms of hydrogen and 2 atoms of oxygen on either side of the equation. The equation is now balanced.

Example 8.2

Sodium metal reacts with water to form sodium hydroxide and hydrogen. Write a balanced chemical equation for this reaction.

Solution

$$\text{Sodium} + \text{Water} \longrightarrow \text{Sodium hydroxide} + \text{hydrogen}$$

$$Na + H_2O \longrightarrow NaOH + H_2 \quad \text{(Unbalanced)}$$

By inspection, Na and O are balanced on both sides of the equation, but hydrogen is not. Balance hydrogen by placing the coefficient 2 in front of H_2O and NaOH.

$$Na + 2\,H_2O \longrightarrow 2\,NaOH + H_2 \quad \text{(Unbalanced)}$$

All the atoms except Na are balanced. Balance Na by placing the coefficient 2 in front of it.

$$2\,Na + 2\,H_2O \longrightarrow 2\,NaOH + H_2 \quad \text{(Balanced)}$$

Final check: There are 4 atoms of H, 2 atoms of O, and 2 atoms of Na on either side of the equation. The equation is balanced.

Example 8.3

Balance the following chemical equations:

A. $Fe + O_2 \longrightarrow Fe_2O_3$
B. $Fe_2O_3 + CO \longrightarrow Fe + CO_2$
C. $H_3PO_4 + Ca(OH)_2 \longrightarrow Ca_3(PO_4)_3 + H_2O$
D. $C_2H_3Cl + O_2 \longrightarrow CO_2 + H_2O + HCl$

Solution

A. $4\ Fe + 3\ O_2 \longrightarrow 2\ Fe_2O_3$
B. $Fe_2O_3 + 3\ CO \longrightarrow 2\ Fe + 3\ CO_2$
C. $2\ H_3PO_4 + 3\ Ca(OH)_2 \longrightarrow Ca_3(PO_4)_2 + 6\ H_2O$
D. $2\ C_2H_3Cl + 5\ O_2 \longrightarrow 4\ CO_2 + 2\ H_2O + 2\ HCl$

8.3. Types of Chemical Reactions

There are numerous types of chemical reactions, which may be classified in several ways. Most chemical reactions conform to one of the following:

- Combination or synthesis
- Decomposition
- Displacement
- Double decomposition or metathesis
- Neutralization
- Oxidation–reduction (dealt with in chapter 22)

This classification is very helpful when writing chemical equations and assists in predicting reaction products.

8.3.1. Combination or synthesis

A combination reaction is one in which two or more reactants (elements or compounds) combine to form a single product. The general equation for a combination is:

$$A + B \longrightarrow AB$$

Some examples involving elements include:

$$H_2 + Br_2 \longrightarrow 2\ HBr$$

$$S + O_2 \longrightarrow SO_2$$

$$2\ Na + S \longrightarrow Na_2S$$

$$4\ Fe + 3\ O_2 \longrightarrow 2\ Fe_2O_3$$

Some examples involving compounds are:

$$2\ NO_2 + H_2O_2 \longrightarrow 2\ HNO_3$$

$$Na_2O + H_2O \longrightarrow 2\ NaOH$$

$$SO_3 + H_2O \longrightarrow H_2SO_4$$

8.3.2. Decomposition

A decomposition reaction is one in which a single reactant is broken down into two or more simpler substances, usually under the action of heat. The general equation for a decomposition reaction is:

$$AB \longrightarrow A + B$$

Representative examples of decomposition reactions include:

$$2\ NaClO_3 \xrightarrow{\Delta} 2\ NaCl + 3\ O_2$$

$$CaCO_3 \xrightarrow{\Delta} CaO + CO_2$$

$$Mg(OH)_2 \xrightarrow{\Delta} MgO + H_2O$$

$$2\ CuO \xrightarrow{\Delta} 2\ Cu + O_2$$

8.3.3. Displacement reactions

A displacement reaction is one in which a free element replaces another element within a compound. The general equation for a displacement replacement is:

$$A + BC \longrightarrow B + AC$$

Representative examples involving metals and nonmetals include:

$$Mg + H_2SO_4 \longrightarrow H_2 + MgSO_4$$

$$Cl_2 + 2\ HBr \longrightarrow Br_2 + 2\ HCl$$

$$Fe + CuCl_2 \longrightarrow FeCl_2 + Cu$$

8.3.4. Double decomposition or metathesis reactions

A double decomposition (or metathesis) reaction is one in which two compounds exchange ions to form two new compounds. The general equation for a double decomposition reaction is:

$$A^+B^- + C^+D^- \longrightarrow A^+D^- + C^+B^-$$

The charges shown here are for clarity. In practice they wouldn't be indicated.

Examples include:

$$AgNO_3 + NaCl \longrightarrow AgCl + NaNO_3$$

$$Ba(NO_3)_2 + Na_2SO_4 \longrightarrow BaSO_4 + 2NaNO_3$$

$$Pb(NO_3)_2 + H_2S \longrightarrow PbS + 2HNO_3$$

8.3.5. Neutralization reactions

A neutralization reaction is one in which an acid reacts with a base to form a salt and water as the only products. The general expression is:

$$\text{Acid} + \text{Base} \longrightarrow \text{Salt} + \text{Water}$$

Some examples include:

$$H_2SO_4 + 2KOH \longrightarrow K_2SO_4 + H_2O$$

$$HCl + NaOH \longrightarrow NaCl + H_2O$$

This is a special form of double displacement reaction. Acid–base reactions will be discussed in chapters 14 and 18.

8.4. Problems

1. Balance the following equations:

   (a) $SO_2 + H_2O \longrightarrow H_2SO_3$
   (b) $H_2O \longrightarrow H_2 + O_2$
   (c) $PCl_3 + Cl_2 \longrightarrow PCl_5$
   (d) $C + O_2 \longrightarrow CO$
   (e) $Mg + O_2 \longrightarrow MgO$

2. Balance the following equations:

   (a) $NH_3 + O_2 \longrightarrow N_2 + H_2O$
   (b) $I_4O_9 \xrightarrow{\Delta} I_2O_5 + I_2 + O_2$
   (c) $K_2S_2O_3 + Cl_2 + H_2O \xrightarrow{\Delta} KHSO_4 + HCl$
   (d) $CS_2 + O_2 \longrightarrow N_2 + SO_2$

3. Write and balance equations for the following:

   (a) Calcium carbonate $\xrightarrow{\Delta}$ Calcium oxide + Carbon dioxide
   (b) Potassium chlorate ($KClO_3$) $\xrightarrow{\Delta}$ Potassium chloride + Oxygen
   (c) Lithium nitride (Li_3N) $\xrightarrow{\Delta}$ Lithium + Nitrogen
   (d) Hydrogen peroxide $\xrightarrow{\Delta}$ Hydrogen + Oxygen

4. The following reactions represent the solid-state synthesis of technologically important materials. Balance the equations:

   (a) $TiCl_4 + H_2S \longrightarrow TiS_2 + HCl$
   (b) $LaCl_3{\cdot}7\,H_2O \longrightarrow LaOCl + HCl + H_2O$
   (c) $CsCl + ScCl_3 \longrightarrow Cs_3Sc_2Cl_9$
   (d) $Li_2CO_3 + Fe_2O_3 \longrightarrow LiFe_5O_8 + CO_2$
   (e) $La_2O_3 + B_2O_3 \longrightarrow LaB_6 + O_2$
   (f) $ScCl_3 + SiO_2 \rightleftarrows Sc_2Si_2O_7 + SiCl_4$

5. Balance the following thermal decomposition equations:

   (a) $NaN_3 \longrightarrow Na + N_2$
   (b) $(NH_4)_2Cr_2O_7 \longrightarrow Cr_2O_3 + N_2 + H_2O$
   (c) $Ag_2CO_3 \longrightarrow Ag + CO_2 + O_2$
   (d) $NaHCO_3 \longrightarrow Na_2CO_3 + CO_2 + H_2O$
   (e) $Al_2(CO_3)_3 \longrightarrow Al_2O_3 + CO_2$

6. Complete and balance the following neutralization reactions:

   (a) $H_2SO_4 + KOH \longrightarrow$
   (b) $H_3PO_4 + Ba(OH)_2 \longrightarrow$
   (c) $HBr + Sn(OH)_4 \longrightarrow$
   (d) $HBr + NaOH \longrightarrow$
   (e) $H_4P_2O_7 + NaOH \longrightarrow$

7. Complete and balance the following double decomposition reactions:

   (a) $Ca(NO_3)_2 + Na_2SO_4 \longrightarrow$
   (b) $Mg(NO_3)_2 + Li_2S \longrightarrow$
   (c) $Pb(NO_3)_2 + Cs_2CrO_4 \longrightarrow$
   (d) $FeCl_3 + CaS \longrightarrow$
   (e) $Na_3PO_4 + Zn(OH)_2 \longrightarrow$

8. The industrial production of phosphorus is represented by the following reaction:

$$Ca_3(PO_4)_2 + SiO_2 + C \longrightarrow P_4 + CaSiO_3 + CO_2$$

   Balance this equation.

9. Hydrocarbons and their oxygen derivatives burn in oxygen to produce carbon dioxide and water. The relative amounts of carbon dioxide and water formed depend on the composition of the parent hydrocarbon. Complete and balance the following equations:

   (a) $C_4H_8 + O_2 \longrightarrow$
   (b) $C_6H_6 + O_2 \longrightarrow$
   (c) $C_6H_{12}O + O_2 \longrightarrow$
   (d) $C_{12}H_{22}O_{11} + O_2 \longrightarrow$
   (e) $C_4H_{10} + O_2 \longrightarrow$

10. Write balanced chemical equations for the following reactions:

    (a) Iron and sulfur react at elevated temperature to form iron (III) sulfide.
    (b) Octane, C_8H_{18}, a component of gasoline, burns in oxygen to produce carbon dioxide and water.
    (c) Sulfur dioxide burns in oxygen to form sulfur trioxide.
    (d) Barium hydroxide reacts with phosphoric acid to produce barium phosphate and water.
    (e) Solid aluminum dissolves in hydrochloric acid to form aluminum chloride and hydrogen gas.

11. Write balanced chemical reactions for the following:

    (a) Aluminum reacts with chlorine to form aluminum chloride.
    (b) Potassium reacts with water to form potassium hydroxide and hydrogen gas.
    (c) Boron trichloride gas reacts with steam (gaseous H_2O) to form boron trihydroxide and hydrogen chloride.
    (d) Silver nitrate solution reacts with sodium chloride to give sodium nitrate and precipitate of silver chloride.
    (e) Silicon tetrafluoride reacts with sodium hydroxide to form sodium silicate (Na_4SiO_4), sodium fluoride, and water.
    (f) Calcium hydroxide reacts with ammonium chloride to form ammonia, calcium chloride, and water.

12. Balance the following reactions:

    (a) $Si_4H_{10} + O_2 \longrightarrow SiO_2 + H_2O$
    (b) $CH_3NO_2 + Cl_2 \longrightarrow CCl_3NO_2 + HCl$
    (c) $NaF + CaO + H_2O \longrightarrow CaF_2 + NaOH$
    (d) $Al_4C_3 + H_2O \longrightarrow Al(OH)_4 + C_2H_2$
    (e) $TiO_2 + B_4C + C \longrightarrow TiB_2 + CO$
    (f) $C_7H_{16}O_4S_2 + O_2 \longrightarrow CO_2 + H_2O + SO_2$

9

Stoichiometry

9.1. Reaction Stoichiometry

Stoichiometry is the study of the quantities of reactants and products involved in a chemical reaction.

9.2. Information from a Balanced Equation

In addition to identifying the reactants and products in a chemical reaction, a balanced equation gives useful information that is helpful in calculations. Consider the equation for the reaction between ammonia and oxygen to produce nitrogen (II) oxide:

$$4\ NH_3(g) + 5\ O_2(g) \longrightarrow 4\ NO\ (g) + 6\ H_2O\ (g)$$

The following information can be obtained:

1. Molecules of reactant and products: 4 molecules of NH_3 react with 5 molecules of O_2 to form 4 molecules of NO and 6 molecules of H_2O.
2. Moles of reactants and products: 4 mol of NH_3 react with 5 mol of O_2 to produce 4 mol of NO and 6 mol of H_2O.
3. Mass of reactants and products: 68 g of NH_3 (4 mol) react with 160 g of O_2 (5 mol) to produce 120 g of NO (4 mol) and 108 g of H_2O (6 mol).
4. Volumes of gases: 4 volumes of NH_3 react with 5 volumes of O_2 to produce 4 volumes of NO and 6 volumes of H_2O at the same temperature and pressure (by Avogadro's law, which will be discussed in detail in section 11.5 of chapter 11).

9.3. Types of Stoichiometric Problems

There are several types of stoichiometric problems. The common types include:

1. Mole – Mole
2. Mass – Mass

3. Mass – Mole (or Mole – Mass)
4. Mole – Volume (or Volume to Mole)
5. Mass – Volume (or Volume to Mass)
6. Volume – Volume

9.3.1. Solving stoichiometric problems

The following general steps can be used to solve many stoichiometric problems:

1. Write the balanced chemical equation for the reaction.
2. Organize your data; determine which quantities you know and which ones you need to find.
3. Write the mole relationship between the given substance (a reactant or a product) and the required substance (a reactant or a product).
4. Calculate molar masses and convert masses, molecules, or volumes of the known substance to moles.
5. Use stoichiometric coefficients or conversion factors (mole ratios) from the equation to determine the moles of the unknown substance.
6. Convert moles of the unknown substance to the desired mass, molecules, or volume.

Figure 9-1 is a summary of the various conversion processes outlined above.

9.3.2. Mole-to-mole stoichiometric problems

In this type of problem, you are given the moles of one component and need to find the moles of another.

Example 9.1

The balanced equation for the Haber process used for the industrial production of ammonia is:

$$N_2(g) + 3\,H_2(g) \longrightarrow 2\,NH_3(g)$$

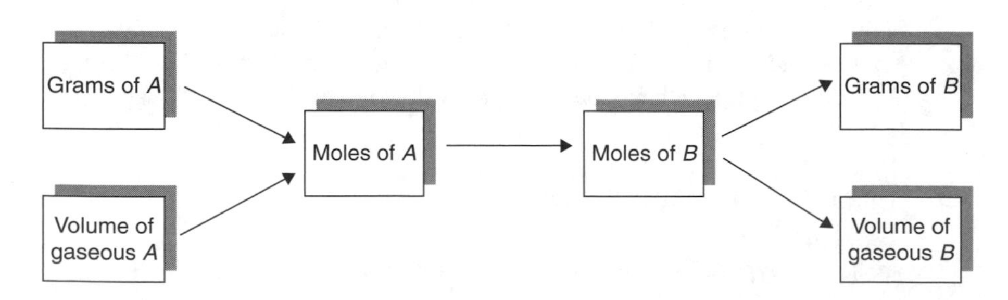

Figure 9-1 A block diagram summarizing the conversion processes used in solving stoichiometric problems.

Calculate the exact moles of nitrogen required to produce 10 mol of ammonia.

Solution

Step 1: We have a balanced equation.
Step 2: Organize your information:
Known: 10 mol of NH_3
Unknown = moles of N_2
Step 3: Write the mole relationship between NH_3 and N_2 and formulate the conversion factor.
1 mol of N_2 corresponds to 2 mol of NH_3.
The conversion factors are:

$$\frac{1 \text{ mol } N_2}{2 \text{ mol } NH_3} \quad \text{and} \quad \frac{2 \text{ mol } NH_3}{1 \text{ mol } N_2}$$

Step 4: Select the conversion factor from step 3 that will cancel the units of the known (mol of NH_3) when multiplied by the given amount of the known (i.c., 10 mol NH_3). Usc this convcrsion factor to calculatc thc moles of N_2:

$$\text{moles } N_2 = (10 \text{ mol } NH_3) \times \left(\frac{1 \text{ mol } N_2}{2 \text{ mol } NH_3}\right) = 5 \text{ mol } N_2$$

Example 9.2

Iron pyrite (FeS_2) burns in air according to the equation:

$$4\ FeS_2 + 11\ O_2 \longrightarrow 2\ Fe_2O_3 + 8\ SO_2$$

How many moles of iron (III) oxide (Fe_2O_3) are obtained from the combustion of 5 mol of iron pyrite?

Solution

Step 1: We have a balanced equation.
Step 2: Organize your information:
Known: 5 mol of FeS_2
Unknown = moles Fe_2O_3
Step 3: Write the mole relationship between FeS_2 and Fe_2O_3 and formulate the conversion factor.
4 mol of FeS_2 corresponds to 2 mol of Fe_2O_3
The conversion factors are:

$$\frac{2 \text{ mol } Fe_2O_3}{4 \text{ mol } FeS_2} \quad \text{and} \quad \frac{4 \text{ mol } FeS_2}{2 \text{ mol } Fe_2O_3}$$

Step 4: Select the conversion factor from step 3 that will cancel the units of the known (mol of FeS_2) when multiplied by the given amount of the known (i.e., 5 mol FeS_2). Use this conversion factor to calculate the moles of Fe_2O_3:

$$\text{mol } Fe_2O_3 = (5 \text{ mol } FeS_2) \times \left(\frac{2 \text{ mol } Fe_2O_3}{4 \text{ mol } FeS_2}\right) = 2.5 \text{ mol } Fe_2O_3$$

9.3.3. Mass-to-mole stoichiometry problems

Here, information about the known quantities is given in mass units, and information about the unknown is required in mole units, or vice-versa.

Example 9.3

The combustion of acetylene gas is represented by the equation:

$$2\ C_2H_2(g) + 5\ O_2(g) \longrightarrow 4\ CO_2(g) + 2\ H_2O(g)$$

Calculate the moles of oxygen (O_2) required for the complete combustion of 25 g of acetylene, (C_2H_2).

Solution

Step 1: We have an equation—in this case, already balanced.

Step 2: Organize your information and calculate molar masses of C_2H_2 and O_2:
$C_2H_2 = 26$ g/mol
$O_2 = 32$ g/mol
Known: 25 g of C_2H_2
Unknown = moles O_2

Step 3: Write the mole relationship between $C_2H_2$and O_2 and formulate the conversion factor.
2 mol of $C_2\,H_2$ ⇒ 5 mol of O_2 or
52 g of $C_2\,H_2$ ⇒ 5 mol of O_2
The conversion factors are:

$$\frac{52 \text{ g } C_2H_2}{5 \text{ mol } O_2} \quad \text{and} \quad \frac{5 \text{ mol } O_2}{52 \text{ g } C_2H_2}$$

Step 4: Select the conversion factor from step 3 that will cancel the units of the known (g of C_2H_2) when multiplied by the given amount of the known (i.e., 25 g C_2H_2). Use this conversion factor to calculate the moles of O_2:

$$\text{moles } O_2 = 25 \text{ g } C_2H_2 \times \left(\frac{5 \text{ mol } O_2}{52 \text{ g } C_2H_2}\right) = 2.404 \text{ mol } O_2$$

9.3.4. Mass-to-mass stoichiometry problems

Here information about the known quantities is given in mass units, and information about the unknown is required in mass units, or vice-versa.

Example 9.4

Pure iron can be produced by reacting hematite ore, Fe_2O_3, with carbon monoxide, CO. Write a balanced equation for the reaction. How many grams of carbon monoxide gas are needed to react with 500 g of hematite?

Solution

Step 1: Write a balanced equation.

$$Fe_2O_3 + 3\,CO \longrightarrow 2\,Fe + 3\,CO_2$$

The sequence of conversions needed for this calculation is:

$$\text{grams } Fe_2O_3 \longrightarrow \text{moles } Fe_2O_3 \longrightarrow \text{moles CO} \longrightarrow \text{grams CO}$$

Step 2: Organize your information and calculate molar masses of Fe_2O_3 and CO:
$Fe_2O_3 = 159.7$ g/mol
$CO = 26$ g/mol
Known: 500 g of Fe_2O_3
Unknown = grams CO

Step 3: Write the mole relationship between Fe_2O_3 and CO and formulate the conversion factors.
1 mol of $Fe_2\,O_3 \Rightarrow$ 3 mol of CO
159.7 g $Fe_2\,O_3 \Rightarrow$ 1 mol $Fe_2\,O_3$
28 g CO $\Rightarrow$ 1 mol CO
The conversion factors are:

$$\frac{1\text{ mol }Fe_2O_3}{159.7\text{ g }Fe_2O_3},\ \frac{1\text{ mol }Fe_2O_3}{3\text{ mol CO}},\ \frac{3\text{ mol CO}}{1\text{ mol }Fe_2O_3},\ \text{and}\ \frac{1\text{ mol CO}}{28\text{ g CO}}$$

Step 4: Select the conversion factors from step 3 that will cancel the units of the known (g and moles of Fe_2O_3) when multiplied by the given amount of the known (i.e., 500 g Fe_2O_3). Use these conversion factors to calculate the mass of CO_2.

$$\text{g CO} = 500\text{ g }Fe_2O_3 \times \frac{1\text{ mol }Fe_2O_3}{159.7\text{g }Fe_2O_3} \times \frac{3\text{ mol CO}}{1\text{ mol }Fe_2O_3} \times \frac{28\text{ g CO}}{1\text{ mol CO}}$$

$$= 262.99\text{ g CO} \approx 263.0\text{ g CO}$$

9.3.5. Mass-to-volume stoichiometry problems

Mass–volume stoichiometric calculations involve reactions in which one or more of the substances is a gas. Here the information about one component is given in moles or grams, and information about the unknown is required in terms of volume of gas at a given temperature and pressure (or vice-versa). The conversion of moles or grams of a gaseous reactant or product to volume at standard conditions can be accomplished by remembering that 1 mol of a gas at standard temperature and pressure (STP) occupies 22.4 L. When nonstandard conditions of temperature and pressure are imposed, the required conversion can be accomplished by the ideal gas equation (as discussed in chapter 11).

Example 9.5

In the laboratory, oxygen can be prepared by heating potassium chlorate. The equation for the reaction is:

$$2\ KClO_3(s) \xrightarrow{\text{heat}} 2\ KCl(s) + 3\ O_2(g)$$

Calculate the volume of oxygen produced at STP by heating 20 g of $KClO_3$.

Solution

Step 1: The balanced equation is known.
Step 2: The sequence of conversions needed for this calculation is:

$$\text{grams } KClO_3 \longrightarrow \text{moles } KClO_3 \longrightarrow \text{moles } O_2 \longrightarrow \text{volume } O_2$$

Calculate molar mass of $KClO_3$:
$KClO_3 = 122$ g/mol
Known: 500 g of Fe_2O_3
Unknown = grams CO

Step 3: Write the mole relationship between $KClO_3$ and O_2 and formulate the conversion factors.
2 mol of $KClO_3$ $\Rightarrow$ 3 mol of O_2
The conversion factors are:

$$\frac{1\ \text{mol}\ KClO_3}{122\ \text{g}\ KClO_3}, \frac{2\ \text{mol}\ KClO_3}{3\ \text{mol}\ O_2}, \frac{3\ \text{mol}\ O_2}{2\ \text{mol}\ KClO_3}, \text{ and } \frac{1\ \text{mol}\ O_2}{22.4\ \text{dm}^3\ (\text{or L})\ O_2}$$

Step 4: Select the conversion factors from step 3 that will cancel the units of the known (g and mol of $KClO_3$) when multiplied by the given amount of the known (i.e., 20 g $KClO_3$). Use these conversion factors to calculate the volume of O_2 as follows:

$$\text{volume } O_2 = 20\text{ g } KClO_3 \times \frac{1\text{ mol } KClO_3}{122\text{ g } KClO_3} \times \frac{3\text{ mol } O_2}{2\text{ mol } KClO_3} \times \frac{22.4\text{ dm}^3 O_2}{1\text{ mol } O_2}$$

$$= 5.51\text{ dm}^3 O_2$$

Example 9.6

What volume of nitrogen at STP will be released by the thermal decomposition of 40 g of ammonium nitrite?

Solution

Step 1: Write the balanced chemical equation:

$$NH_4NO_2(aq) \xrightarrow{\text{heat}} N_2(g) + 2\, H_2O(g)$$

Step 2: Organize information and calculate molar mass of NH_4NO_2:
$NH_4NO_2 = 64$ g/mol
Known: 40 g of NH_4NO_2
Unknown = volume of N_2 at 700 mmHg and 30°C.

Step 3: Write the mole relationship between NH_4, NO_2, and N_2, and formulate the conversion factors.

$$1\text{ mol } NH_4NO_2 \Leftrightarrow 1\text{ mol } N_2 \Leftrightarrow 22.4\text{ L } N_2 \text{ at STP}$$

$$\frac{1\text{ mol } N_2}{1\text{ mol } NH_4NO_2}, \frac{1\text{ mol } NH_4NO_2}{64\text{ g } NH_4NO_2}, \frac{22.4\text{ L } N_2}{1\text{ mol } N_2}$$

Step 4: Calculate the volume at STP:

$$\text{volume } N_2 = 40\text{ g } NH_4NO_2 \times \frac{1\text{ mol } NH_4NO_2}{64\text{ g } NH_4NO_2} \times \frac{1\text{ mol } N_2}{1\text{ mol } NH_4NO_2}$$

$$\times \frac{22.4\text{ L } N_2}{1\text{ mol } N_2} = 14.00\text{ L}$$

9.3.6. Volume-to-volume stoichiometry problems

Volume–volume stoichiometry problems are based on the law of combining volumes, also known as Gay-Lussac's law, which states that at the same temperature and pressure, when gases react or are formed, they do so in simple whole-number ratios of volume. This volume ratio is directly proportional to the values of the corresponding stoichiometric coefficients in the balanced equation. Hence, we can use the stoichiometric coefficients in the balanced equation to form volume relationships as in other types of stoichiometry problems. For example, consider the following gaseous reaction, in which all substances are at the same temperature and pressure.

$$N_2(g) + 3\,H_2(g) \longrightarrow 2\,NH_3(g)$$

In terms of reacting moles, 1 mol of nitrogen gas reacts with 3 mol of hydrogen gas to form 2 mol of ammonia gas. Now, in terms of reacting volumes, the equation indicates that 1 volume of nitrogen gas reacts with 3 volumes of hydrogen gas to form 2 volumes of ammonia gas. If the volumes of reactants and products were measured at STP, then we would see 22.4 L of nitrogen reacting with 67.2 L (3 × 22.4 L) of hydrogen to form 44.8 L (2 × 22.4 L) of ammonia. The ratio remains the same as in the balanced equation, i.e., $N_2 : H_2 : NH_3$ equals 22.4 L : 67.2 L : 44.8 L (or simply 1 : 3 : 2).

Example 9.7

Calculate the volume in liters at STP of oxygen gas needed, and the volume of nitrogen (II) oxide gas and water vapor produced, from the reaction of 1.12 L of ammonia gas. The equation for the reaction is:

$$4\,NH_3(g) + 5\,O_2(g) \longrightarrow 4\,NO(g) + 6\,H_2O(g)$$

Solution

All the reactants and products are gases at the same temperature and pressure. Therefore, their volumes are directly proportional to their stoichiometric coefficients in the balanced equation.

Step 1: The balanced equation is known.
Step 2: Organize the information given:
Known: 1.12 L of NH_3
Unknown = liters of O_2, NO, and H_2O
Step 3: Write the mole and volume relationship between NH_3 and each of the unknowns (O_2, NO, and H_2O)and formulate the conversion factors.

$$4 \text{ mol } NH_3 \quad \Rightarrow \quad 5 \text{ mol } O_2 \quad \Rightarrow \quad 4 \text{ mol NO} \quad \Rightarrow \quad 6 \text{ mol } H_2O$$

or

$$4\text{ L } NH_3 \quad \Rightarrow \quad 5\text{ L } O_2 \quad \Rightarrow \quad 4\text{ L NO} \quad \Rightarrow \quad 6\text{ L } H_2O$$

This corresponds to the following conversion factors in volumes:

$$\frac{5\text{ L } O_2}{4\text{ L } NH_3};\ \frac{4\text{ L NO}}{4\text{ L } NH_3};\ \frac{6\text{ L } H_2O}{4\text{ L } NH_3}$$

Step 4: Use the conversion factors to calculate the volumes of the unknowns:

$$\text{volume } O_2\text{: } 1.12\ \text{L}\ NH_3 \times \frac{5\ \text{L}\ O_2}{4\ \text{L}\ NH_3} = 1.4\ \text{L}\ O_2$$

$$\text{volume NO: } 1.12\ \text{L}\ NH_3 \times \frac{4\ \text{L NO}}{4\ \text{L}\ NH_3} = 1.12\ \text{L NO}$$

$$\text{volume } H_2O\text{: } 1.12\ \text{L}\ NH_3 \times \frac{6\ \text{L}\ H_2O}{4\ \text{L}\ NH_3} = 1.68\ \text{L}\ H_2O$$

9.4. Limiting Reagents

When chemical reactions are carried out in the laboratory, we do not add reactants in the exact molar ratios indicated by the balanced equation. For various reasons, one or more reactants is usually present in large excess. In most cases, only one reactant is completely consumed at the end of the reaction, and this reactant determines the amount of products that can be formed. This is called the *limiting reagent* or *reactant*. Once the limiting reagent is entirely consumed, the reaction stops; the other excess reactants will still be present in some amount.

9.4.1. Limiting reagent calculations

For chemical reactions involving two or more reactants, it is necessary to determine which one is the limiting reagent. The following procedure is helpful in determining the limiting reagent:

1. Make sure the chemical equation is balanced.
2. Calculate the number of moles of each of the reactants present from their given amount.
3. Calculate the amount of product that can be formed from the complete reaction of each reactant.
4. Determine which of the reactants would produce the least amount of the product. This is the limiting reagent.

Example 9.8

A mixture of 50 g of CaO and 50 g of H_2O were reacted in an autoclave to form calcium hydroxide according to the equation.

$$CaO + H_2O \longrightarrow Ca(OH)_2$$

What is the limiting reagent?

Solution

1. We have a balanced equation.
   $CaO + H_2O \longrightarrow Ca(OH)_2$
   50 g   50 g   ? g
2. Determine the number of moles of each reactant:
   Moles of $CaO = 50\ g \times \frac{1\ mol\ CaO}{56\ g} = 0.893\ mol$
   Moles of $H_2O = 50\ g \times \frac{1\ mol\ CaO}{18\ g} = 2.778\ mol$
3. Calculate the moles of $Ca(OH)_2$ formed from each reactant:
   Moles of $Ca(OH)_2$produced from 0.893 mol CaO supplied
   Moles $Ca(OH)_2 = 0.893\ mol\ CaO \times \frac{1\ mol\ Ca(OH)_2}{1\ mol\ CaO} = 0.893\ mol$
   Moles of $Ca(OH)_2$produced from 2.778 mol H_2O supplied
   Moles $Ca(OH)_2 = 2.778\ mol\ H_2O \times \frac{1\ mol\ Ca(OH)_2}{1\ mol\ H_2O} = 2.778\ mol$
4. Now determine the limiting reagent. Since CaO produces the least amount of $Ca(OH)_2$, it is the limiting reagent.

9.5. Reaction Yields: Theoretical, Actual, and Percent Yields

In laboratory reactions, the amount of product isolated is always less than the amount predicted from the balanced equation. There are several reasons for this. For example, some of the products may be lost during isolation and purification; some undesired products may be formed due to side reactions, or the reactant may not undergo complete reaction. The *theoretical yield* is the maximum amount of a product that can be formed, as calculated from a chemical equation representing the reaction. The *actual yield* of a product is the amount actually formed when the experiment is performed. The *percent yield* is the ratio of the actual yield to the theoretical yield multiplied by 100. That is:

$$\text{Percent yield } (\%) = \frac{\text{Actual yield}}{\text{Theoretical yield}} \times 100$$

Example 9.9

Phosphoric acid is formed from the combustion of 70 g of PH_3 in excess oxygen according to the following equation:

$$PH_3(g) + 2\ O_2(g) \longrightarrow H_3PO_4(s)$$

If 188 g of phosphoric acid were produced in the reaction, calculate the theoretical yield and the percent yield.

Solution

1. We have a balanced equation.
2. Calculate the number of moles of PH_3:

$$\text{moles } PH_3 = 70 \text{ g} \times \frac{1 \text{ mol } PH_3}{33.97 \text{ g } PH_3} = 2.061 \text{ mol}$$

3. Determine the limiting reagent. Since we know that oxygen is in excess, PH_3 becomes the limiting reagent.
4. Calculate the theoretical yield of H_3PO_4 using moles of the limiting reactant. Convert moles of H_3PO_4 to grams of H_3PO_4:

$$\text{g } H_3PO_4 = 2.061 \text{ mol } PH_3 \times \frac{1 \text{ mol } H_3PO_4}{1 \text{ mol } PH_3} \times \frac{98 \text{ g } H_3PO_4}{1 \text{ mol } H_3PO_4}$$

$$= 201.98 \text{ g } H_3PO_4$$

5. Calculate the percent yield:

$$\text{Percent yield (\%)} = \frac{\text{Actual yield}}{\text{Theoretical yield}} \times 100 = \frac{188 \text{ g}}{201.98 \text{ g}} \times 100 = 93.1\%$$

Example 9.10

In a certain experiment, 70.0 g Al reacted with excess oxygen according to the equation:

$$4 \text{ Al(s)} + 3 \text{ O}_2\text{(g)} \longrightarrow 2 \text{ Al}_2\text{O}_3\text{(s)}$$

If a yield of 82.5% was obtained, what was the actual yield of Al_2O_3, in grams, from the experiment?

Solution

1. We have a balanced equation.
2. Calculate the number of moles of Al:

$$\text{moles Al} = 75 \text{ g} \times \frac{1 \text{ mol Al}}{27 \text{ g Al}} = 2.778 \text{ mol}$$

3. Determine the limiting reagent. Since we know that oxygen is in excess, Al becomes the limiting reagent.
4. Calculate the theoretical yield of Al_2O_3 using moles of the limiting reactant. Convert moles of Al_2O_3 to grams of Al_2O_3:

$$\text{grams } Al_2O_3 = 2.778 \text{ mol Al} \times \frac{2 \text{ mol } Al_2O_3}{4 \text{ mol Al}} \times \frac{102 \text{ g } Al_2O_3}{1 \text{ mol } Al_2O_3} = 141.68 \text{ g } Al_2O_3$$

5. Calculate the actual yield:

$$\text{Percent yield (\%)} = \frac{\text{Actual yield}}{\text{Theoretical yield}} \times 100$$

$$82.5\ \% = \frac{\text{Actual yield}}{141.68} \times 100$$

$$\text{Actual yield} = \frac{141.68 \times 82.5}{100} = 116.88\ \text{g}\ Al_2O_3$$

9.6. Problems

1. Consider the formation of ethanol by the fermentation of glucose, given by the following equation:

$$C_6H_{12}O_6 \longrightarrow 2\ C_2H_5OH + 2\ CO_2$$

Calculate the maximum masses of ethanol and gaseous carbon dioxide that can be produced from the fermentation of 250 g of glucose.

2. What mass of oxygen will react with 10 g of hydrogen to form water?
3. The thermal decomposition of potassium chlorate is represented by the equation

$$2\ KClO_3 \longrightarrow 2\ KCl + 3\ O_2$$

Calculate (a) the mass of $KClO_3$ needed to generate 6.25 mol of oxygen; (b) the moles of KCl produced from the decomposition of 5.125 g of $KClO_3$; (c) the mass and number of moles of oxygen produced from the decomposition of 5.125 g of $KClO_3$.

4. Chlorine is prepared in the laboratory by the action of hydrochloric acid on manganese dioxide, MnO_2, according to:

$$4\ HCl + MnO_2 \longrightarrow MnCl_2 + 2\ H_2O + Cl_2$$

What mass of MnO_2 is required to completely react with 10.25 g of HCl?

5. Very pure silicon (used in computer chips) is manufactured by heating silicon tetrachloride, $SiCl_4$, with zinc according to the equation:

$$SiCl_4 + 2\ Zn \longrightarrow Si + 2\ Zn\ Cl_2$$

If 0.25 mol of Zn is added to excess $SiCl_4$, how many grams of silicon are produced?

6. Tungsten metal, W, is used widely in the industrial production of incandescent-bulb filaments. The metal is produced by the action of hydrogen on tungsten (VI) oxide according to the following equation:

$$WO_3 + 3\,H_2 \longrightarrow W + 3\,H_2O$$

If a sample of WO_3 produces 25.825 g of water, how many grams of W are formed?

7. When steam is passed over iron filings at red heat, tri-iron tetraoxide (Fe_3O_4) and hydrogen are produced. The balanced equation for the reaction is:

$$3\,Fe(s) + 4\,H_2O(g) \xrightleftharpoons{\text{red hot}} Fe_3O_4(s) + 4\,H_2(g)$$

Calculate the volume of hydrogen produced at STP by reacting 0.25 mol of Fe with steam.

8. Calculate the volume of carbon dioxide produced at STP when 3.25 g propane burns in a rich supply of oxygen.

$$C_3H_8 + 5\,O_2 \longrightarrow 3\,CO_3 + 4\,H_2O$$

9. Large quantities of nitrogen gas can be produced from the thermal decomposition of sodium azide (NaN_3):

$$2\,NaN_3 \longrightarrow 2\,Na + 3\,N_2$$

Calculate the volume of nitrogen produced at STP by the decomposition of 85.25 g of NaN_3.

10. Nickel tetracarbonyl, $Ni(CO)_4$, decomposes at elevated temperature according to the following equation:

$$Ni(CO)_4(g) \rightleftharpoons Ni(s) + 4\,CO(g)$$

Calculate the volume of carbon monoxide produced at STP by the decomposition of 0.050 mol of $Ni(CO)_4$.

11. Acetylene gas, which is used in welding, burns in oxygen according to the following equation:

$$C_2H_2 + O_2 \longrightarrow H_2O + CO_2$$

(a) Balance the equation for the reaction.
(b) How many moles of oxygen are needed for the complete combustion of 5.25 g of acetylene?
(c) Calculate the volume of carbon dioxide obtained at STP from 5.25 g of acetylene.

12. Aluminum sulfate is used as a leather-tanning agent and as an antiperspirant. The compound can be obtained, along with water, by the reaction of aluminum hydroxide with sulfuric acid.

    (a) Write a balanced equation for the reaction.
    (b) Assume that 100 g of aluminum hydroxide reacts with 155 g of sulfuric acid. What is the limiting reagent?
    (c) How many grams of aluminum sulfate are produced?
    (d) How many grams of the excess reactant are left at the end of the reaction?

13. One of the chemical reactions involved in photographic film development is the reaction of AgBr with sodium thiosulfate, $Na_2S_2O_3$:

$$Na_2S_2O_3 + AgBr \longrightarrow Na_3Ag(S_2O_3)_2 + NaBr$$

    (a) Balance the equation for the reaction.
    (b) If 15 g of $Na_2S_2O_3$ and 18 g of AgBr are present in a reaction, which is the limiting reagent for the formation of $Na_3Ag(S_2O_3)_2$?
    (c) How many moles of the excess reagent are left at the end of the reaction?

14. The explosive trinitrotoluene (TNT) can be prepared by reacting toluene and nitric acid according to the equation:

$$C_7H_8 + HNO_3 \longrightarrow \underset{\text{(TNT)}}{C_7H_5N_3O_6} + H_2O$$

    (a) Balance the equation.
    (b) Determine the limiting reagent if 65 g C_7H_8 reacts with 50 g of HNO_3 .
    (c) What is the theoretical yield of TNT that can be obtained from a reaction mixture that contains 65 g of C_7H_8 and 50 g of HNO_3?
    (d) If the actual yield of TNT for the reaction mixture in part (c) is 55 g, calculate the percent yield of TNT for the reaction.

15. Phosphine, PH_3, can be prepared by the hydrolysis of calcium phosphide, Ca_3P_2:

$$Ca_3P_2 + H_2O \longrightarrow Ca(OH)_2 + PH_3$$

    (a) Balance the equation.
    (b) What is the maximum mass of PH_3 that could be prepared by mixing 44.25 g of Ca_3P_2 and 125.0 mL of water? (Density of water at 25°C is 1.000 g/mL.)

16. Acetylsalicylic acid (the chemical name for aspirin) helps to relieve pain and lowers body temperature. Industrially, it is produced from the reaction of salicylic acid and acetic anhydride:

$$\underset{\text{Salicylic acid}}{HOC_6H_4COOH} + \underset{\text{Acetic anhydride}}{(CH_3CO)_2O} \longrightarrow \underset{\text{Acetysalicylic acid}}{CH_3COOC_6H_4COOH} + \underset{\text{Acetic acid}}{CH_3COOH}$$

Starting with 12.00 g of salicylic acid and excess acetic anhydride, a chemistry student synthesized aspirin in the laboratory and reported a yield of 91%. What was the student's actual yield of aspirin?

17. The main chemical component responsible for artificial banana flavor is isopentyl acetate. An industrial chemist wants to prepare 500 g of isopentyl acetate by the reaction of acetic acid with isopentyl alcohol:

$$CH_3COOH + HO(CH_2)_2CH(CH_3)_2 \longrightarrow CH_3COO(CH_2)_2CH(CH_3)_2 + H_2O$$

Acetic acid   Isopentyl alcohol   Isopentyl acetate

How many grams of acetic acid are needed if the chemist expects a yield of only 78%? Assume that isopentyl alcohol is present in excess.

18. An analytical chemist wanted to determine the composition of copper-bearing steel. He placed 5.00 g of the sample in excess aqueous HCl solution. All the iron in the steel dissolved as iron (III) chloride, and 0.15 g of hydrogen was produced. What is the percent composition of the alloy? Assume the alloy is composed only of Cu and Fe, and that the Cu did not react with the HCl.

19. A drug enforcement agent seized 250.00 g of heroin, $C_{21}H_{23}O_5N$, and burned a 25.00-g sample in the presence of 35.25 g of oxygen. The end products were CO_2, NO_2, and H_2O. (a) Write a balanced equation for the combustion reaction. (b) What is the limiting reagent? (c) How many grams of the excess reactant remain unreacted? (d) What was the percentage completion of the combustion of heroin? (e) Determine the total mass of CO_2, NO_2, and H_2O produced in the combustion process. (1 mol of gas at STP occupies a volume of 22.4 L.)

20. Iron is produced by the carbon-monoxide reduction of iron ore (usually hematite, Fe_2O_3), in a blast furnace. The following reactions occur in various temperature regions in the reactor, resulting in the formation of molten iron.

$$3\ Fe_2O_3 + CO \longrightarrow 2\ Fe_3O_4 + CO_2 \quad (1)$$

$$Fe_3O_4 + CO \longrightarrow 3\ FeO + CO_2 \quad (2)$$

$$FeO + CO \longrightarrow Fe + CO_2 \quad (3)$$

How many kilograms of iron can be produced from 25.00 kg of the hematite ore? Assume that carbon monoxide is present in excess. (*Hint*: Manipulate these equations such that FeO and Fe_3O_4 cancel out of the final equation.)

10

Structure of the Atom

10.1. Electronic Structure of the Atom

The arrangement of electrons around the nucleus of an atom is known as its *electronic structure*. Since electrons determine all the chemical and most physical properties of an atomic system, it is important to understand the electronic structure. Much of our understanding has come from spectroscopy, the analysis of the light absorbed or emitted by a substance.

10.2. Electromagnetic Radiation

Electromagnetic radiation is a form of energy; light is the most familiar type of electromagnetic radiation. But radio waves, microwaves, X-rays, and many other similar phenomena are also types of electromagnetic radiation. All these exhibit wavelike properties, and all travel through a vacuum at the speed of light (see figure 10-1). The wavelike propagation of electromagnetic radiation can be described by its frequency (ν), wavelength (λ), and speed (c).

Wavelength (lambda, λ): The wavelength of a wave is the distance between two successive peaks or troughs.

Frequency (nu, ν): The frequency of a wave is the number of waves (or cycles) that pass a given point in space in one second. The unit is expressed as the reciprocal of seconds (s^{-1}) or as hertz (Hz). A hertz is one cycle per second ($1\ Hz = 1\ s^{-1}$).

Speed of light (c): The speed of light in a vacuum is one of the fundamental constants of nature, and does not vary with the wavelength. It has a numerical value of 2.9979×10^8 m/s, but for convenience we use 3.0×10^8 m/s.

These measurements are related by the equation:

$$\text{Speed of light} = \text{Wavelength} \times \text{Frequency}$$

$$c = \lambda\nu$$

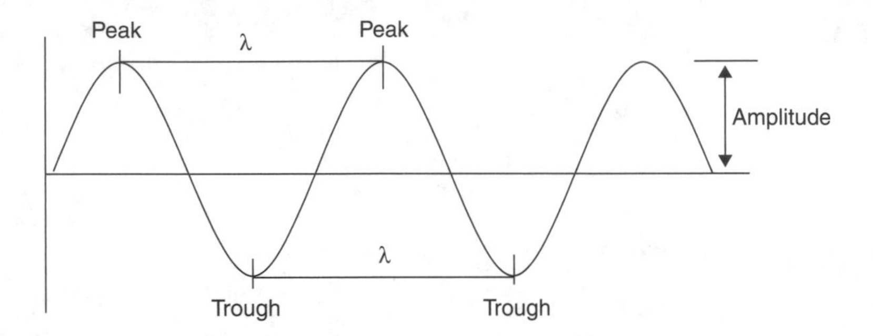

Figure 10-1 The wavelike properties of electromagnetic radiation, illustrating the wavelength and frequency of an electromagnetic wave.

This expression can be rearranged to give:

$$\lambda = \frac{c}{\nu}, \quad \text{or} \quad \nu = \frac{c}{\lambda}$$

Wave number ($\overline{\nu}$): The wave number is a characteristic of a wave that is proportional to energy. It is defined as the number of wavelengths per unit of length (usually in centimeter, cm). Wave number may be expressed as

$$\overline{\nu} = \frac{1}{\lambda}$$

Example 10.1

Calculate the frequency of red light with a wavelength of 700 nm. Note: 1 nm = 10^{-9}m.

Solution

Use the expression:

$$\nu = \frac{c}{\lambda}$$

$$c = 3.0 \times 10^{8}\ \text{m/s}$$

$$\lambda = 700\ \text{nm} = 700 \times 10^{-9}\ \text{m} = 7 \times 10^{-7}\ \text{m}$$

$$\nu = \frac{3.0 \times 10^{8}\ \text{m/s}}{7.0 \times 10^{-7}\ \text{m}} = 4 \times 10^{14}\ \text{s}^{-1} \quad \text{or} \quad 4 \times 10^{14}\ \text{Hz}$$

Example 10.2

What is the wavelength of a yellow light that has a frequency of 5.2×10^{14} Hz?

Solution

Use the expression:

$$\lambda = \frac{c}{\nu}$$

$$c = 3.0 \times 10^{8}\,\text{m/s}$$

$$\nu = 5.2 \times 10^{14}\,\text{Hz}$$

$$\lambda = \frac{3.0 \times 10^{8}\,\text{m/s}}{5.2 \times 10^{14}\,\text{Hz}} = 5.8 \times 10^{-7}\text{m} \quad \text{or} \quad 580\,\text{nm}.$$

10.3. The Nature of Matter and Quantum Theory

While electromagnetic radiation behaves like a wave, with characteristic frequency and wavelength, experiment has shown that electromagnetic radiation also behaves as a continuous stream of particles or energy packets. At the end of the 19th century the general belief among scientists was that matter and energy were unrelated. In 1901, Max Planck investigated the so-called blackbody radiation—the radiation emitted by hot solids at different temperatures. He observed that the intensity of the blackbody radiation varied with the wavelength. To explain his results, Planck suggested that the emission and absorption of radiant energy is quantized and cannot vary continuously. That is, energy can be gained or lost only in small packets or discrete units called quanta. The smallest amount of radiant energy, E, that can be emitted is called a quantum of energy, and is given by Planck's equation:

$$E = h\nu$$

where h is Planck's constant, with a value of 6.63×10^{-34} J · s, and ν is the frequency of the radiated light. Planck concluded that radiant energy from a given source must be an exact integral multiple of the simplest quantum $h\nu$:

$$E = nh\nu$$

where $n = 1, 2, 3, \ldots$ etc.

10.3.1. Photoelectric effect

In 1905, Albert Einstein used Planck's results to explain the photoelectric effect. Here, light shining on a metal surface causes electrons to be ejected; but once again, the process occurs only in discrete quantities, rather than continuously. Einstein suggested that electromagnetic radiation is quantized, and that light could be thought of as "particles" called photons. Planck referred to these "particles" as quanta. The energy of a photon is given as:

$$E_{\text{photon}} = h\nu = \frac{hc}{\lambda}$$

Example 10.3

What is the energy of a photon of infrared radiation that has a frequency of 9.8×10^{13} Hz?

Solution

$$\begin{aligned} E &= h\nu \\ &= 6.63 \times 10^{-34}\,\text{J}\cdot\text{s photon}^{-1} \times 9.8 \times 10^{13}\,\text{s}^{-1} \\ &= 9.60 \times 10^{-20}\,\frac{\text{J}}{\text{photon}} \end{aligned}$$

10.4. The Hydrogen Atom

When a beam of light passes through a prism, it produces a continuous spectrum containing light of all different colors or wavelengths. But when we observe the light given off by a heated gas—whether element or compound—we see only certain colors of light, which are characteristic of the substance.

The emission spectrum of atomic hydrogen consists of several series of isolated lines (four lines are observed in the visible region). The frequencies of the emission lines for hydrogen fit the Balmer equation:

$$\nu = C\left(\frac{1}{2} - \frac{1}{n^2}\right), \quad n = 3, 4, 5, 6$$

C is called the Rydberg constant for hydrogen, whose value is $3.29 \times 10^{15}\ \text{s}^{-1}$, and n is an integer greater than 2. For atoms other than the hydrogen atom, the emission-line frequencies follow a more complex relation.

10.4.1. The Bohr model

To account for the line spectrum of the hydrogen atom, Niels Bohr proposed what is commonly referred to as Bohr's model for the atom:

- Electrons move about the central nucleus in circular orbits of fixed radius called energy levels.

- Electrons can occupy only certain (permitted) discrete energy level in atoms. Electrons do not radiate energy when they are in one of the allowed energy levels.
- The angular momentum of the electron is quantized, and is a whole-number multiple of $h/2\pi$:

$$\text{Angular momentum} = mur = n\left(\frac{h}{2\pi}\right)$$

where m is the mass of the electron, u is velocity, r is radius of Bohr's orbit, n is an integer called the *principal quantum number* (n is discussed later in the chapter), and h is Planck's constant.
- Electrons absorb or emit energy in discrete amounts as they move from one energy level to another. The change in energy for an electron dropping from an excited state E_2 to lower (the ground) state E_1 is proportional to the frequency of the radiation emitted and is given by Planck's relationship:

$$\Delta E = E_2 - E_1 = h\nu$$

From his model, Bohr derived the following formula for the electron energy levels in the hydrogen atom:

$$E_n = -\left(\frac{R_H}{n^2}\right) \quad \text{or} \quad -\left(\frac{2.18 \times 10^{-18}\,\text{J}}{n^2}\right), \qquad n = 1, 2, 3, \ldots \infty$$

where R_H is the Rydberg constant, with a value of 2.18×10^{-18} J, and n is an integer called the *principal quantum number*, which describes the size of the allowed orbits (or shells). The minus sign indicates that E decreases as the electron moves closer to the nucleus. When n is equal to one, the orbit lies closest to the nucleus; an electron in this orbit is said to be in the *ground state*. When n is equal to 2, the electron is at a higher energy level. As n becomes infinitely large, the energy level becomes zero:

$$E_{n=\infty} = -\left(\frac{R_H}{\infty^2}\right) \quad \text{or} \quad -\left(\frac{2.18 \times 10^{-18}\,\text{J}}{\infty^2}\right) = 0$$

This is the reference or zero-energy state in which an electron is removed from the nucleus of the atom. Note that the zero energy state is higher in energy than the states closer to the nucleus (with negative energy).

10.4.2. Emission and absorption spectra

Absorption and emission spectra arise from electrons moving between energy levels. An absorption spectrum is produced when an atom absorbs light of a particular wavelength. The light energy absorbed excites or promotes an electron from a lower

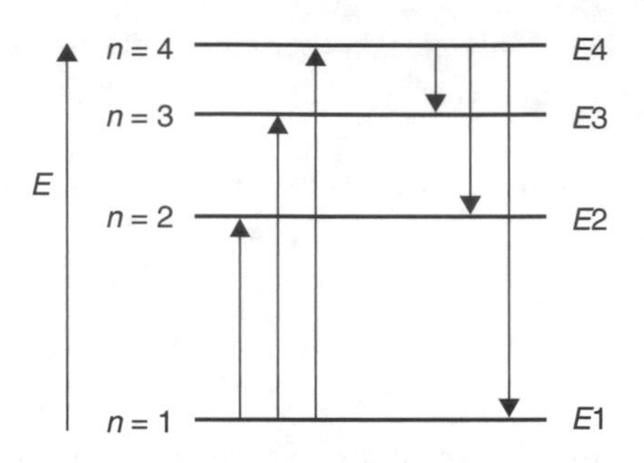

Figure 10-2 Energy levels in the hydrogen atom, from the Bohr model. The arrows indicate the transitions of the electron from one energy level to another.

energy level to a higher energy level. The wavelength of the absorbed light is obtained from the difference between the two energy levels.

$$\Delta E = \frac{hc}{\lambda} \quad \text{and} \quad \lambda = \frac{hc}{\Delta E}$$

where $\Delta E = E_2 - E_1$.

Similarly, an emission spectrum arises when an electron previously excited to a higher energy level falls back to a lower energy. The wavelength of the emitted light is obtained from the difference between the two energy levels, as before.

Figure 10-2 shows the absorption and emission of photons by an electron in an atom. The upward arrows represent absorption and the downward arrows represent emission. If an electron jumps from a state with an initial energy level E_i to a state with a final energy level E_f, the change in energy, ΔE, can be obtained from Bohr's equation:

$$\Delta E = \left[\left(\frac{-R_H}{n_f^2}\right) - \left(\frac{-R_H}{n_i^2}\right)\right] = -R_H\left(\frac{1}{n_f^2} - \frac{1}{n_i^2}\right) = R_H\left(\frac{1}{n_i^2} - \frac{1}{n_f^2}\right)$$

Since $\Delta E = E_f - E_i = hv$, the frequency of emitted radiation can be obtained from the expression:

$$hv = R_H\left(\frac{1}{n_i^2} - \frac{1}{n_f^2}\right) \quad \text{or} \quad v = \frac{R_H}{h}\left(\frac{1}{n_i^2} - \frac{1}{n_f^2}\right)$$

Note:

- For energy absorption, n_f is greater than n_i and ΔE is positive as electrons move from a lower-energy state to a higher energy.
- For radiant energy emission, n_f is less than n_i and ΔE is negative as electrons move from a higher-energy state to a lower energy.

Example 10.4

What is the amount of energy emitted when an electron in an excited hydrogen atom falls from energy level $n = 4$ to energy level $n = 1$?

Solution

This problem represents an electronic transition from energy level $n = 4$ to level $n = 1$. Radiant energy is emitted and ΔE will be negative. Use $R_H = 2.18 \times 10^{-18}$ J.

$$\Delta E = R_H \left(\frac{1}{n_i^2} - \frac{1}{n_f^2} \right) = 2.18 \times 10^{-18}\,\mathrm{J} \left(\frac{1}{4^2} - \frac{1}{1^2} \right) = -2.04 \times 10^{-18}\,\mathrm{J}$$

Example 10.5

Calculate (a) the frequency and (b) the wavelength of the radiation related to an electron in an excited hydrogen atom falling from $n = 3$ to $n = 1$. Is light emitted or absorbed?

Solution

This is an electronic transition from $n = 3$ to $n = 1$ state of the hydrogen atom. Therefore we expect light to be emitted as ΔE will be negative.

(a) To calculate the frequency, use the following expression:

$$\nu = \frac{R_H}{h} \left(\frac{1}{n_i^2} - \frac{1}{n_f^2} \right) = \left(\frac{2.18 \times 10^{-18}\mathrm{J}}{6.63 \times 10^{-34}\mathrm{J \cdot s}} \right) \left(\frac{1}{3^2} - \frac{1}{1^2} \right) = 2.92 \times 10^{15}\mathrm{s}^{-1}$$

(b) To calculate the wavelength, use the value of the frequency obtained above in the expression:

$$\lambda = \frac{c}{\nu} = \frac{3.0 \times 10^{8}\,\mathrm{ms}^{-1}}{2.92 \times 10^{15}\mathrm{s}^{-1}} = 1.03 \times 10^{-7}\mathrm{m} \quad \text{or} \quad 103\ \mathrm{nm}$$

10.5. The Quantum-Mechanical Description of the Hydrogen Atom

Bohr's model of the hydrogen atom significantly advanced our knowledge of atomic structure but it could not explain the spectra of heavier atoms. This led to further studies on atomic structure by many scientists.

10.5.1. The wave nature of the electron

The wave-particle dual nature of light led Louis de Broglie to propose in 1924 that any particle with mass m moving with velocity v should have a wavelength (λ) associated with it. The relationship between the velocity and wavelength of the particle can be expressed mathematically by the de Broglie equation:

$$\lambda = \frac{h}{mv}$$

where λ is the wavelength, h is Planck's constant, m is the mass of the particle, and v is the velocity. This is particularly relevant to particles as small as electrons; the wavelength associated with their motion tends to be significant in comparison to the size of the particle, which is not the case for large objects.

Example 10.6

Calculate the wavelength of an electron moving with a velocity of 6.5×10^6 m/s. The mass of an electron is 9.1×10^{-31} kg.

Solution

$$\lambda = \frac{h}{mv}$$

$$\lambda = \frac{(6.63 \times 10^{-34}\,\text{J}\cdot\text{s})}{(9.1 \times 10^{-31}\,\text{kg})(6.5 \times 10^{6}\text{m/s})}$$

$$= 1.12 \times 10^{-10}\,\text{m} \quad \text{or} \quad 0.112\ \text{nm}$$

10.5.2. The Heisenberg uncertainty principle

The *uncertainty principle* states that it is impossible to determine accurately both the momentum and position of an individual electron simultaneously. Mathematically, the law states that the uncertainty in the speed (Δv) times the uncertainty in position (Δx) multiplied by the mass (m) of the particle must be greater than or equal to Planck's constant (h) divided by 4π:

$$(\Delta x)(m \times \Delta v) \geq \frac{h}{4\pi}$$

The more accurately we can determine the position of a moving electron, the less accurately we can measure its momentum. Hence we can speak only of the probability of finding an electron at any given region in space within the atom.

10.6. Quantum Mechanics and Atomic Orbitals

Erwin Schrodinger's work on the wave-mechanical model of the atom in the mid-1920s greatly improved our understanding of atomic structure. The model is based on de Broglie's postulate and other fundamental assumptions. Schrodinger formulated what is known as the *Schrodinger wave equation*, which is the fundamental assumption of quantum theory. This equation successfully predicts every aspect of the hydrogen spectrum, and much of what is seen for heavier elements. Mathematically, the equation treats the electron as though it were a wave, specifically a standing wave. Though it is very complex, and often cannot be solved exactly, it gives an explicit mathematical description of the functions which determine the electron's distribution around the nucleus. These take the form of a set of *wave functions* (also known as *orbitals*) and their energies (which correspond to the energy levels in the Bohr model).

A wave function is represented by the symbol ψ (psi). The square of the wave function, ψ^2, is called the *probability density*; it measures the probability of finding an electron in a region of space. The probability distributions generated by these functions are essentially maps of the electron density around the nucleus, which define the shapes of the various orbitals.

10.6.1. Orbitals and quantum numbers

The solutions to Schrodinger's equation for an electron in a hydrogen atom (also applicable to other atoms) give rise to a series of wave functions or atomic orbitals, and also tells us that the energy of the hydrogen atom is quantized. This means that only certain energies, and hence energy levels that meet certain quantum conditions, are possible. Each orbital is defined by four *quantum numbers*, which describe the various properties of atomic orbitals, such as the number of possible orbitals within a given main energy level and the orientation of the electrons within each orbital.

1. *Principal quantum number (n)*
   The principal quantum number, n, indicates the size and energy of the main orbital (or level) an electron occupies in an atom. It can have any positive whole number value; 1, 2, 3, ... As n increases, the overall size of the orbital it describes increases, and the orbital can accommodate more electrons. Note that each main orbital, or value of n, is referred to as a *shell*. In the past, energy levels (or values) of n were represented by the letters $K, L, M, N, \ldots$.
   The number of electrons that can be accommodated in each energy level n is given by $2n^2$. Thus the electron capacities of energy levels 1, 2, 3, and 4, are 2, 8, 18, and 32, respectively.

Shell	K	L	M	N
Value of n	1	2	3	4
Electron capacity	2	8	18	32

Table 10-1 Relationship Between Shells, Subshells, and Their Relative Energies

Shell	*Subshells*	*Relative energies*
$n = 1$	1s	
$n = 2$	$2s$, $2p$	$2p > 2s$
$n = 3$	$3s$, $3p$, $3d$	$3d > 3p > 3s$
$n = 4$	$4s$, $4p$, $4d$, $4f$	$4f > 4d > 4p > 4s$

2. *Angular-momentum (or azimuthal) quantum number (l)*
   The angular-momentum quantum number, l, determines a region—a *subshell*—within the main shell (n) and defines its shape. Only certain values of l are allowed. For each value of n, l can have integral values from 0 to $n-1$. This implies that there are n different shapes of orbitals within each shell. For example, when $n = 4$, $l = 0$, 1, 2, or 3. The first four subshells are generally designated by the letters s, p, d, and f, corresponding to $l = 0$, 1, 2, and 3. This is summarized below:

Valule of l	0	1	2	3
Letter used	s	p	d	f

   The number of subshells within a given shell is shown in table 10-1, along with their relative energies.
3. *Magnetic (or orbital) quantum number (m_l)*
   The *magnetic quantum number* defines the spatial orientation of the subshell. Only certain values of m_l are allowed, and these depend upon l. Values of m_l are whole numbers varying from $-l$ through zero to $+l$, thus the number of allowed values of m_l is given by $2l + 1$. Therefore, there is one s orbital, three p orbitals, five d orbitals, and seven d orbitals.
4. *Spin quantum number (m_s)*
   The *spin quantum number*, m_s, specifies the spin of an electron and so the orientation of the magnetic field generated by this spin. The spin can be clockwise or counterclockwise. Therefore, for every set of n, l, and m_l values, m_s can take the value $+\frac{1}{2}$ or $-\frac{1}{2}$. Table 10-2 summarizes the permissible values for each quantum number.

Example 10.7

Calculate the maximum number of electrons that can be placed in

(a) A shell with $n = 3$
(b) A shell with $n = 4$
(c) A shell with $n = 6$

Table 10-2 Allowed Values of Quantum Numbers Through $n = 4$

Shell	n	t	m_l	m_s	*Designation of subshells*	*No. of orbitals in subshell*	*No. of electrons in subshell*	*Total electrons in main shell*
K	1	0	0	$+\frac{1}{2}, -\frac{1}{2}$	1*s*	1	2	2
L	2	0	0	$+\frac{1}{2}, -\frac{1}{2}$	2*s*	1	2	8
		1	$-1, 0, +1$	$\pm\frac{1}{2}$ for each value of m_l	2*p*	3	6	
M	3	0	0	$+\frac{1}{2}, -\frac{1}{2}$	3*s*	1	2	
		1	$-1, 0, -1$	$\pm\frac{1}{2}$ for each value of m_l	3*p*	3	6	18
		2	$-2, -1, 0, -1, +2$	$\pm\frac{1}{2}$ for each value of m_l	3*d*	5	10	
N	4	0	0	$+\frac{1}{2}, -\frac{1}{2}$	4*s*	1	2	
		1	$-1, 0, +1$	$\pm\frac{1}{2}$ for each value of m_l	4*p*	3	6	32
		2	$-2, -1, 0, +1, +2$	$\pm\frac{1}{2}$ for each value of m_l	4*d*	5	10	
		3	$-3, -2, -1, 0, +1, +2, +3$	$\pm\frac{1}{2}$ for each value of m_l	4*f*	7	14	

Solution

Use the expression $2n^2$.

(a) For $n = 3$, the maximum number of electrons in the shell will be $2n^2 = 2(3^2) = 18$.
(b) For $n = 4$, the maximum number of electrons in the shell will be $2n^2 = 2(4^2) = 32$.
(c) For $n = 6$, the maximum number of electrons in the shell will be $2n^2 = 2(6^2) = 72$.

Example 10.8

Describe the sublevels in the $n = 4$ energy level.

Solution

For $n = 4$, l can have values of 0, 1, 2, and 3. Therefore the number of sublevels and their designations include:

$$l = 0 \quad (4s)$$
$$l = 1 \quad (4p)$$
$$l = 2 \quad (4d)$$
$$l = 3 \quad (4f)$$

Example 10.9

What quantum numbers are permissible for a $5p$ orbital?

Solution

For a $5p$ orbital, $n = 5$. The fact that it is a p orbital tells us that $l = 1$. For $l = 1$, m_l can have any of the values -1, 0, or $+1$, hence there are three sets of permissible quantum numbers for a $5p$ orbital. They are:

$$n = 5, \quad l = 1, \quad m_l = -1$$
$$n = 5, \quad l = 1, \quad m_l = 0$$
$$n = 5, \quad l = 1, \quad m_l = +1$$

Note that m_s is not counted here. It describes an electron, not an orbital.

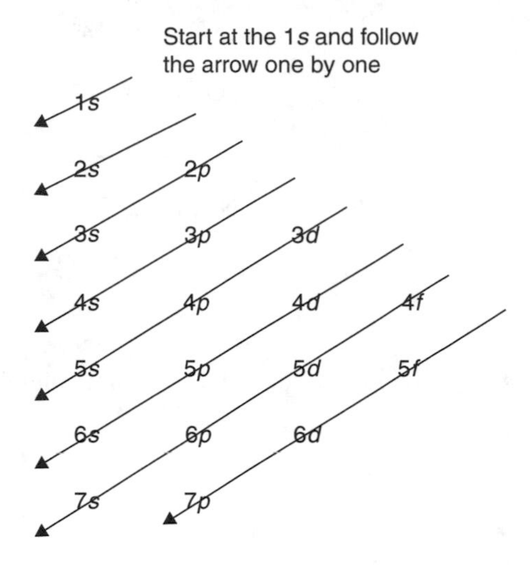

Figure 10-3 An aid to remembering the building-up or Aufbau order of atomic orbitals.

10.7. Electronic Configuration of Multielectron Atoms

An electronic configuration indicates how many electrons an atom has in each of its subshells or energy levels. Knowing the relative energies of the various orbitals allows one to predict for each element or ion the orbitals that are occupied by electrons. The ground state electronic configurations are obtained by placing electrons in the atomic orbitals of lowest possible energy, provided no single orbital holds more than two electrons. In building up the electronic configuration of an atom, three principles are used. These include the Aufbau principle, the Pauli exclusion principle, and Hund's principle.

10.7.1. Aufbau principle

Electrons occupy the lowest-energy orbitals available to them, entering the higher-energy orbitals only after the lowest-energy orbitals are filled.

The Aufbau diagram (see figure 10-3) can assist you in writing electron configurations for any element in the periodic table as long as you know the atomic number.

In a compressed form, electrons fill the sub-shells available in each main energy level in the following order:

$$1s2s2p3s3p4s3d4p5s4d5p6s4f5d6p7s5f6d$$

For example, the electron configurations of the first ten elements in the periodic table are:

$$_1\text{H} : 1s^1 \quad _6\text{C} : 1s^2 2s^2 2p^2$$

$$_{2}He : 1s^2 \qquad _{7}N : 1s^2 2s^2 2p^3$$
$$_{3}Li : 1s^2 2s^1 \qquad _{8}O : 1s^2 2s^2 2p^4$$
$$_{4}Be : 1s^2 2s^2 \qquad _{9}F : 1s^2 2s^2 2p^5$$
$$_{5}B : 1s^2 2s^2 2p^1 \qquad _{10}Ne : 1s^2 2s^2 2p^6$$

Example 10.10

Write the electron configuration for (a) Ca (atomic number, $Z = 20$), and (b) Mn ($Z = 25$).

Solution

(a) Ca: $1s^2\, 2s^2\, 2p^6\, 3s^2\, 3p^6\, 4s^2$
(b) Mn: $1s^2\, 2s^2\, 2p^6\, 3s^2\, 3p^6\, 4s^2\, 3d^5$

Example 10.11

Predict the electron configuration of Ca^{2+}.

Solution

Calcium has atomic number 20, i.e., 20 protons and 20 electrons. But Ca^{2+} has lost 2 electrons. Therefore 18 electrons must be placed in the various sub-shells of Ca orbitals in order of increasing energy:

$$_{20}Ca^{2+}(18 \text{ electrons}) : 1s^2\, 2s^2\, 2p^6\, 3s^2\, 3p^6$$

10.7.2. Shorthand notation for electron configuration

Electron configurations can also be written using a shorthand method which shows only the outer electrons, those surrounding the inert noble-gas core. This involves using $[_{2}He]$ for $1s^2$, $[_{10}Ne]$ for $1s^2 2s^2 2p^6$, $[_{18}Ar]$ for $1s^2 2s^2 2p^6 3s^2 3p^6$, etc. Thus we may represent the electron configurations for Si and K as:

$$_{14}Si : [_{10}Ne]3s^2 3p^2 \quad \text{and} \quad _{19}K : [_{18}Ar]4s^1$$

Example 10.12

Using the shorthand notation, write the electron configuration for Mn.

Solution

The atomic number of Mn is 25. Therefore it has 25 electrons, which will fill its orbitals from lowest energy to highest.

$$_{25}\mathrm{Mn} : [_{18}\mathrm{Ar}]4s^2 3d^5$$

10.7.3. The Pauli exclusion principle

The Pauli exclusion principle states that no two electrons in an atom can have the same four quantum numbers. Since there are only two possible values for m_s $+\frac{1}{2}$ and $-\frac{1}{2}$, an orbital can accommodate only two electrons, which must have opposite spins.

10.7.4. Hund's principle (or rule)

Hund's principle states that if two or more orbitals have equal energy (degenerate orbitals), a single electron goes into each until all orbitals are half-filled; only then can a second electron (with an opposite spin quantum number) enter any of the orbitals. In other words, the orbitals in a given subshell will each gain one electron before any of them gain a second one.

10.7.5. Summary of building-up principles

1. The lowest-energy orbitals are filled first (Aufbau principle).
2. Only two electrons of opposite spin go into any one orbital (Pauli exclusion principle).
3. If two or more orbitals have the same energy (degenerate orbitals), each is half-filled before any one of them is completely filled (Hund's rule).

10.7.6. Orbital diagrams

Orbital diagrams are used to show the electron occupancy in each orbital around the nucleus. They serve to illustrate both Hund's rule and Pauli's principle. Each atomic orbital is represented by a box, which contains up to two electrons. An electron is represented by a half-arrow, which points upward (for $m_s = \frac{1}{2}$) or downward (for $m_s = \frac{1}{2}$). If two electrons are in the same orbital (box), the two half-arrows representing them will be opposed as illustrated below.

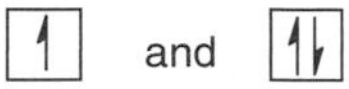

[↿] and [↿⇂]

An *s* orbital is represented by one box, a *p* orbital by three merged boxes, and a *d* orbital by five merged boxes. That is:

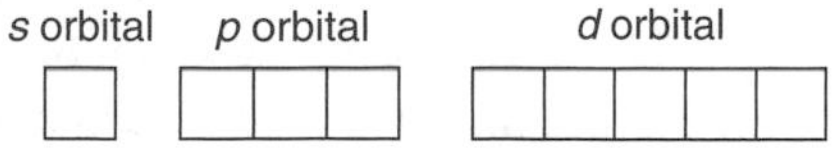

s orbital *p* orbital *d* orbital

Example 10.13

Draw the orbital diagrams for (a) fluorine and (b) magnesium.

Solution

(a)

(b)

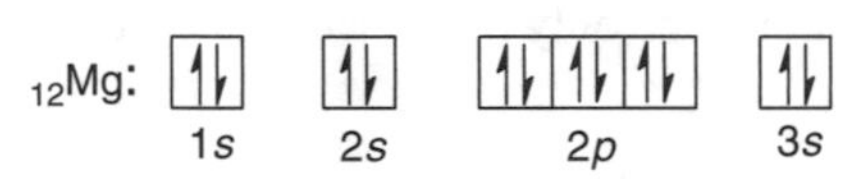

10.8. Problems

1. Express the following in nanometers:
   (a) 2.34×10^{-12} cm
   (b) 1.73×10^{-8} cm
   (c) 5.44×10^{3} m
   (d) 0.575 m
2. Express the following in centimeters:
   (a) 5.34×10^{5}nm
   (b) 265 nm
   (c) 6.64×10^{12}Å
   (d) 879 Å
3. What is the frequency of yellow light with a wavelength of 589 nm?
4. What is the wavelength of light having a frequency of $9.29 \times 10^{16}\ s^{-1}$?
5. Calculate the wave number, $\bar{\nu}$, of light having a wavelength of 589 nm.
6. A laser emits light having a frequency of $6.94 \times 10^{14}\ s^{-1}$. Calculate (a) the wavelength, (b) the wave number of this light.
7. What is the energy of photons in kJ that correspond to light with a frequency of $3.70 \times 10^{12}\ s^{-1}$?
8. What is the energy of a light with a wavelength of 408 nm?
9. The blue-green light emitted by a particular atom has a wavelength of 525 nm (in the visible range) of the electromagnetic spectrum. Find (a) the frequency, (b) the wave number, and (c) the energy of a photon of this radiation.
10. Nuclear magnetic resonance (NMR) spectrometers allow chemists to measure the absorption of energy by some atomic nuclei such as hydrogen. NMR instruments

are available that operate at 60 MHz, 200 MHz, 400 MHz, 600 MHz, and more recently, 700 MHz. At what energy do the 60 MHz, 200 MHz, 400 MHz, 600 MHz, and 700 MHz NMR instruments operate, in kJ?

11. What is the energy of an electron in the third energy level of a hydrogen atom?
12. The Lyman series of lines in the hydrogen emission spectrum corresponds to $n = 1$ and is observed in the ultraviolet region. Calculate (a) the wave number of the spectral line in this series for a value of m equal to 5 (note $n = 1$); (b) the energy emitted by an excited hydrogen atom having a spectral line with value of $n = 1$, and $m = 5$.
13. For a hydrogen atom in the ground state ($n = 1$), calculate the energy necessary to cause ionization ($m = \infty$).
14. Calculate the energy change and the wavelength that occur when an electron falls from (a) $n = 4$ to $n = 1$, (b) $n = 5$ to $n = 2$ in the hydrogen atom.
15. Calculate the wavelength, in nanometers, of a proton moving at a velocity of 4.56×10^4 m s^{-1}. The mass of a proton is 1.67×10^{-27} kg.
16. If the wavelength of an electron is 25 nm, what is the velocity of the electron? The mass of an electron is $9.11\,4 \times 10^{-28}$ g.
17. In an electron diffraction experiment, a beam of electrons is accelerated by a potential difference of 50 kV. What is the characteristic de Broglie wavelength of the electrons in the beam?
(*Hint*: $KE = \frac{1}{2}mv^2$; mass of an electron $= 9.11 \times 10^{-31}$ kg; 1 J = 1 kg m^2 s^{-2})
18. What is the maximum number of electrons permissible in the following energy levels:

(a) $n = 1$ energy level
(b) $n = 2$ energy level
(c) $n = 3$ energy level
(d) $n = 4$ energy level

19. For the energy level $n = 3$, what sublevels are possible?
20. What is the shorthand notation for describing the orbitals having the following quantum numbers?

(a) $n = 3, l = 1, m_l = 0$
(b) $n = 3, l = 2, m_l = 1$
(c) $n = 4, l = 3, m_l = 3$

21. Give the sets of quantum numbers that describe electrons in the following orbitals:

(a) $2p$ orbital
(b) $3d$ orbital
(c) $4p$ orbital
(d) $4f$ orbital

22. Write the electronic configuration of the following elements:
(a) Na (b) P (c) Ca (d) Rb (e) Fe

23. Write the electronic configuration of each of the following ions:
    (a) S^{2-} (b) Ni^{2+} (c)Cu^{2+} (d) Ti^{4+} (e) Br^-
24. Write the electronic configuration for the following atoms or ions, using the relevant noble-gas inner-core abbreviation:
    (a) O (b) Zn^{2+} (c) Mg (d) Pb^{2+} (e) V (f) Te^{2-} (g) Mn^{7+}
25. Identify the elements corresponding to the following electronic configurations:
    (a) $[He]2s^22p^1$
    (b) $[Ar]3d^{10}4s^24p^2$
    (c) $[Kr]4d^{10}5s^24p^5$
    (d) $[Xe]4f^{14}5d^{10}6s^26p^3$
    (e) $[Ar]3d^54s^2$
    (f) $[Ne]3s^23p^3$
26. Identify the elements corresponding to the following electronic configurations:
    (a) $1s^22s^22p^5$
    (b) $1s^22s^22p^63s^23p^3$
    (c) $1s^22s^22p^63s^23p^64s^23d^1$
    (d) $1s^22s^22p^63s^23p^64s^23d^8$
    (e) $1s^22s^22p^2$
    (f) $1s^22s^22p^63s^23p^64s^23d^2$
27. Draw the orbital diagram for:
    (a) oxygen (b) sodium (c) titanium (d) zinc.
28. Two elements, *A* and *B*, have the following isotopic masses and electronic configuration:

    A = 17.99916 amu and $1s^22s^22p^4$
    B = 25.98259 amu and $1s^22s^22p^63s^2$
    (a) Give the chemical identity of *A* and *B*.
    (b) How many subatomic particles are present in *A* and *B*?

11

Gas Laws

11.1. Standard Temperature and Pressure

Volumes and densities of gases vary significantly with changes in pressure and temperature. This means that measurements of the volumes of gases will likely vary from one laboratory to another. To correct for this, scientists have adopted a set of standard conditions of temperature and pressure (STP) as a reference point in reporting all measurements involving gases. They are 0°C (or 273 K) and 760 mm Hg or 1 atm (or 1.013×10^5 N m^{-2} in S.I. units). Therefore *standard temperature and pressure*, as used in calculations involving gases, are defined as 0°C (or 273 K) and 1 atmosphere (or 760 torr). (Note: For calculations involving the gas laws, temperature must be in K.)

11.2. Boyle's Law: Pressure vs. Volume

Boyle's law states that *the volume of a given mass of gas at constant temperature is inversely proportional to the pressure*. The law can be expressed in mathematical terms:

$$V \propto \frac{1}{P} \quad \text{or} \quad PV = k \text{ at constant } n \text{ and } T$$

Since $P \times V =$ constant, problems dealing with P–V relationships can be solved by using the simplified equation:

$$P_1V_1 = P_2V_2$$

Here P_1, V_1 represent one set of conditions and P_2, V_2 represent another set of conditions for a given mass of gas.

Example 11.1

A sample of gas occupies 5.0 L at 760 mmHg at 25°C. Calculate the volume of the gas if the pressure is reduced to 650 mmHg at 25°C.

Solution

First, list the known quantities and the unknown quantity:

$$\text{Known: } V_1 = 5.0 \text{ L}$$
$$P_1 = 760 \text{ mmHg}$$
$$P_2 = 650 \text{ mmHg}$$
$$\text{Unknown: } V_2 = ?$$

Then solve for the unknown quantity:

$$P_1V_1 = P_2V_2$$

$$V_2 = V_1 \times \frac{P_1}{P_2} = 5.0 \text{ L} \times \frac{760 \text{ mm}}{650 \text{ mm}} = 5.85 \text{ L}$$

Example 11.2

A given mass of gas occupies 300 mL at an atmospheric pressure of 1.0×10^5 N m^{-2}. What will be the pressure if the volume is reduced to 50 mL, assuming no change in temperature?

Solution

First, list the known quantities and the unknown quantity.

$$\text{Known: } V_1 = 300.0 \text{ mL}$$
$$P_1 = 10^5 \text{ N m}^{-2}$$
$$V_2 = 50 \text{ mL}$$
$$\text{Unknown: } P_2 = ?$$

Then solve for the unknown quantity:

$$P_1V_1 = P_2V_2$$

$$P_2 = P_1 \times \frac{V_1}{V_2} = 10^5 \text{ N m}^{-2} \times \frac{300 \text{ mL}}{50 \text{ mL}} = 6 \times 10^5 \text{ N m}^{-2}$$

The final pressure of the gas is 6.0×10^5 N m^{-2}.

11.3. Charles's Law: Temperature vs. Volume

Charles's law states that *the volume of a given mass of gas is directly proportional to its absolute temperature*. So if the absolute temperature is doubled, say from 300 K to 600 K, the volume of the gas will also double. A plot of the volume of a gas versus its temperature (K) gives a straight line. A notable feature of such a plot is that the volume of all gases extrapolates to zero at the same temperature, −273.2°C. This point is defined as 0 K, and is called *absolute zero*. Thus, the relationship between the Kelvin and Celsius temperature scales is given as: K = 0°C + 273. Scientists believe that the absolute zero temperature, 0 K, cannot be attained, although some laboratories have reported producing 0.0001 K.

In mathematical terms, Charles's law can be stated as:

$$V \propto T \quad \text{or} \quad \frac{V}{T} = k, \text{ where } k \text{ is a constant}$$

Since V/T = constant, problems dealing with V–T relationships can be solved by using the simplified equation:

$$\frac{V_1}{T_1} = \frac{V_2}{T_2}$$

Here V_1, T_1 represent one set of conditions and V_2, T_2 represent another set of conditions for a given mass of gas.

Example 11.3

A 6.0-L sample of a gas was heated from −5°C to 80°C at constant pressure. Calculate the final volume of the gas.

Solution

1. Identify the known quantities and the unknown quantity:

$$\begin{aligned} \text{Known: } V_1 &= 6.0 \text{ L} \\ T_1 &= -5 + 273 = 268 \text{ K} \\ T_2 &= 80 + 273 = 353 \text{ K} \\ \text{Unknown: } V_2 &= ? \end{aligned}$$

2. Then solve for the unknown quantity using the expression:

$$\frac{V_1}{T_1} = \frac{V_2}{T_2}$$

$$V_2 = V_1 \times \frac{T_2}{T_1} = 6.0 \text{ L} \times \frac{353 \text{ K}}{268 \text{ K}} = 7.90 \text{ L}$$

The final volume of the gas is 7.9 L.

Example 11.4

A given mass of gas occupies 500 mL at 15°C in a cylinder. If the maximum capacity of the cylinder is 950 mL, what is the highest temperature to which the cylinder can be heated at constant pressure?

Solution

1. Identify the known quantities and the unknown quantity:

$$\begin{aligned} \text{Known: } V_1 &= 500 \text{ mL} \\ T_1 &= 15 + 273 = 288 \text{ K} \\ V_2 &= 950 \text{ mL} \\ \text{Unknown: } T_2 &= ? \end{aligned}$$

2. Then solve for the unknown quantity:

$$\frac{V_1}{T_1} = \frac{V_2}{T_2}$$

$$T_2 = T_1 \times \frac{V_2}{V_1} = 288 \text{ K} \times \frac{950 \text{ mL}}{500 \text{ mL}} = 547.2 \text{ K} \quad \text{or} \quad 274.2°\text{C}$$

The maximum temperature to which the cylinder can be heated is 547 K or 274°C.

11.4. The Combined Gas Law

For a fixed mass of gas we can combine Boyle's law and Charles's law to express the relationships among pressure, volume, and temperature. This combined gas law may be stated as: *For a given mass of gas, the volume is directly proportional to the absolute temperature and inversely proportional to the pressure*. Mathematically, we have:

$$\frac{P_1V_1}{T_1} = \frac{P_2V_2}{T_2}$$

Remember that temperature must be in K.

Example 11.5

A 250 mL sample of nitrogen gas was collected at −24°C and 735 torr. Calculate the volume at STP.

Solution

1. Identify the known quantities and the unknown quantity:

$$\begin{aligned} \text{Known: } V_1 &= 250 \text{ mL} \\ T_1 &= -24 + 273 = 249 \text{ K} \\ T_2 &= 273 \text{ K} \\ P_1 &= 735 \text{ torr} \\ P_2 &= 760 \text{ torr} \\ \text{Unknown: } V_2 &= ? \end{aligned}$$

2. Then solve for the unknown quantity using the combined gas law:

$$\frac{P_1 V_1}{T_1} = \frac{P_2 V_2}{T_2}$$

$$V_2 = (V_1)\left(\frac{T_2}{T_1}\right)\left(\frac{P_1}{P_2}\right) = 250 \text{ mL} \times \left(\frac{273 \text{ K}}{249 \text{ K}}\right)\left(\frac{735 \text{ torr}}{760 \text{ torr}}\right) = 265.1 \text{ mL}$$

The final volume of the gas is 265 mL.

Example 11.6

At 300 K and 2.25 atm, a sample of helium occupies 3.5 L. Determine the temperature at which the gas will occupy 4.50 L at 1.50 atm.

Solution

1. Identify the known quantities and the unknown quantity.

$$\begin{aligned} \text{Known: } V_1 &= 3.5 \text{ L};\; V_2 = 4.5 \text{ L} \\ T_1 &= 300 \text{ K} \\ P_1 &= 2.25 \text{ atm};\; P_2 = 1.5 \text{ atm} \\ \text{Unknown: } T_2 &= ? \end{aligned}$$

2. Then solve for the unknown quantity:

$$\frac{P_1 V_1}{T_1} = \frac{P_2 V_2}{T_2}$$

$$T_2 = (T_1)\left(\frac{P_2}{P_1}\right)\left(\frac{V_2}{V_1}\right) = 300 \text{ K} \times \left(\frac{1.5 \text{ atm}}{2.25 \text{ atm}}\right)\left(\frac{4.5 \text{ L}}{3.5 \text{ L}}\right) = 257.1 \text{ K}.$$

The final temperature of the gas is 257.1 K.

11.5. Gay-Lussac's Law and Reactions Involving Gases

Chemical reactions involving gases occur in definite and simple proportions by volume. In 1808, Gay-Lussac published his observation on the volume of gases involved in chemical reactions; his results are known as Gay-Lussac's law of combining volumes. This law states: *When gases react, they do so in volumes which are in simple ratios to one another, and to the volume of the gaseous products, provided the volumes are measured at the same temperature and pressure.*

For example, it has been determined experimentally that one volume of hydrogen and one volume of chlorine react to give two volumes of hydrogen chloride:

1 volume of H_2 + 1 volume of Cl_2				⟶	2 volumes of HCl
Ratio	1	:	1	:	2

Similarly, one volume of nitrogen will react with three volumes of hydrogen to form two volumes of ammonia:

1 volume of N_2 + 3 volumes of H_2				⟶	2 volumes of NH_3
Ratio	1	:	3	:	2

Example 11.7

50 mL of CH_4 are ignited with 100 mL of oxygen at 100°C and 760 torr, according to the following equation:

$$CH_4(g) + 2O_2(g) \longrightarrow CO_2(g) + 2H_2O(g)$$

The reaction produces 50 mL of CO_2 and 100 mL of H_2O. Show that the data illustrate Gay-Lussac's law.

Solution

Check the volume relationships between reactants and products to see if they are in simple whole-number ratios.

$$\frac{\text{Volume of } CH_4}{\text{Volume of } O_2} = \frac{50}{100} = \frac{1}{2}$$

$$\frac{\text{Volume of } CH_4}{\text{Volume of } CO_2} = \frac{50}{50} = \frac{1}{1}$$

$$\frac{\text{Volume of } CH_4}{\text{Volume of } H_2O} = \frac{50}{100} = \frac{1}{2}$$

$$\frac{\text{Volume of } O_2}{\text{Volume of } CO_2} = \frac{100}{50} = \frac{2}{1}$$

$$\frac{\text{Volume of } O_2}{\text{Volume of } H_2O} = \frac{100}{100} = \frac{1}{1}$$

These ratios fit Gay-Lussac's law.

Example 11.8

100 cm^3 of acetylene, C_2H_2, are mixed with 500 cm^3 of oxygen and exploded. If all the volumes are measured at STP, calculate the volume of the resulting gases. What is the percentage of steam in the resulting gas mixture?

Solution

1. First write a balanced equation for the reaction:

$$2\,C_2H_2(g) + 5O_2(g) \longrightarrow 4\,CO_2(g) + 2H_2O(g)$$

2. Determine the volume of O_2 required to react completely with the acetylene. Applying Gay-Lussac's law, C_2H_2 and O_2 react in the ratio of 2 to 5 by volume. The volume of O_2 required is calculated as:

$$\text{vol of } O_2 = 100\ cm^3 C_2H_2 \times \frac{5\ cm^3 O_2}{2\ cm^3 C_2H_2} = 250\ cm^3 O_2$$

3. Determine the excess oxygen.
   Since only 250 cm^3 were used out of the initial 500 cm^3, the excess O_2 is $(500 - 250) = 250\ cm^3$.
4. Determine the volume of the products.
   2 volumes of C_2H_2 produce 4 volumes of CO_2 and 2 volumes of H_2O:

$$\text{vol of } CO_2 = 100\ cm^3 C_2H_2 \times \frac{4\ cm^3 CO_2}{2\ cm^3 C_2H_2} = 200\ cm^3 CO_2$$

$$\text{vol of } H_2O = 100\ cm^3 C_2H_2 \times \frac{2\ cm^3 H_2O}{2\ cm^3 C_2H_2} = 100\ cm^3 H_2O$$

5. Determine the volume of the resulting gases:

$$\text{Total volume of resulting gases} = \text{vol of } CO_2 + \text{vol of } H_2O + \text{excess } O_2$$

$$= 200\ cm^3 + 100\ cm^3 + 250\ cm^3 = 550\ cm^3$$

6. Calculate the percentage of steam in the gas mixture:

$$\%\ H_2O = \frac{\text{Vol of } H_2O}{\text{Total vol of resulting gases}} \times 100 = \frac{100}{550} \times 100 = 18.2\%$$

11.6. Avogadro's Law

Initially, scientists (including Gay-Lussac himself) could not explain the law of combining volumes of gases. In 1811, Amadeo Avogadro, an Italian professor of physics, put forward a hypothesis that:

Equal volumes of all gases at the same temperature and pressure contain the same number of molecules.

This became known as Avogadro's law after it was experimentally proved to be correct with ±2% margin of error. For example, it can be shown experimentally that nitrogen reacts with hydrogen to form ammonia in a volume ratio of one to three.

$$1 \text{ volume of } N_2 + 3 \text{ volumes of } H_2 \longrightarrow 2 \text{ volumes of } NH_3$$

Furthermore, a very similar experiment indicates that they react in a ratio of 1 mol to 3 mol (as shown by measuring the masses which react). Thus the reacting volumes are proportional to the number of moles involved—which we can take as support for Avogadro's law and Gay-Lussac's law.

Mathematically, the law can be represented as:

$$V \propto n, \quad \text{or} \quad V = kn$$

where V = volume, n = number of moles, and k = a proportionality constant. Essentially, this equation means that the volume of a sample of gas at constant temperature and pressure is directly proportional to the number of moles of molecules in the gas. Hence:

$$\frac{V_1}{n_1} = \frac{V_2}{n_2}$$

Example 11.9

A balloon filled with 5.0 mol of He gas has a volume of 2.5 L at a given temperature and pressure. If an additional 1.0 mol of He is introduced into the balloon without altering the temperature and pressure, what will be the new volume of the balloon?

Solution

1. Identify the known quantities and the unknown quantity:

$$\begin{aligned} \text{Known: } V_1 &= 2.5 \text{ L} \\ n_1 &= 5.0 \text{ mol} \\ n_2 &= n_1 + 1.0 = 5.0 + 1.0 = 6 \text{ mol.} \\ \text{Unknown: } V_2 &= ? \end{aligned}$$

2. Then solve for the unknown:

$$\frac{V_1}{n_1} = \frac{V_2}{n_2}$$

$$V_2 = V_1 \times \frac{n_2}{n_1} = 2.5 \text{ L} \times \frac{6.0 \text{ mol}}{5.0 \text{ mol}} = 3.0 \text{ L}$$

The new volume of the balloon is 3.0 L.

11.7. The Ideal Gas Law

Boyle's law, Charles's law and Avogadro's law show that the volume of a given mass of gas depends on pressure, temperature, and number of moles of the gas.

Boyle's law: $V \propto \frac{1}{P}$ (constant n, T)

Charles's law: $V \propto T$ (constant n, P)

Avogadro's law: $V \propto n$ (constant P, T)

Mathematically, if we combine these laws we have:

$$V \propto \frac{nT}{P} \quad \text{or} \quad V = R\left(\frac{nT}{P}\right)$$

R is a proportionality constant, known as the *ideal gas constant*. Depending on the units of volume and pressure, R can have a value of 0.082 atm L/K-mol or 8.31 J/K-mol.

This equation can be rearranged to give what is generally known as the *ideal gas law*:

$$PV = nRT$$

Example 11.10

Calculate the number of moles of an ideal gas which occupies a volume of 15 L at 25°C and a pressure of 3 atm.

Solution

1. Identify the known quantities and the unknown quantity:

Known: $V = 15 \text{ L}$

$T = 25 + 273 = 298 \text{ K}$

$$P = 3 \text{ atm}$$

$$R = 0.08206 \text{ L-atm/K-mol}$$

$$\text{Unknown: } n = ?$$

2. Then solve for the unknown:

$$PV = nRT$$

$$n = \frac{PV}{RT} = \frac{(3 \text{ atm})(15 \text{ L})}{(0.08206 \text{ L-atm/K-mol})(298 \text{ K})} = 1.84 \text{ mol}$$

Example 11.11

A 0.25-mol sample of an ideal gas occupies 56 L at 19 atm. Calculate the temperature of this gas.

Solution

1. Identify the known quantities and the unknown quantity:

$$\text{Known: } V = 56 \text{ L}$$

$$n = 0.25 \text{ mol}$$

$$P = 19 \text{ atm}$$

$$R = 0.08206 \text{ L-atm/K-mol}$$

$$\text{Unknown: } T = ?$$

2. Then solve for the unknown quantity using the ideal gas law:

$$PV = nRT$$

$$T = \frac{PV}{nR} = \frac{(19 \text{ atm})(56 \text{ L})}{(0.25 \text{ mol})(0.08206 \text{ L-atm/K-mol})} = 5.19 \times 10^4 \text{ K}$$

11.8. Density and Molecular Mass of a Gas

Density is defined as mass per unit volume and can be used to convert volume into mass and vice-versa. By applying the ideal gas equation, the molecular mass of a gas can be calculated from its density. The relationship between the density and molecular weight of a gas is derived as follows.

From the ideal gas equation:

$$V = \frac{nRT}{P}$$

But:

$$n = \frac{\text{mass(g)}}{\text{molecular weight} \cdot \text{(g/mol)}} = \frac{m}{M_w}$$

Substituting for n in the ideal gas equation gives:

$$P = \frac{nRT}{V} = \frac{mRT}{VM_w}$$

This equation is the basis of the use of the Dumas method for finding the molar mass of volatile compounds.

Recall that density, ρ, is equal to m/V (see section 2.5), hence:

$$P = \rho \frac{RT}{M_w}$$

And:

$$M_w = \rho \frac{RT}{P}$$

So if you know the density of a gas at a given pressure and temperature, you can calculate its molecular weight.

Example 11.12

What is the density of nitrogen gas at STP?

Solution

1. Identify the known quantities and the unknown quantity:

$$\begin{aligned} \text{Known: } P &= 1.0 \text{ atm} \\ T &= 273 \text{ K} \\ M_w &= 28 \text{ g/mol} \\ R \text{ is a constant} &= 0.08206 \text{ L-atm/K-mol} \\ \text{Unknown: } \rho &= ? \end{aligned}$$

2. Then solve for the unknown quantity:

$$\rho = \frac{PM_w}{RT} = \frac{(1.0 \text{ atm})(28 \text{ g/mol})}{(0.08206 \text{ L-atm/K-mol})(273 \text{ K})} = 1.25 \text{ g/L}$$

Example 11.13

An organic vapor suspected to be CCl_4 has a density of 4.43 g/L at 714 mmHg and 125°C. What is the molecular mass of the gas?

Solution

1. Identify the known quantities and the unknown quantity:

$$\begin{aligned} \text{Known: } P &= 714 \text{ mmHg} \\ T &= 125 + 273 = 398 \text{ K} \\ \rho &= 4.43 \text{ g/L} \\ R &= 0.08206 \text{ L-atm/K-mol} \end{aligned}$$

Unknown: $M_w = ?$

2. Then solve for the unknown quantity:

$$M_w = \frac{\rho RT}{P} = \frac{(4.43 \text{ g/L})(0.08206 \text{ L-atm/K-mol})(398 \text{ K})}{\left(\dfrac{714 \text{ mmHg}}{760 \text{ mmHg/atm}}\right)} = 154.0 \text{ g/mol}$$

The molar mass is 154.0 g/mol, and this corresponds to the molar mass of CCl_4.

11.9. Molar Volume of an Ideal Gas

Molar volume is defined as the volume occupied by 1 mol of any gas at STP. Using the ideal gas equation, the molar volume of a gas has been determined experimentally to be 22.4 L. In other words, 1 mol of hydrogen gas will occupy 22.4 L, 1 mol of nitrogen gas will occupy 22.4 L, and 1 mol of oxygen will occupy 22.4 L at STP. However, the molar volumes of many real gases do show deviations from the ideal value, for reasons to be discussed later.

We can rearrange the ideal gas law:

$$PV = nRT$$

to calculate volume:

$$V = \frac{nRT}{P}$$

At STP, $T = 273.2$ K and $P = 1$ atm. R has the usual value of 0.08206 L-atm/K-mol. Since we want to find the volume of 1 mol of gas, $n = 1.0$ mol. Thus:

$$V = \frac{(1.000\ \text{mol})(0.08206\ \text{L-atm/K-mol})(273.2\ \text{K})}{1.000\ \text{atm}} = 22.42\ \text{L}$$

Example 11.14

Quicklime (CaO) is produced by the thermal decomposition of limestone ($CaCO_3$). Calculate the volume of carbon dioxide, CO_2, at 30°C and 700 mmHg given off by the decomposition of 300 g of $CaCO_3$.

Solution

1. Write a balanced equation for the reaction.

$$CaCO_3(s) \xrightarrow{\text{heat}} CaO(s) + CO_2(g)$$

2. Calculate molar masses of $CaCO_3$ and CO_2.

$$CaCO_3 = 100\ \text{g/mol}$$

$$CO_2 = 44\ \text{g/mol}$$

$$\text{Known: } 300\ \text{g of } CaCO_3$$

$$\text{Unknown} = \text{vol of } CO_2$$

3. Formulate appropriate conversion factors and calculate the volume of CO_2 produced at STP.

$$\text{vol of } CO_2 = 300\ \text{g } CaCO_3 \times \frac{1\ \text{mol } CaCO_3}{100\ \text{g } CaCO_3} \times \frac{1\ \text{mol } CO_2}{1\ \text{mol } CaCO_3} \times \frac{22.4\ \text{L } CO_2}{1\ \text{mol } CO_2} = 67.2\ \text{L}$$

4. Convert 67.2 L of CO_2 at STP to the volume at 30°C and 700 mmHg using the combined gas equation:

$$\frac{P_1V_1}{T_1} = \frac{P_2V_2}{T_2}$$

$$P_1 = 760\ \text{mmHg}, \quad V_1 = 67.2\ \text{L}, \quad T_1 = 273\ \text{K}$$

$$P_2 = 700 \text{ mmHg}, \quad T_2 = 30 + 273 = 303 \text{ K}$$

$$V_2 = (V_1)\left(\frac{T_2}{T_1}\right)\left(\frac{P_1}{P_2}\right) = 67.2 \text{ L} \times \left(\frac{273 \text{ K}}{303 \text{ K}}\right)\left(\frac{700 \text{ mm}}{760 \text{ mm}}\right) = 55.77 \text{ L}$$

Thus the decomposition of 300 g of $CaCO_3$ will liberate 55.8 L of CO_2 at 30°C and 700 mmHg.

11.10. Dalton's Law of Partial Pressures

John Dalton experimented with the pressure exerted by a mixture of gases, and in 1803 he summarized his observations in terms of the partial pressures of gases. The *partial pressure* of a gas is the pressure that it would exert on the container if it were present alone. Dalton's law of partial pressure can be stated as follows: *In a mixture of gases, which do not react chemically with one another, the total pressure exerted is the sum of the pressures that each gas would exert if it were alone.*

In mathematical terms, the law can be expressed as:

$$P_{\text{total}} = P_1 + P_2 + P_3 + \cdots,$$

where P_{total} is the total pressure of the mixture, and P_1, P_2, P_2, ..., are the partial pressures exerted by the individual gases 1, 2, 3, ..., that constitute the mixture.

If each gas obeys the ideal gas equation, then:

$$P_1 = n_1 \frac{RT}{V}, \quad P_2 = n_2 \frac{RT}{V}, \quad P_3 = n_3 \frac{RT}{V}, \quad \text{etc.}$$

Substituting for P_1, P_2, P_3, etc. in Dalton's law and factoring gives:

$$P_{\text{total}} = \frac{RT}{V}(n_1 + n_2 + n_3 + \cdots) = n_t \left(\frac{RT}{V}\right)$$

where $n_t = n_1 + n_2 + n_3 + \cdots$.

In other words, the total gas pressure depends on how many moles of gas are present, but not on their individual identities.

Example 11.15

A tank containing a mixture of He, Ne, Ar, and N_2 has a total pressure of 6 atm. The pressure exerted by He is 0.5 atm, by Ne is 1.0 atm, and by Ar, 2.5 atm. Calculate the partial pressure of N_2 in the tank.

Solution

Use the expression for total pressure and solve for the unknown (N_2):

$$P_t = P_{He} + P_{Ne} + P_{Ar} + P_{N_2}$$
$$P_{N_2} = P_t - (P_{He} + P_{Ne} + P_{Ar}) = 6.0 - 4.0 = 2.0 \text{ atm}$$

11.10.1. Collecting gases over water

In the laboratory, many gases such as oxygen and hydrogen are collected by allowing them to displace water from a container (downward displacement of water). The gas collected will be mixed with water vapor. The total pressure of the sample is the sum of the pressures of the gas collected and the water vapor present in the sample:

$$P_t = P_{gas} + P_{H_2O}$$

Example 11.16

What is the partial pressure of hydrogen gas collected over water at 25°C at a pressure of 745 torr? Water's vapor pressure at 25°C is 23.8 torr.

Solution

Applying Dalton's law we have:

$$P_t = P_{gas} + P_{H_2O}$$
$$P_{gas} = P_t - P_{H_2O} = 745 \text{ torr} - 23.8 \text{ torr} = 721.2 \text{ torr}$$

11.11. Partial Pressure and Mole Fraction

The mole fraction of a substance in a mixture is defined as the ratio of the number of moles of that substance to the total number of moles in the mixture. In a mixture containing three components, 1, 2, and 3, the mole fractions X_1, X_2, and X_3 of the components 1, 2, and 3 are expressed as:

$$X_1 = \frac{n_1}{n_1 + n_2 + n_3} = \frac{n_1}{n_{total}}$$

where $n_t = n_1 + n_2 + n_3$ and similarly for X_2 and X_3. From our discussion of Dalton's law, we can express P_t and P_1 as follows:

$$P_1 = n_1 \frac{RT}{V}, \quad P_{\text{total}} = n_{\text{total}} \frac{RT}{V}$$

$$\frac{P_1}{P_{\text{total}}} = \frac{n_1(RT/V)}{n_{\text{total}}(RT/V)} = \frac{n_1}{n_{\text{total}}}$$

From our definition of mole fraction, we can rewrite the above equation as:

$$\frac{P_1}{P_{\text{total}}} = \frac{n_1}{n_{\text{total}}} = X_1$$

Hence:

$$P_1 = X_1 P_{\text{total}}$$

In other words, the partial pressure of a gas in a mixture is directly related to its mole fraction.

Example 11.17

A flask contains 2.5 mol He, 1.5 mol Ar, and 3.0 mol O_2. If the total pressure is 700 mmHg, what is the mole fraction and partial pressure of each gas?

Solution

First we calculate the mole fraction of each gas using the definition or expression for mole fraction.

$$X_{\text{He}} = \frac{n_{\text{He}}}{n_{\text{He}} + n_{\text{Ar}} + n_{O_2}} = \frac{2.5 \text{ mol}}{2.5 \text{ mol} + 1.5 \text{ mol} + 3.0 \text{ mol}} = 0.357$$

$$X_{\text{Ar}} = \frac{n_{\text{Ar}}}{n_{\text{He}} + n_{\text{Ar}} + n_{O_2}} = \frac{1.5}{2.5} + 1.5 + 3.0 = 0.214$$

$$X_{O_2} = \frac{n_{O_2}}{n_{\text{He}} + n_{\text{Ar}} + n_{O_2}} = \frac{1.5}{2.5} + 1.5 + 3.0 = 0.429$$

We know the total pressure. Therefore we can calculate the partial pressure of each gas using the mole fraction:

$$P_{\text{He}} = X_{\text{He}} P_{\text{total}}$$

$$P_{\text{He}} = 0.357 \times 700 \text{ mmHg} = 249.9 \text{ mmHg}$$

$$P_{\text{Ar}} = X_{\text{Ar}} P_{\text{total}}$$

$$P_{Ar} = 0.214 \times 700 \text{ mmHg} = 149.8 \text{ mmHg}$$

$$P_{O_2} = \chi_{O_2} P_{total}$$

$$P_{O_2} = 0.429 \times 700 \text{ mmHg} = 300.3 \text{ mmHg}$$

11.12. Real Gases and Deviation from the Gas Laws

Real gases obey gas laws (i.e., behave like ideal gases) only at high temperatures and low pressures. At low temperatures or high pressures, real gases show marked deviations from the behavior described by the ideal gas law. There are two reasons for the deviation from the ideal behavior.

1. Real gas molecules have small but measurable volumes; the kinetic theory for ideal gases assumes that the particles have zero volume.
2. The force of attraction and repulsion between real gas molecules (at high concentrations) is not negligible. Gas molecules do attract each other, contrary to the assumption of the kinetic theory.

In 1873, Johannes van der Waals, a professor of physics at the University of Amsterdam, developed an equation for real gases. He modified the ideal gas law by adding correction factors to account for the effects of intermolecular attraction and repulsions.

The resulting van der Waals equation is:

$$P_{obs} = \frac{nRT}{V - nb} - a\left(\frac{n}{V}\right)^2$$

For easy comparison with the ideal gas law, this equation can be rearranged to give:

$$\left[P_{obs} + a\left(\frac{n}{V}\right)^2\right](\text{V} - \text{nb}) = nRT$$

The values $a(n/V)^2$ and nb correct for the attractive forces and volume of the gas molecules. The values for the van der Waals constants, a and b, are different for each gas and are determined experimentally. Values of a and b for some gases are represented in table 11-1. Professor van der Waals was awarded the Nobel Prize in 1910 for his work.

Example 11.18

Using the van der Waals equation, calculate the pressure for a 2.5-mol sample of ammonia gas confined to 1.12 L at 25°C; a = 4.17 atm-L^2/mol^2 and

Table 11-1 van der Waals Constants for Some Gases

Gas	a (atm-L^2/mol^2)	b (L/mol)
He	0.034	0.0237
Ne	0.211	0.0171
Ar	1.35	0.0322
Kr	2.32	0.0398
Xe	4.19	0.0511
H_2	0.244	0.0266
N_2	1.39	0.0391
O_2	1.36	0.0318
Cl_2	6.49	0.0562
H_2O	5.46	0.0305
CH_4	2.25	0.0428
CO_2	3.59	0.0427
NH_3	4.17	0.0371

$b = 0.0371$ L/mol. How does the result compare with predictions of the ideal gas law?

Solution

The following data are given:

$$V = 1.12 \text{ L}$$

$$n = 2.5 \text{ mol}$$

$$T = 25 + 273 = 298 \text{ K}$$

$$R = 0.08206 \frac{\text{L-atm}}{\text{mol-K}}$$

$$a = 4.17 \frac{\text{atm-L}^2}{\text{mol}^2}, \quad \text{and} \quad b = 0.0371 \frac{\text{L}}{\text{mol}}$$

Substitute the values given directly into the van der Waals equation:

$$P_{\text{obs}} = \frac{nRT}{V - nb} - a\left(\frac{n}{V}\right)^2$$

$$P_{\text{obs}} = \frac{(2.5 \text{ mol})(0.08206 \text{ L-atm/mol-K})(298 \text{ K})}{1.12 \text{ L} - (2.5 \text{ mol})(0.0371 \text{ L/mol})} - \left(4.17 \frac{\text{atm-L}^2}{\text{mol}^2}\right)\left(\frac{2.5 \text{ mol}}{1.12 \text{ L}}\right)^2$$

$$= 59.5101 \text{ atm} - 20.7769 \text{ atm} = 38.7332 \text{ atm} \approx 38.7 \text{ atm}$$

The prediction of the ideal gas would be:

$$P = \frac{nRT}{V}$$

$$P = \frac{(2.5 \text{ mol})(0.08206 \text{ L-atm/K-mol})(298 \text{ K})}{1.12 \text{ L}} = 54.5846 \text{ atm} \approx 54.6 \text{ atm}$$

The pressure predicted by the ideal gas law is 41% higher than that by the van der Waals equation.

11.13. Graham's Law of Diffusion

Graham's law of diffusion states that *at constant pressure and temperature, the rates of diffusion of gases are inversely proportional to the square roots of their vapor densities or molecular weights*. In mathematical terms, this can be stated as:

$$\text{Rate} \propto \frac{1}{\sqrt{d}}, \quad \text{or} \quad \text{Rate} = \frac{k}{\sqrt{d}}$$

For two gases A and B, diffusing at rates R_A and R_B at the same temperature and pressure:

$$\frac{R_A}{R_B} = \frac{\sqrt{d_B}}{\sqrt{d_A}} = \sqrt{\frac{d_B}{d_A}}$$

where d_A and d_B are the densities of gases A and B.

But density is proportional to the relative molecular mass M of a gas. Hence:

$$\frac{R_A}{R_B} = \sqrt{\frac{M_B}{M_A}}$$

Furthermore, rate is also inversely proportional to time of diffusion, t, i.e.:

$$R \propto \frac{1}{t}$$

Therefore:

$$\frac{R_A}{R_B} = \sqrt{\frac{M_B}{M_A}} = \frac{t_B}{t_A}$$

Example 11.19

An unknown gas X diffuses one-third as fast as ammonia, NH_3. What is the molecular mass of X?

Solution

Use the following expression:

$$\frac{R_X}{R_{NH_3}} = \sqrt{\frac{M_{NH_3}}{M_X}}$$

$$R_X = 1; \quad R_{NH_3} = 3; \quad M_{NH_3} = 17 \text{ g/mol}; \quad M_X = ?$$

$$\frac{1}{3} = \sqrt{\frac{17}{M_X}}$$

$$\left(\frac{1}{3}\right)^2 = \left(\sqrt{\frac{17}{M_X}}\right)^2 = \frac{17}{M_X}$$

$$\frac{1}{9} = \frac{17}{M_X}$$

$$M_x = 9 \times 17 \text{ g/mol}$$

$$M_X = 153 \text{ g/mol}$$

Example 11.20

In 45 s, 100 mL of oxygen diffused through a small hole, while 150 mL of an unknown gas *A* diffused through the same hole in 60 s. Calculate (a) the rates of diffusion of the two gases, (b) the molecular mass of gas *A*.

Solution

(a) Rate of diffusion $= \dfrac{\text{Volume}}{\text{Time}}$

Rate of diffusion of $O_2 = \dfrac{100 \text{ mL}}{45 \text{ s}} = 2.22 \text{ mL/s}$

Rate of diffusion of gas $A = \dfrac{250 \text{ mL}}{60 \text{ s}} = 4.17 \text{ mL/s}$

(b) Use Graham's law:

$$\frac{R_A}{R_{O_2}} = \sqrt{\frac{M_{O_2}}{M_A}}$$

$$\frac{4.17 \text{ mL/s}}{2.22 \text{ mL/s}} = \sqrt{\frac{32}{M_A}}$$

$$\left(\frac{4.17\ \text{mL/s}}{2.22\ \text{mL/s}}\right)^2 = \frac{32}{M_A}$$

$$3.528 = \frac{32}{M_A}$$

$$M_A = \frac{32}{3.528} = 9.09\ \text{g/mol}$$

11.14. Problems

1. A certain mass of gas occupies a volume of 500 mL at a pressure of 700 mmHg. If the temperature remains constant, what would be the pressure reading when the volume is increased to 1500 mmHg?
2. A sample of nitrogen gas occupies a volume of 1.5 L at 25°C at a pressure of 0.79 atm. Calculate the volume of the sample at 25°C and a pressure of 5.5 atm.
3. A 25-g sample of argon gas occupies 650 cm^3 at a pressure of 1.0×10^5 N m^{-2}. What will be its pressure if the volume is increased to 1950 cm^3, assuming there is no change in temperature?
4. A fixed mass of gas occupies a volume of 300 cm^3 at 25°C and at 101.3 kPa pressure. What is the volume of the gas at 5°C and a pressure of 101.3 kPa?
5. A fixed mass of gas initially occupying a volume of 430 cm^3 is heated from 0°C to 120°C at a constant pressure of 740 mmHg. What is the final volume?
6. A sample of gas occupies a volume of 15.0 L at 1.15 atm and 30°C. Find the temperature in °C at which its volume becomes 7.5 L, assuming that its pressure remains constant.
7. A 2.50-L container is filled with helium gas to a pressure of 700 mmHg at 225°C. Determine the volume of a new container that can be used to store the same gas at STP.
8. Find the temperature to which 20.0 L of CO_2 at 20°C and 650 mmHg must be brought in order to decrease the volume to 12 L at a pressure of 760 mmHg.
9. The volume of some SO_2 gas initially at 100°C is to be decreased by 25%. Calculate the final temperature to which the gas must be heated if the pressure of the gas is to be tripled at the same time.
10. A 3.0-L steel tank contains nitrogen gas at 23.5°C and at a pressure of 7.5 atm. What is the internal gas pressure when the temperature of the tank and its contents is raised to 95°C?
11. A 350 cm^3 sample of hydrogen gas was collected over water at 55°C and 745 mmHg. What would be the volume of the dry gas at STP? The vapor pressure of water at 55°C is 118.04 mmHg.

12. Butane (C_4H_{10}), which is used as fuel in some cigarette lighters, reacts with oxygen in air to produce steam and carbon dioxide.

    (a) Write a balanced chemical equation for this reaction.
    (b) If 1.50 cm^3 of C_4H_{10}, initially at 760 mmHg and 27°C, are used in the combustion process in a carefully controlled experiment, how many cm^3 of steam measured at 110°C and 800 mmHg are formed?

13. An ideal gas, originally at 1.25 atm and 45°C, was allowed to expand to a volume of 25 L at 0.75 atm and 95°C. What was the initial volume of this gas?
14. The production of chlorine trifluoride is given by the equation:

$$Cl_2(g) + 3\,F_2(g) \longrightarrow 2\,ClF_3(g)$$

    What volume of ClF_3 is produced if 10 L of F_2 are allowed to react completely with Cl?

15. A 1.25-mol sample of NO_2 gas at 1.25 atm and 50°C occupies a volume of 2.50 L. What volume would 1.25 mol of N_2O_5 occupy at the same temperature and pressure?
16. Using the standard condition values for P (atm), T (K), and V (L), evaluate the constant R in the ideal gas equation for 1 mol of an ideal gas.
17. What is the volume occupied by 10.25 g of CO_2 at STP?
18. What volume will 3.125 g of NH_3 gas occupy at 33°C and 0.85 atm?
19. A 50.0-L cylinder contains fluorine gas at 24°C and 2,650 mmHg. Find the number of moles of fluorine in the cylinder.
20. At STP, 0.50 L of a gas weighs 2.0 g. What is the molecular weight of the gas?
21. At 725 mmHg and 30°C, 0.725 g of a gas occupies 0.85 L. What is the molecular weight of the gas?
22. What is the density of sulfur trioxide (SO_3) in grams per liter at 700 mmHg and 27°C?
23. The molecular weight of an unknown gas was determined to be 88 g/mol at 25°C and 1 atmosphere pressure using the Dumas method. What is the density of the unknown gas?
24. One of the reactions involved in the production of elemental sulfur from the hydrogen sulfide content of sour natural gas is burning the gas in oxygen in the front-end furnace. The equation for the reaction is:

$$2\,H_2S(g) + 3O_2(g) \longrightarrow 2H_2O(g) + 2SO_2(g)$$

    If all gases are present at the same temperature and pressure, calculate:

    (a) The volume of oxygen needed to burn 1000 L of H_2S.
    (b) The volume of SO_2 formed.

25. A 19.51-g mixture of K_2CO_3 and $CaHCO_3$ was treated with excess hydrochloric acid (HCl). The CO_2 gas liberated occupied 4.0 L at 28°C and 0.95 atm. Calculate the percentage by mass of $CaHCO_3$ in the mixture.

26. 10 cm^3 of a gaseous hydrocarbon required 50 cm^3 of oxygen for complete combustion. If 30 cm^3 of CO_2 gas were evolved, what is the molecular formula of the hydrocarbon?

27. An organic compound containing only carbon, hydrogen, and oxygen was burned in a rich supply of oxygen. A 51 cm^3 sample of the organic compound needed 204 cm^3 of oxygen for complete combustion. If 153 cm^3 of CO_2 and 153 cm^3 of steam (H_2O) were collected:

    (a) Determine the molecular formula of the compound, and
    (b) Write a balanced equation to show the complete combustion in oxygen.

    All measurements are at the same temperature and pressure. *Hint*: the combustion process may be represented by the equation:

$$C_xH_yO_z + \left(x + \frac{y}{4} - \frac{z}{2}\right) O_2 \longrightarrow xCO_2 + \frac{y}{2} H_2O$$

28. Oxygen gas may be produced in the laboratory by decomposing solid $KClO_3$ at elevated temperature (products are KCl and O_2).

    (a) Write a balanced equation for the reaction.
    (b) Calculate the mass of $KClO_3$ needed to produce 10.5 L of oxygen at STP.

29. Acetylene gas may be prepared by the hydrolysis of calcium carbide according to the equation:

$$CaC_2 + 2H_2O \longrightarrow C_2H_2 + Ca(OH)_2$$

    Calculate the volume of acetylene gas collected at 25°C and 1.3 atm if 55 g of water were added to excess solid calcium carbide.

30. Dry air is composed of 78.08% N_2, 20.95% O_2, 0.93% Ar, and 0.036% CO_2 by volume. What is the partial pressure of each gas (N_2, O_2, Ar, and CO_2) if the total pressure is 1900 mmHg?

31. A mixture of gases contains 1.5 mol helium, 2.0 mol neon, and 2.5 mol argon.

    (a) What is the mole fraction of each gas in the mixture?
    (b) Calculate the partial pressure of each gas if the total pressure is 2.5 atm.

32. At STP, a 25.0-L flask contains 8.25 g of oxygen, 12.50 g of nitrogen, and 15.75 g of carbon dioxide. Calculate:

    (a) The mole percent of each gas in the mixture.
    (b) The volume percent of nitrogen in the mixture.

33. What are the relative rates of diffusion of the following pairs of gases?

    (a) N_2 and CO_2
    (b) H_2 and O_2
    (c) Ne and Ar

34. At STP the density of helium is 0.18 g/L, and of oxygen is 1.43 g/L. Under the same conditions of temperature and pressure, determine which gas effuses faster through a small orifice in a cylinder and by how much.
35. In 25 minutes, 300 mL of sulfur dioxide diffused through a porous partition. Under the same conditions of temperature and pressure, 250 mL of an unknown gas Y diffused through the same partition in 9.11 minutes. Calculate: (a) the rates of diffusion of the two gases, (b) the molecular mass of gas Y.
36. The van der Waals constants for steam, H_2O, are $a = 5.464\ L^2atm/mol^2$, $b = 0.03049$ L/mol. Calculate the pressure of 2.50 mol of steam contained in a volume of 5.0 L at 100°C, (a) assuming H_2O obeys the ideal gas law, (b) using the van der Waals equation.
37. A 4.5-L pressurized gas cylinder was filled with 12.25 mol of hydrogen gas at 25°C. Find the pressure of the gas (a) using the ideal gas law, (b) using the van der Waals equation. The van der Waals constants for hydrogen, H_2, are $a = 0.244\ L^2atm/mol^2$, $b = 0.0266$ L/mol.
38. A 1.0-mol sample of dry air contains 0.7808 mol N_2, 0.2095 mol O_2, 0.0093 mol Ar, and 0.00036 mol CO_2 confined to 0.750 L at – 25°C. (a) Using the van der Waals equation of state, calculate the partial pressures of the individual component gases. (b) Use these partial pressures to predict the pressure of a 1.00-mol air sample under these conditions. Select the appropriate van der Waals constants from table 11-1.

12

Liquids and Solids

12.1. The Liquid State

The atoms or molecules in a liquid have enough kinetic energy to partially overcome the forces of attraction between them. Therefore, they are in constant random motion (as in a gas) but they are still relatively close together. However, they are not as tightly packed, or as well ordered, as in a solid. There is not as much free space in a liquid as in a gas. The atoms or molecules may aggregate together to form chains or rings that readily move relative to one another; this gives a liquid its fluid (flow) properties.

Liquids generally occur as compounds. For example, water, ethanol, and carbon tetrachloride are liquids at room temperature. However, a few elements are also liquids at room temperature: bromine, cesium, gallium, mercury, and rubidium.

12.1.1. Properties of liquids

A liquid is characterized by the following physical properties: boiling point and freezing point, density, compressibility, surface tension, and viscosity. These properties of a liquid are greatly influenced by the strength of its intermolecular forces.

In summary:

- Liquids have definite volume but no definite shape. They take on the shape of their containers.
- Liquids are characterized by low compressibility, low rigidity, and high density relative to gases.
- Liquids diffuse through other liquids.
- Liquids can vaporize into the space above them and produce a vapor pressure.

12.2. Polar Covalent Bonds and Dipole Moment

Polar molecules possess an electric dipole moment, μ, defined as the product of the magnitude of the partial charges Q^+ and Q^- on the molecule and the distance r separating the charges. In mathematical terms, it is given by the equation:

$$\mu = Qr$$

The unit for μ is debyes (D), and 1 D = 3.336×10^{-30} coulomb meter (C-m).

No interatomic bonds are completely ionic. Knowing the dipole moment of a compound, though, lets us differentiate ionic from covalent bonds by calculating the percent ionic character for the bonds. The percent ionic character of a bond is found by comparing the measured dipole moment of the molecule of the type $A-B$ with the calculated dipole moment for the 100% ionized compound A^+B^-. The expression for percent ionic character is:

$$\%\text{ Ionic character} = \left(\frac{\mu_{\text{measured for } A-B}}{\mu_{\text{calculated for } A^+B^-}}\right) \times 100\%$$

Example 12.1

The dipole moment of HBr is $\mu = 0.82$ D and the bond length is 1.41Å. Calculate the magnitude of charge in coulombs that would results in this dipole moment. Note that 1 D = 3.34×10^{-30}C-m and 1 Å = 1×10^{-10} m.

Solution

First write the expression for calculating the dipole moment. Then solve for Q.

$$\mu = Qr$$

$$Q = \frac{\mu}{r}$$

Now calculate Q by substituting the values of μ and r in the equation.

$$Q = \frac{\mu}{r} = \frac{(0.82\text{ D})\left(\dfrac{3.34 \times 10^{-30}\text{ C-m}}{1\text{ D}}\right)}{(1.41\text{ Å})\left(\dfrac{1 \times 10^{-10}\text{ m}}{1\text{ Å}}\right)} = 1.94 \times 10^{-20}\text{ C}$$

Example 12.2

The experimentally determined dipole moment of HI is 0.44 D and the interatomic distance between H and I is 1.61 Å. Calculate the percent ionic character of HI.

Solution

First, assume that HI is 100% ionic, with a positively charged H ion separated from the negatively charged I ion by 1.61 Å. Then calculate the theoretical dipole moment for the ionized HI:

$$\mu(\text{theoretical}) = (\text{Charge on 1 electron}) \times (\text{interatomic distance})$$

$$= \left(1.60 \times 10^{-19}\ \text{C}\right)(1.61\ \text{Å})\left(\frac{10^{-10}\ \text{m}}{1\ \text{Å}}\right)\left(\frac{1\ \text{D}}{3.34 \times 10^{-30}\ \text{C-m}}\right)$$

$$= 7.713\ \text{D}$$

To calculate the percent ionic character, use the following expression:

$$\%\ \text{ionic character} = \left(\frac{\mu_{(\text{observed})}}{\mu_{(\text{theoretical})}}\right) \times 100\%$$

$$= \left(\frac{0.44\ \text{D}}{7.713\ \text{D}}\right) \times 100\%$$

$$= 5.70\%$$

With respect to the measured dipole moment, this result suggests that the $H-I$ bond is only about 5.7% ionic in character.

12.3. Vapor Pressure and the Clausius-Clapeyron Equation

The vapor pressures of all substances increase with increasing temperature. The quantitative dependence of a liquid's vapor pressure on the absolute temperature is known as the Clausius-Clapeyron equation:

$$\ln P = \frac{-\Delta H_{\text{vap}}}{RT} + C$$

where P is the vapor pressure, ΔH_{vap} is the molar heat of vaporization, and R is the gas constant (8.314 J/K-mol), T is the absolute temperature, and C is a constant. The equation can be expressed in a linear form as:

$$\ln P = \left(\frac{-\Delta H_{\text{vap}}}{R}\right)\left(\frac{1}{T}\right) + C$$

$$\updownarrow \qquad \updownarrow \qquad \updownarrow \qquad \updownarrow$$

$$y = \quad (m) \quad . \quad (x) + b$$

The above equation predicts that a graph of ln P against 1/T should give a straight line with a slope equal to $-\Delta H_{\text{vap}}/R$. From the slope, the heat of vaporization, which is assumed to be independent of temperature, can be determined:

$$\Delta H_{vap} = -\text{slope} \times R$$

The Clausius-Clapeyron equation can be used to calculate the vapor pressure of a liquid at various temperatures if ΔH_{vap} and the vapor pressure at one temperature are known. Suppose the vapor pressure of a liquid at temperature T_1 is P_1 and at temperature T_2 is P_2; then we can write the corresponding Clausius-Clapeyron equations as:

$$\ln P_1 = \frac{-\Delta H_{vap}}{RT_1} + C$$

$$\ln P_2 = \frac{-\Delta H_{vap}}{RT_2} + C$$

Subtracting the first equation from the second equation:

$$\ln P_2 - \ln P_1 = \frac{\Delta H_{vap}}{R}\left(\frac{1}{T_1} - \frac{1}{T_2}\right)$$

On rearrangement, the equation becomes:

$$\ln \frac{P_2}{P_1} = \frac{\Delta H_{vap}}{R}\left(\frac{1}{T_1} - \frac{1}{T_2}\right) \quad \text{or} \quad \frac{\Delta H_{vap}}{R}\left(\frac{T_2 - T_1}{T_1 T_2}\right)$$

Converting this equation to ordinary logarithms gives:

$$\log \frac{P_2}{P_1} = \frac{\Delta H_{vap}}{2.303\ R}\left(\frac{1}{T_1} - \frac{1}{T_2}\right) \quad \text{or} \quad \frac{\Delta H_{vap}}{2.303\ R}\left(\frac{T_2 - T_1}{T_1 T_2}\right)$$

which lets us calculate the vapor pressure P_2 at any temperature T_2 if we know the vapor pressure at any other pressure.

Example 12.3

The vapor pressure of water at 50°C is 92.4 mmHg and its molar heat of vaporization is 42.21 kJ/mol. Calculate the vapor pressure at the boiling point of water (100°C).

Solution

To begin solving this problem, organize the data, convert ΔH_{vap} from kJ/mol to J/mol, and change temperatures in degrees Celsius to kelvins.

$$\Delta H_{vap} = 42.21\ \text{kJ/mol} = 42210\ \text{J/mol}$$

$$T_1 = 50°\text{C} = 323\ \text{K}; \quad T_2 = 100°\text{C} = 373\ \text{K}$$

$$P_1 = 92.4\ \text{mmHg}; \quad P_2 = ?$$

Next, write the modified Clausius-Clapeyron equation and solve for P_2 using the given data.

$$\ln\frac{P_2}{P_1} = \frac{\Delta H_{\text{vap}}}{R}\left(\frac{1}{T_1}-\frac{1}{T_2}\right) \quad \text{or} \quad \frac{\Delta H_{\text{vap}}}{R}\left(\frac{T_2-T_1}{T_1T_2}\right)$$

$$\ln\frac{P_2}{92.4} = \frac{42210\ \text{J/mol}}{8.314\ \text{J/K}}\left(\frac{373\ \text{K}-323\ \text{K}}{373\ \text{K}\times 323\ \text{K}}\right)$$

$$\ln\frac{P_2}{92.4} = 2.107$$

$$\frac{P_2}{92.4} = e^{2.107} = 8.22$$

$$\frac{P_2}{92.4} = 8.22, \quad \Rightarrow P_2 = 759.9\ \text{mmHg} \approx 760\ \text{mmHg}$$

The calculated value agrees with the value known by definition.

Example 12.4

The vapor pressure of ethanol at 30°C is 98.5 mmHg and the molar heat of vaporization is 39.3 kJ/mol. Determine the approximate normal boiling point of ethanol from these data.

Solution

We are required to solve for the normal boiling point, T_2. At this temperature, the vapor pressure is equal to atmospheric pressure, i.e., 760 mmHg. To begin, we organize the data, and change temperatures in degrees Celsius to kelvins.

$$T_1 = 30°\text{C} = 300\ \text{K}; \quad T_2 = ?$$

$$P_1 = 98.5\ \text{mmHg}; \quad P_2 = 760\ \text{mmHg}$$

$$\Delta H_{\text{vap}} = 39.3\ \text{kJ/mol}$$

Next, write the modified Clausius-Clapeyron equation and solve for T_2.

$$\ln\frac{P_2}{P_1} = \frac{\Delta H_{\text{vap}}}{R}\left(\frac{1}{T_1}-\frac{1}{T_2}\right)$$

$$\ln\frac{760}{98.5} = \frac{39300}{8.314}\left(\frac{1}{300}-\frac{1}{T_2}\right)$$

$$4.322\times 10^{-4} = \left(\frac{1}{300}-\frac{1}{T_2}\right)$$

$$\frac{1}{T_2} = \frac{1}{300} - 4.322 \times 10^{-4} = 0.0029$$

$$T_2 = 344.7 \text{ K} \quad \text{or} \quad 71.7°\text{C}$$

Example 12.5

The vapor pressure of diethyl ether at 19°C is 400 mmHg and its normal boiling point is 34.6°C. Calculate the molar enthalpy of vaporization of diethyl ether. (The pressure at the boiling point is 1 atm = 760 mmHg.)

Solution

To begin solving this problem, organize the data, and change temperature in degrees Celsius to kelvins.

$$T_1 = 19°\text{C} = 292 \text{ K}; \quad T_2 = 34.6°\text{C} = 307.6 \text{ K}$$

$$P_1 = 400 \text{ mmHg}; \quad P_2 = 760 \text{ mmHg}$$

$$\Delta H_{\text{vap}} = ?$$

Next, write the modified Clausius-Clapeyron equation and solve for ΔH_{vap} using the given data.

$$\ln\frac{P_2}{P_1} = \frac{\Delta H_{\text{vap}}}{R}\left(\frac{T_2 - T_1}{T_1 T_2}\right)$$

$$\ln\frac{760}{400} = \frac{\Delta H_{\text{vap}}}{8.314 \text{ J/K.mol}}\left(\frac{307.6 \text{ K} - 292 \text{ K}}{307.6 \text{ K} \times 292 \text{ K}}\right)$$

$$0.642 = 2.08 \times 10^{-5} \Delta H_{\text{vap}}$$

$$\Delta H_{\text{vap}} = 30{,}732 \text{ J/mol} = 30.7 \text{ kJ/mol}$$

12.4. The Solid State

Unlike liquids and gases, solids have a definite volume and shape. This is because the structural units (atoms, ions, or molecules) that make up the solid are held in close proximity and (usually) rigid order by chemical bonds, electrostatic attraction, or intermolecular forces.

12.4.1. Types of solids

Solids may be crystalline or amorphous. A *crystalline solid* is composed of one or more crystals characterized by a well-defined three-dimensional ordering of the basic structural units. An *amorphous solid* is one that has a random and disordered three-dimensional arrangement of basic structural units. An amorphous solid lacks long-range molecular order.

There are four different types of crystalline solids, distinguished by the forces holding the structural units together: covalent, ionic, metallic, and molecular solids.

Covalent solids: A covalent solid is a solid made up of atoms joined together by covalent bonds. Examples include diamond, graphite, quartz (SiO_2), and silicon carbide (SiC).

Ionic solids: An ionic solid consists of positive and negative ions held together in an ordered three-dimensional structure by electrostatic attraction between oppositely charged nearest neighbors. The crystal is therefore held together by ionic bonds. Examples include NaCl, CsI, and $CaSO_4$.

Metallic solids: A metallic solid consists of cations held together by a "sea" of delocalized electrons. Examples include K, Fe, Zn, and Cu.

Molecular solids: A molecular solid is composed of atoms and molecules held together by intermolecular forces such as dispersion forces, dipole–dipole forces and hydrogen bonds. Examples include solid water (ice), solid carbon dioxide (dry ice), solid argon, and sucrose.

12.4.2. Crystal lattices

A *crystal* is a solid with a regular, three-dimensional arrangement of atoms, molecules, or ions. Crystals often have plane surfaces, sharp edges, and regular polyhedral shapes, but the defining characteristic is the regular internal arrangement.

The constituents of a crystal are placed according to a geometrical pattern known as the *crystal lattice* or *space lattice*, which is a system of points representing the repeating pattern. The *crystal structure* of a solid is generated by an indefinite (or infinite) three-dimensional repetition of identical structural units called the unit cell. Crystal lattices may be *primitive*, with one lattice point per unit cell, or *centered*, having two or four points per cell.

12.4.3. Unit cells

A *unit cell* is the fundamental repeating structural unit of a crystalline solid. It is the smallest fraction of the crystal lattice that contains a representative portion of the crystal structure. Thus when a unit cell is stacked repeatedly in three dimensions without gaps, it reproduces the entire crystal structure. In general the unit cell is described by the lengths of its edges (a, b, c) and the angles between the edges (α, β, γ).

12.5. The Crystal System

There are seven basic shapes of unit cells, commonly referred to as the seven crystal systems. They are described in table 12-1. From symmetry considerations,

Table 12-1 The Unit Cell Relationships for the Seven Crystal Systems

Crystal system	Bravais lattices (symbol)	Unit cell		Example
		Lengths	Angles	
Cubic	P, I, F	$a = b = c$	$\alpha = \beta = \gamma = 90°$	NaCl (rock salt)
Tetragonal	P, I	$a = b \neq c$	$\alpha = \beta = \gamma = 90°$	TiO_2 (rutile)
Orthorhombic	P, C, I, F	$a \neq b \neq c$	$\alpha = \beta = \gamma = 90°$	$MgSO_4 \cdot 7H_2O$ (epsomite)
Monoclinic	P, C	$a \neq b \neq c$	$\alpha = \beta = 90°; \gamma \neq 90°$	$KClO_4$ (potassium chlorate)
Triclinic	P	$a \neq b \neq c$	$\alpha \neq \beta \neq \gamma \neq 90°$	$K_2Cr_2O_7$ (potassium dichromate)
Hexagonal	P	$a = b \neq c$	$\alpha = \beta = 90°; \gamma \neq 120°$	SiO_2 (Silica)
Rhombohedral	R	$a = b = c$	$\alpha = \beta = \gamma \neq 90°$	$CaCO_3$ (calcite)

variations of the seven types of unit cells are possible. If we combine the crystal systems (primitive translational symmetry) and the centering translations, this will give rise to the 14 possible unique crystal lattices, also known as the Bravais Lattices (see table 12-1). Hence 14 different unit cell geometries can occur in a crystalline solid.

In this book, however, we will consider only unit cells with cubic symmetry. There are three types of cubic unit cells. They are primitive or simple cubic, body-centered cubic, and face-centered cubic.

12.5.1. Close-packed structures

Most metals and several molecular substances have close-packed structures. These can be thought of as consisting of identical objects packed together as closely as possible. There are two possible close-packed arrangements:

1. Cubic close-packed (ccp)
2. Hexagonal close-packed (hcp)

These two arrangements represent the most efficient way of packing spheres. In this book, we will focus only on cubic close-packed structures.

12.5.2. Cubic unit cells

There are three types of cubic unit cells. They are primitive or simple cubic, body-centered cubic, and face-centered cubic. These are shown in figure 12-1.

Guidelines for determining the number of atoms in a unit cell

1. An atom occupying a corner is shared equally among 8 adjacent unit cells. Therefore it contributes only 1/8 of an atom to each unit cell.
2. An atom occupying an edge is shared equally among 4 adjacent unit cells. Therefore it contributes only 1/4 of an atom to each unit cell.

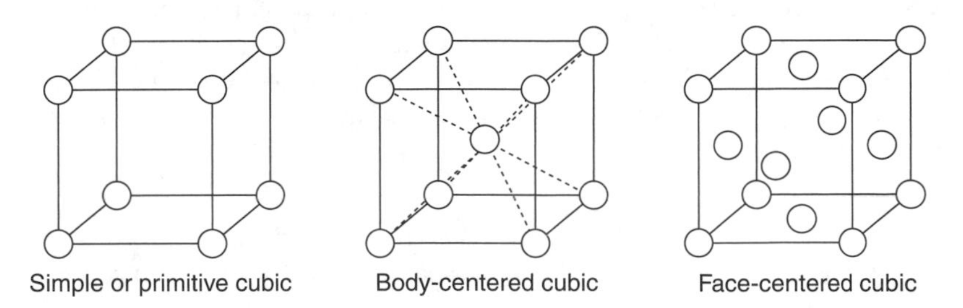

Figure 12-1 Types of cubic cells.

3. An atom occupying a face is shared equally between 2 adjacent unit cells. Therefore it contributes only 1/2 of an atom to each unit cell.
4. An atom completely inside the unit cell, i.e., a body-centered atom, is not shared with neighboring unit cells. Therefore, a body-centered atom contributes 1 atom to the unit cell.

12.5.3. Simple or primitive cubic unit cell

A simple cubic (SC) unit cell has one lattice point (or atom) at each of its 8 corners; each corner atom is shared with neighboring cells such that only 1/8 of the atom is in each unit cell. Therefore the number of atoms per unit cell is:

$$\left(\frac{\frac{1}{8}\text{ atom}}{\text{corner}}\right) \times \frac{8\text{ corners}}{\text{unit cell}} = 1\frac{\text{atom}}{\text{unit cell}}$$

In this book, the number of atoms in a unit cell is denoted by the letter "Z," so that $Z = 1$ for the primitive unit cell.

12.5.4. Body-centered cubic unit cell

A body-centered cubic (BCC) unit cell has one lattice point or atom in the center of each unit cell and one atom at each of its eight corners. Each corner atom is shared with neighboring cells such that only 1/8 of the atom is in each unit cell. There is one body center per unit cell and one atom per body center. Therefore the number of atoms per unit cell is:

$$\left(\frac{\frac{1}{8}\text{ atom}}{\text{corner}}\right) \times \left(\frac{8\text{ corners}}{\text{unit cell}}\right) + \left(\frac{1\text{ body center}}{\text{unit cell}}\right)\left(\frac{1\text{ atom}}{\text{body center}}\right) = 2\frac{\text{atoms}}{\text{unit cell}}$$

Thus $Z = 2$ for the BCC unit cell.

12.5.5. Face-centered cubic unit cell

A face-centered cubic (FCC) unit cell has one atom in each face of the unit cell and one atom at each corner. There are 8 corners and 6 faces per unit cell. Each corner atom is shared with neighboring cells such that only 1/8 of the atom is in each unit cell. Also, each face contributes 1/2 face atom to each unit cell. Therefore the number of atoms per unit cell is:

$$\left(\frac{\frac{1}{8}\text{ atom}}{\text{corner}}\right) \times \left(\frac{8\text{ corners}}{\text{unit cell}}\right) + \left(\frac{6\text{ faces}}{\text{unit cell}}\right)\left(\frac{\frac{1}{2}\text{ atom}}{\text{face}}\right) = 4\frac{\text{atoms}}{\text{unit cell}}$$

Thus $Z = 4$ for the FCC unit cell.

Example 12.6

Tungsten (W) crystallizes in a body-centered cubic lattice. What is the number of W atoms per unit cell?

Solution

The BCC system contains one atom in the center of each unit cell and one atom at each corner. The number of atoms per unit cell is:

$$\left(\frac{\frac{1}{8}\text{ W atom}}{\text{corner}}\right) \times \left(\frac{8\text{ corners}}{\text{unit cell}}\right) + \left(\frac{1\text{ body center}}{\text{unit cell}}\right)\left(\frac{1\text{ W atom}}{\text{body center}}\right) = 2\,\frac{\text{W atoms}}{\text{unit cell}}$$

Example 12.7

Copper crystallizes in a face-centered cubic system with a unit cell edge of 3.62 Å at 20°C. Calculate the number of copper atoms in one of the unit cells.

Solution

The FCC system contains one atom in each face of each unit cell and one atom at each corner. The number of atoms per unit cell is:

$$\left(\frac{\frac{1}{8}\text{ Cu atom}}{\text{corner}}\right) \times \left(\frac{8\text{ corners}}{\text{unit cell}}\right) + \left(\frac{6\text{ faces}}{\text{unit cell}}\right)\left(\frac{\frac{1}{2}\text{ Cu atom}}{\text{face}}\right) = 4\,\frac{\text{Cu atoms}}{\text{unit cell}}$$

12.5.6. Coordination number

Coordination number, as used here, is defined as the number of nearest neighbors of an atom or an ion in a crystal structure.

12.6. Calculations Involving Unit Cell Dimensions

The structure and dimensions of a unit cell can be determined by X-ray diffraction studies (described in section 12.9). Once these are known, we can calculate several physical parameters such as volume, density, atomic mass, and Avogadro's number.

The theoretical density of a crystalline solid can be computed from the following equation:

$$\rho = \frac{ZM}{N_A V}$$

where ρ is the density, Z is the total number of atoms in the unit cell, M the molecular mass, N_A Avogadro's number, and V is the volume of the unit cell. For a cubic unit cell, $V = a^3$ where a is the unit cell edge.

Example 12.8

Calcium crystallizes in a cubic lattice system with a unit cell edge of 5.6 Å. The density of Cu is 1.55 g/cm^3.

(a) Calculate the number of Ca atoms per unit cell.
(b) Use the answer in part (a) to determine what cubic crystal system Ca belongs to.

Solution

(a) Step 1: Determine the mass of Ca in one unit cell using the density and unit cell dimensions.

$$\text{Density} = \frac{\text{Mass of unit cell}}{\text{Volume of unit cell}}$$

$$\text{Mass of unit cell} = \text{Density} \times \text{Volume of unit cell}$$

For a cubic unit cell,

$$V = a^3 = \left(5.6\ \text{Å} \times \frac{1\ \text{cm}}{10^8\ \text{Å}}\right)^3 = 1.756 \times 10^{-22}\ \text{cm}^3$$

$$\text{Mass of unit cell} = \left(1.55\ \frac{\text{g Ca}}{\text{cm}^3}\right)\left(1.756 \times 10^{-22}\frac{\text{cm}^3}{1\ \text{unit cell}}\right)$$

$$= 2.722 \times 10^{-22}\frac{\text{g Ca}}{\text{unit cell}}$$

Step 2: Calculate the mass of a Ca atom from its atomic mass and N_A (i.e., 6.02×10^{23} atoms/mol).

$$\text{Mass of 1 Ca atom} = \left(\frac{40.08\ \text{g}}{1\ \text{mol Ca}}\right)\left(\frac{1\ \text{mol Ca}}{6.02 \times 10^{23}\ \text{atoms Ca}}\right)$$

$$= \frac{6.66 \times 10^{-23}\text{g Ca}}{\text{atom Ca}}$$

Step 3: Determine the number of Ca atoms per unit cell from the mass of Ca per unit cell and the mass of a Ca atom.

$$\text{Number of Ca atoms per unit cell} = \left(2.722 \times 10^{-22} \frac{\text{g Ca}}{\text{unit cell}}\right) \times \left(\frac{1 \text{ atom Ca}}{6.66 \times 10^{-23} \text{g Ca}}\right) = 4 \frac{\text{atoms Ca}}{\text{unit cell}}$$

(b) The number of atoms per unit cell corresponds to a face-centered cubic system. That is:

$$\left(\frac{\frac{1}{8} \text{ Cu atom}}{\text{corner}}\right) \times \left(\frac{8 \text{ corners}}{\text{unit cell}}\right) + \left(\frac{6 \text{ faces}}{\text{unit cell}}\right)\left(\frac{\frac{1}{2} \text{ Cu atom}}{\text{face}}\right) = 4 \frac{\text{Cu atoms}}{\text{unit cell}}$$

Example 12.9

Elemental silver, Ag, crystallizes in a face-centered cubic lattice with a unit cell edge of 4.10 Å. The density of Ag is 10.51 g/cm^3. Determine the number of atoms in 108.0 g of Ag.

Solution

Step 1: Calculate the volume of the unit cell:

$$V = a^3 = \left(4.10 \text{ Å} \times \frac{1 \text{ cm}}{10^8 \text{ Å}}\right)^3 = 6.89 \times 10^{-23} \frac{\text{cm}^3}{\text{unit cell}}$$

Step 2: Calculate the volume of 108 g of Ag:

$$\text{Volume} = \frac{\text{Mass}}{\text{Density}} = \left(\frac{108 \text{ g}}{10.51 \text{ g/cm}^3}\right) = 10.28 \text{ cm}^3$$

Step 3: Calculate the number of unit cells in this volume and hence the total number of atoms:

$$\text{Number of unit cells} = \left(10.28 \text{ cm}^3\right)\left(\frac{1 \text{ unit cell}}{6.89 \times 10^{-23} \text{cm}^3}\right) = 1.50 \times 10^{23} \text{ unit cells}$$

There are 4 atoms per unit cell in a face-centered cubic system. Thus:

$$\text{Total number of atoms} = \left(4\ \frac{\text{atoms}}{\text{unit cell}}\right)\left(1.50 \times 10^{23}\ \text{unit cells}\right)$$
$$= 6.0 \times 10^{23}\ \text{atoms}$$

Example 12.10

Nickel crystallizes in a face-centered cubic structure with a unit cell edge length of 352.5 pm. (a) Calculate the radius of a Ni atom in pm. (b) Calculate the density of Ni in g/cm^3.

Solution

(a) In an fcc unit cell, the face atoms touch the corner atoms along the diagonal of each face (see figure 12-2). However, the corner atoms do not touch each other along the edges. Therefore, the length of the diagonal is equal to 4 × the radius ($4r$). Note that the diagonal and the edges of the cube form a right-angled triangle. Hence we can solve for r using the Pythagorean theorem.

$$a^2 + a^2 = (4\text{r})^2$$

$$2a^2 = 16r^2$$

$$r = \sqrt{\frac{2a^2}{16}} = \sqrt{\frac{a^2}{8}} = \sqrt{\frac{352.5^2}{8}} = 124.6\ \text{pm}$$

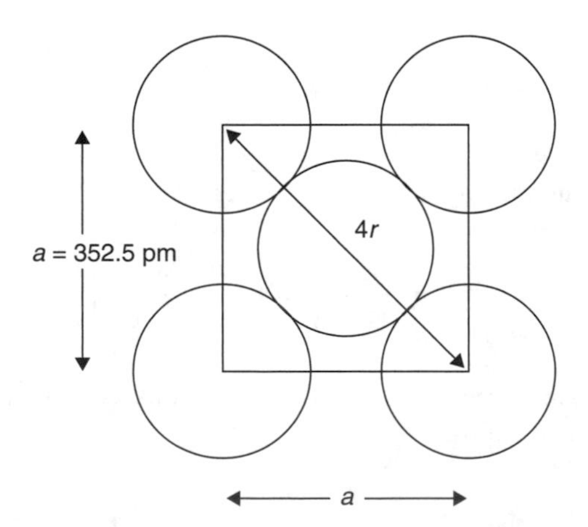

Figure 12-2 A representation of the unit cell of nickel, indicating the relative sizes and positions of the atoms.

(b) Calculate the density of Ni.

$$\text{Density } (d) = \frac{\text{mass of a single unit cell } (m)}{\text{volume of a single unit cell } (v)}$$

$$\text{Volume} = a^3 = (352.5 \text{ pm})^3 = \left(\frac{4.38 \times 10^7 \text{ pm}^3}{\text{unit cell}}\right)\left(\frac{10^{-10} \text{ cm}}{1 \text{ pm}}\right)^3$$

$$= 4.38 \times 10^{-23} \text{ cm}^3$$

To calculate the mass of a unit cell, we will need to find the number of atoms per unit and then convert the number of atoms to mass in grams using the atomic mass and Avogadro's number, N_A.

In an FCC cubic system, there are 4 atoms per unit cell, i.e., $\left[\left(\frac{1}{8} \times 8\right) + \left(\frac{1}{2} \times 6\right) = 4\right]$. The atomic mass of Ni is 58.7 g/mol and $N_A = 6.022 \times 10^{23}$ atoms/mol.

$$\text{Mass of a Ni unit cell} = (4 \text{ atoms})\left(\frac{58.7 \text{ g/mol}}{6.022 \times 10^{23} \text{ atom/mol}}\right)$$

$$= 3.90 \times 10^{-22} \text{ g}$$

$$\text{Density} = \frac{3.90 \times 10^{-22} \text{ g}}{4.38 \times 10^{-23} \text{ cm}^3} = 8.90 \text{ g/cm}^3$$

Example 12.11

An unknown metal M crystallizes in a body-centered cubic structure with a unit cell dimension of 315 pm. (a) Calculate the number of M atoms per unit cell. (b) If the measured density of M is 10.15 g/cm^3, determine the relative atomic mass and use the periodic table to identify the unknown metal.

Solution

(a) In a BCC unit cell, there are 8 corner atoms and 1 center atom. Therefore the number of M atoms will be

$$\left(\frac{\frac{1}{8} \text{ atom}}{\text{corner}}\right) \times \left(\frac{8 \text{ corners}}{\text{unit cell}}\right) + \left(\frac{1 \text{ body center}}{\text{unit cell}}\right)\left(\frac{1 \text{ atom}}{\text{body center}}\right) = 2 \frac{M \text{ atoms}}{\text{unit cell}}$$

(b) To calculate the atomic mass of M, we will need to know the volume of the unit cell. This can be calculated from the cell dimensions.

$$\text{Volume} = d^3 = (315 \text{ pm})^3 = \left(\frac{3.126 \times 10^7 \text{ pm}^3}{\text{unit cell}}\right)\left(\frac{10^{-10} \text{ cm}}{1 \text{ pm}}\right)^3$$

$$= 3.126 \times 10^{-23} \text{ cm}^3$$

We can now calculate the atomic mass of M from the mass of the unit cell.

$$\text{Density} = \frac{\text{Mass}}{\text{Volume}}$$

$$10.15\ \frac{\text{g}}{\text{cm}^3} = \frac{\text{Mass}}{3.126 \times 10^{-23}\text{cm}^3}$$

$$\text{Mass of an } M \text{ unit cell} = 3.172 \times 10^{-22}\ \text{g}$$

$$3.172 \times 10^{-22}\ \text{g} = \frac{(2\ \text{atoms})(\text{Relative atomic mass})}{6.022 \times 10^{23}\text{atom/mol}}$$

$$\text{Relative atomic mass} = 95.5\frac{\text{g}}{\text{mol}}$$

From the periodic table M is most likely Mo, which has a relative atomic mass of 95.9.

Example 12.12

Metallic gold (Au) crystallizes in a face-centered cubic system with a unit cell dimension of 4.07 Å. Determine (a) the distance between centers of nearest neighbors, (b) the radius of a gold atom, (c) the volume of a gold atom in cm^3, given that for a spherical atom, $V = \frac{4}{3}\pi r^3$, (d) the per cent volume of a unit cell occupied by Au atoms, (e) the density of Au.

Solution

(a) Figure 12-3 represents the face of an fcc unit cell. Each sphere is a gold atom. Therefore, the distance between nearest neighbors is one-half the distance

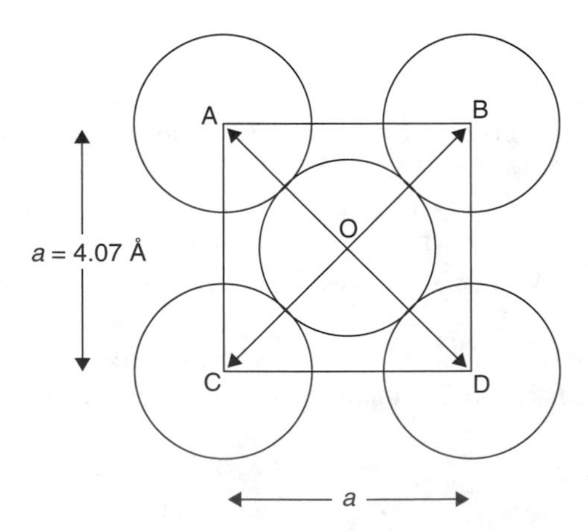

Figure 12-3 One face of the face-centered cubic unit cell of gold.

of the diagonal AD (for example) of the inscribed square. Using Pythagoras' theorem, we can calculate the distance AD, and $AD/2$ is the distance between centers of neighbors.

$$(AD)^2 = a^2 + a^2 = 2a^2$$

$$AD = \sqrt{2a^2} = a\sqrt{2} = 1.414a$$

$$AD = 1.414 \times 4.07\ \text{Å} = 5.76\ \text{Å}$$

Distance between centers of nearest neighbors $= \frac{AD}{2} = \frac{5.76}{2} = 2.88$ Å

(b) Take a close look at the diagram in part (a). Notice that the atoms touch along the diagonals of each face but not on the edges of the cube. The length of the diagonal is $4r$ (or $r + 2r + r$). Once again we can use Pythagoras' theorem to show the relationship between the unit cell dimension and the radius of the atom.

$$a^2 + a^2 = (4r)^2$$

$$2a^2 = (4r)^2 = 16r^2$$

$$a^2 = 8r^2; \quad a = r\sqrt{8}$$

$$r = \frac{a}{\sqrt{8}} = \frac{4.07\ \text{Å}}{\sqrt{8}} = 1.44\ \text{Å}$$

(c) To calculate the volume of a gold atom, substitute the radius of Au in the volume equation:

$$V_{\text{Au atom}} = \frac{4}{3}\pi r^3 = \left(\frac{4}{3}\right)(\pi)\left(1.44\ \text{Å} \times \frac{10^{-8}}{\text{Å}}\right)^3 = 1.25 \times 10^{-23}\ \text{cm}^3$$

(d) There are 4 atoms in an fcc crystal lattice unit cell. Each Au atom occupies a volume of 1.25×10^{-23} cm^3. Therefore, 4 Au atoms will occupy $V_{\text{Au atoms}} = (4)\left(1.25 \times 10^{-23}\ \text{cm}^3\right) = 5.00 \times 10^{-23}$ cm^3. This is the volume of the Au atoms themselves.

Now calculate the volume of the unit cell using $V = a^3$ (since the unit cell is cubic):

$$V_{\text{unit cell}} = a^3 = \left(4.07\ \text{Å} \times \frac{10^{-8}\ \text{cm}}{\text{Å}}\right)^3 = 6.74 \times 10^{-23}\ \text{cm}^3$$

$$\%\ \text{Volume of unit cell occupied by Au atoms} = \frac{V_{\text{Au atoms}}}{V_{\text{unit cell}}} \times 100$$

$$= \frac{5.00 \times 10^{-23}\ \text{cm}^3}{6.74 \times 10^{-23}\ \text{cm}^3} \times 100$$

$$= 74.2\%$$

(e) To calculate the density of Au, we need to determine the mass of the unit cell.

$$\text{Mass of a Au unit cell} = (4\ \text{atoms})\left(\frac{197\ \text{g/mol}}{6.022 \times 10^{23}\ \text{atom/mol}}\right)$$

$$= 1.31 \times 10^{-21}\ \text{g}$$

$$\text{Density} = \frac{\text{Mass of unit cell}}{\text{Volume of unit cell}} = \frac{1.31 \times 10^{-21}\ \text{g}}{6.74 \times 10^{-23}\ \text{cm}^3} = 19.42\ \text{g/cm}^3$$

12.7. Ionic Crystal Structure

Most salts crystallize as ionic solids with anions and cations occupying the unit cell instead of atoms. The sizes of cations are usually different from those of anions. Hence ionic crystal structures are different from those of metals. The structure, properties, and stability of ionic compounds depend on the relative radii of the constituent ions. Ionic compounds of the general formula $M^{n+}X^{x-}$ usually crystallize in the NaCl structure (face-centered cubic lattice), CsCl structure (simple cubic lattice), or zincblende (ZnS) structure (face-centered cubic lattice).

12.7.1. The sodium chloride (NaCl) or "rock-salt" structure

The NaCl structure is based on a face-centered cubic lattice (see figure 12-4). It consists of cubic-close-packed chloride anions with sodium cations occupying all the available octahedral holes to achieve the required 1:1 stoichiometry. Each sodium ion is surrounded by six chloride ions located at the corners of a regular octahedron. Similarly, each chloride ion is surrounded octahedrally by six sodium ions, resulting in a coordination number of 6:6. The NaCl structure is generally adopted when the cation to anion radius ratio is in the range 0.41–0.73. Examples of compounds with the NaCl structure include halides of Li, Na, K, and Rb; as well as divalent metal oxides and sulfides such as CaO, MgO, MnO, and CaS.

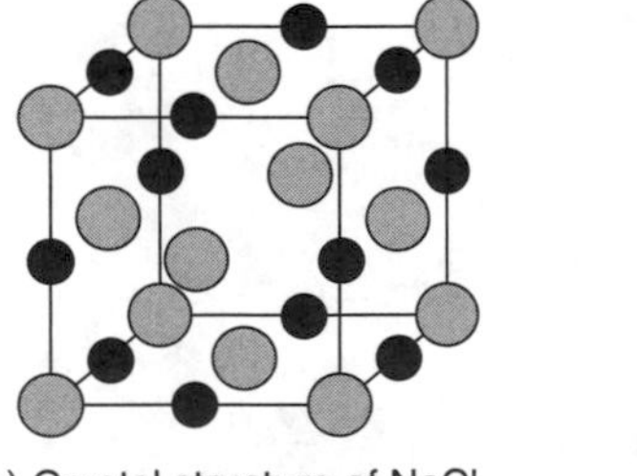

(a) Crystal structure of NaCl

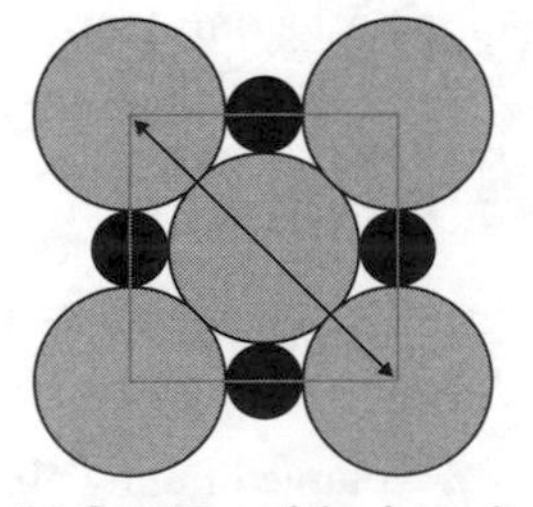

(b) One face of the fcc unit

Figure 12-4 The unit cell of NaCl showing (a) skeletal view and (b) space-filling view. The larger chloride anions adopt an fcc unit cell with the smaller sodium cations occupying the holes between adjacent anions.

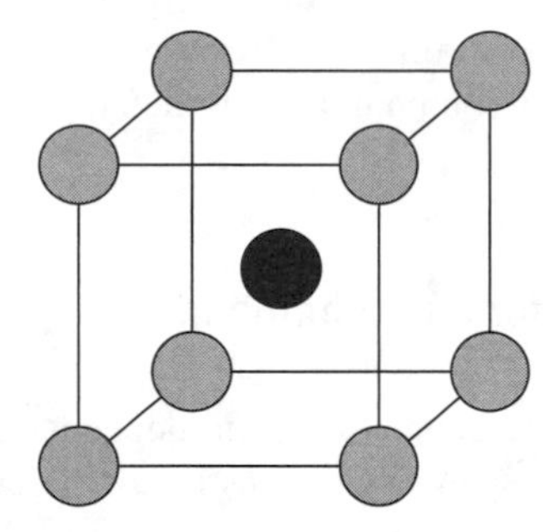

Figure 12-5 The unit cell of CsCl.

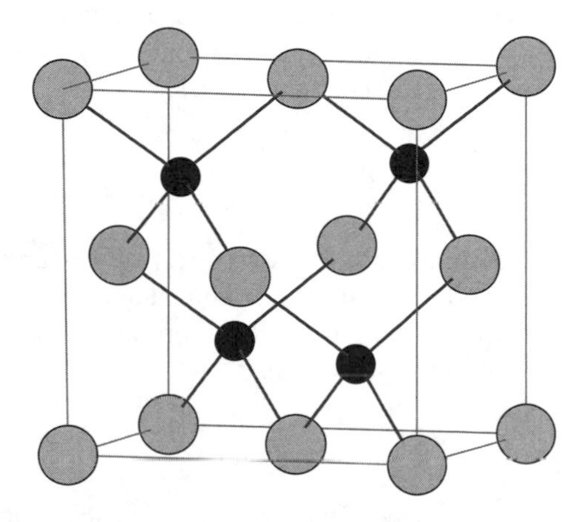

Figure 12-6 The structure of zincblende (ZnS).

12.7.2. The cesium chloride structure

In the CsCl structure, a Cs cation is located at the center of the cube with eight chloride ions occupying the eight corners (see figure 12-5). Similarly, each chloride ion is at the center of a cube, surrounded by eight nearest-neighbor cesium ions. The coordination number of Cs and Cl is 8:8, with one Cs ion and one Cl ion per unit cell. The CsCl structure is preferred when the cation/anion radius ratio is 0.73 or greater. Examples of compounds with the CsCl structure include the halides of monovalent Cs, Tl, and NH_4^+.

12.7.3. The Structure of zincblende

Zinc sulfide occurs in two crystalline forms, zincblende and wurtzite. The zincblende structure is based on the face-centered cubic lattice (see figure 12-6). This structure is adopted when the cation to anion radius ratio is between 0.22 and 0.41. The structure can be viewed as a cubic close-packed (fcc) array of S atoms with Zn ions occupying half of the tetrahedral sites, resulting in a ZnS stoichiometry with 8 Zn and 16 S ions in the unit cell. Examples of compounds with the zincblende structure include BeS, CuCl, CdS, HgS, and SiC.

12.8. Radius Ratio Rule for Ionic Compounds

The structure of a crystalline ionic compound depends on the coordination number of each ion and the relative sizes of the constituent cations and anions. Each ion in an ionic crystal always makes an effort to surround itself with as many oppositely charged ions as possible. This results in a stable and neutral structure. The radius ratio is defined as the ratio of cation radius to anion radius; its value is useful in determining the coordination number of the smaller ion, which in most cases is the cation. Table 12-2 shows the coordination number, crystal structure, and limiting values of the radius ratio for several common cases.

Example 12.13

The unit cell of potassium chloride, KBr, is a face-centered cube with a unit cell edge length of 6.60 Å. Assuming anion–anion contact and anion–cation contact, calculate (a) the ionic radius of the bromide ion, (b) the ionic radius of the potassium ion, (c) the radius ratio.

Solution

(a) Figure 12-7 shows one face of the face-centered cubic unit cell of KBr. Notice that the bromide ions touch along the diagonal d, which is also the hypotenuse of the two right isosceles triangles, and is equal to four times the radius of

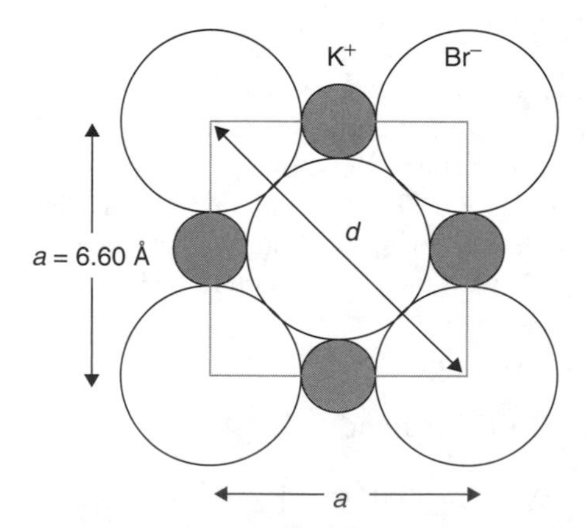

Figure 12-7 One face of the face-centered cubic unit cell of KBr. Notice that the bromide ions touch along the diagonal d.

Table 12-2 Radius Ratio and Crystal Types of Selected Ionic Compounds

Formula	*Crystal structure type*	*Limiting radius ratio* (r^+/r^-)	*n number of* M^-	*Coordination number of* X^-	*Examples with radius ratio in parentheses*
MX	Cesium chloride	>0.73	8	8	CsCl (0.93) and TICL (0.83)
MX_2	Fluoride	>0.73	8	4	CaF_2 (0.73) and PbF_2 (0.88)
MX	Sodium chloride	0.41–0.73	6	6	Nacl (0.52) and AgCL (0.70)
MX_2	Rutile	0.41–0.73	6	3	MgF_2 (0.48) and ZnF_2 (0.54)
MX	Zincblende	<0.41	4	4	ZnS (0.40) and BeO (0.22)

the bromide ion. That is, $d = r_{Br^-} + 2r_{Br^-} + r_{Br^-} = 4r_{Br^-}$. We can calculate d using Pythagoras' theorem, since we know the sides of the right-angled triangle ($a = 6.60$ Å).

$$d^2 = a^2 + a^2 = 2a^2$$

$$d = \sqrt{2a^2} = \sqrt{(2)(6.60)^2} = 9.33 \text{ Å}$$

$$\text{But } d = 4r_{Br^-} = 9.33 \text{ Å}$$

$$r_{Br^-} = \frac{8.91 \text{ Å}}{4} = 2.33 \text{ Å}$$

(b) The unit cell edge length, a, is two times the radius of the bromide ion plus two times the radius of the potassium ion.

$$a = 2r_{K^+} + 2r_{Br^-} = 6.60 \text{ Å}$$

From part (a), $r_{Br^-} = 2.33$ Å

$$2r_{K^+} = 6.60 \text{ Å} - 2r_{Br^-} = 6.60 - 2 \times 2.33$$

$$2r_{K^+} = 1.93 \text{ Å}$$

$$r_{K^+} = 0.97 \text{ Å}$$

(c) $\text{Radius ratio} = \dfrac{\text{Radius of cation}}{\text{Radius of anion}} = \dfrac{r_{K^+}}{r_{Br^-}} = \dfrac{0.97}{2.33} = 0.41$

Example 12.14

CsBr crystallizes in the CsCl structure with a bromide ion at each corner and a cesium ion at the center of the unit cell. The unit cell edge length is 4.29 Å. Calculate (a) the number of ions per unit cell, (b) the distance between the center of a cesium ion and the center of the nearest bromide ion, (c) the radius of the bromide ion, if the radius of the cesium ion is 1.75 Å, (d) the density of CsBr in g/cm^3.

Solution

(a) One Cs^+ ion is at the center of the cell and is in contact with eight corner Br^- ions. Therefore the number of Br^- ions in the unit cell is:

$$8 \text{ corners} \times \frac{1}{8}\left(\frac{Br^-}{\text{corner}}\right) = 1\ Br^-$$

Thus the unit cell of CsBr contains the equivalent of 1 Cs^+ion and 1 Br^- ion. This is consistent with one formula unit per unit cell.

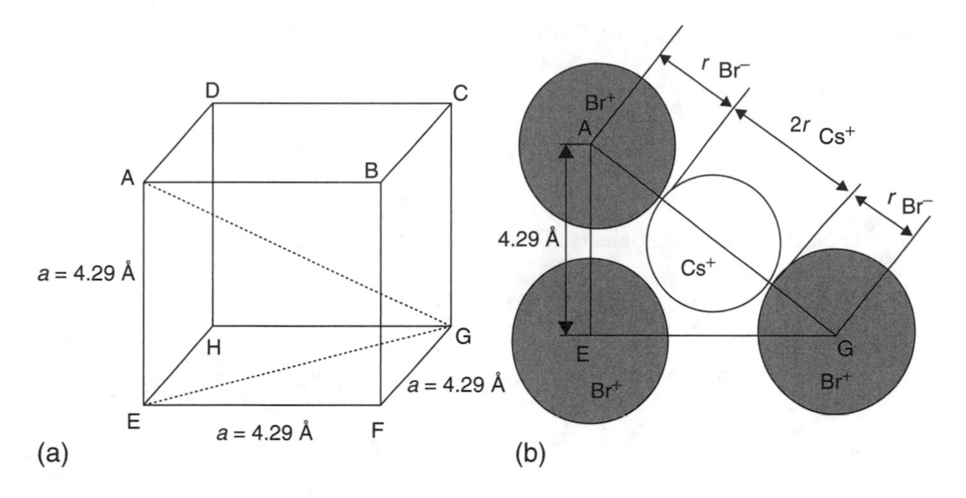

Figure 12-8 CsBr unit cell geometry.

(b) In a body-centered cubic unit cell, all the atoms or ions touch along a diagonal passing through the center of the cube, as illustrated in figure 12-8(b). So the shortest distance between the center of a cesium ion and the center of the nearest bromide ion is along the diagonal of the unit cell and is $(AG)/2$ from figure 12-8(a). Therefore we need to calculate the length of the diagonal AG from figure 12-4(a). By Pythagoras' theorem:

$$(AG)^2 = (AE)^2 + (EG)^2$$

$$\text{But } (EG)^2 = (EF)^2 + (FG)^2$$

$$(AG)^2 = (AE)^2 + (EF)^2 + (FG)^2 \quad (\text{or } 3a^2)$$

$$(AG)^2 = 4.29^2 + 4.29^2 + 4.29^2 = 55.2\ \text{Å}^2$$

$$AG = 7.43\ \text{Å}$$

The distance between the centers of Cs^- and Br^- is 7.43/2 or 3.72 Å.

(c) Since we know the Cs–Br distance and the radius of the cesium ion (1.75 Å), we can easily find the radius of Br^-.

$$2r_{Cs^+} + 2r_{Br^-} = AG$$

$$r_{Cs^+} + r_{Br^-} = \frac{AG}{2} = 3.72\ \text{Å}$$

$$1.75\ \text{Å} + r_{Br^-} = 3.72\ \text{Å}$$

$$r_{Br^-} = 1.97\ \text{Å}$$

(d) To find the density of CsBr, first we need to calculate the volume of the unit cell.

$$V = a^3 = \left(4.29\ \text{Å} \times \frac{10^{-8}\ \text{cm}}{1\ \text{Å}}\right)^3 = 7.90 \times 10^{-23}\ \text{cm}^3$$

Next we calculate density. There are two ions per unit cell, 1 Cs^+ and 1 Br^-:

$$\text{MW of CsBr} = 213\ \text{g/mol}$$

$$\text{Mass of per unit cell} = \left(213\ \frac{\text{g CsBr}}{\text{mol CsBr}}\right)\left(\frac{1\ \text{mol CsBr}}{6.02 \times 10^{23}\ \text{CsBr}}\right)$$

$$= 3.538 \times 10^{-22}\ \text{g}$$

$$\text{Density} = \frac{\text{Mass}}{\text{Volume}} = \frac{3.54 \times 10^{-22}\ \text{g}}{7.90 \times 10^{-23}\ \text{cm}^3} = 4.47\ \text{g/cm}^3$$

Example 12.15

Predict the unit cell structure of CaS given that the ionic radii of Ca^{2+} and S^{2-} are 0.99 Å and 1.84 Å respectively.

Solution

$$\text{Radius ratio} = \frac{r_{Ca^{2+}}}{r_{S^{2-}}} = \frac{0.99}{1.84} = 0.54$$

CaS should crystallize in the NaCl unit cell structure.

Example 12.16

The cubic unit cell of TlCl is similar to that of CsCl and has a radius ratio of 0.83. Estimate the ionic radius of Tl^+ given that the ionic radius of Cl^- is 1.81 Å.

Solution

$$\text{Radius ratio} = \frac{r_{Tl^+}}{r_{Cl^-}}$$

The radius ratio for TlCl is 0.83

$$0.83 = \frac{r_{Tl^+}}{1.84} \quad \Rightarrow \quad r_{Tl^+} = 1.53\ \text{Å}$$

The ionic radius of Tl^+ is 1.53 Å.

12.9. Determination of Crystal Structure by X-ray Diffraction

X-rays are electromagnetic radiation of very short wavelength, comparable to the distance between atoms or (lattice points) in a crystal. When passed through crystals, X-rays are scattered—diffracted—in ways which depend on the arrangement of atoms, molecules, or ions within the crystals. Thus, X-rays can be used to determine the important parameters of crystal structure.

In 1913, W. H. Bragg and W. L. Bragg (father and son) formulated a fundamental geometrical equation to explain X-ray diffraction. To understand how a diffraction pattern may be produced, consider the reflections of monochromatic X-rays of wavelength λ incident on a pair of atoms in two different layers of a crystalline solid, as shown in figure 12-9. Let the lattice-point spacing between the two layers be d, and let θ be the grazing angle. The lower wave travels an extra distance equal to the sum of $BC + CD$. This is equal to the path difference between the two reflected waves. The two waves will be in phase after reflection and reinforce each other as they exit the crystal if the path difference (i.e., $BC + CD$) is an integral number of wavelengths ($n = 1, 2, 3 \ldots$).

In mathematical terms,

$$BC + CD = n\lambda$$

Applying trigonometry, we can show that

$$BC + CD = 2d \sin\theta$$

where θ is the angle between the X-rays and the plane of the crystal, and d is the separation between adjacent planes.

Combining these two equations yields

$$2d \sin\theta = n\lambda \qquad \text{(where } n = 1, 2, 3 \ldots)$$

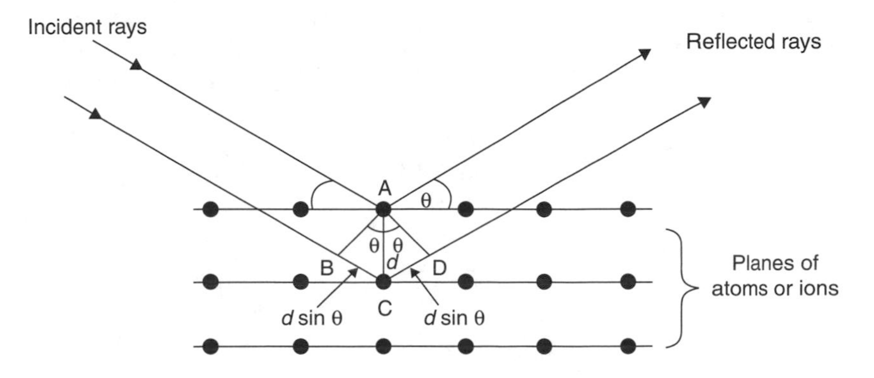

Figure 12-9 Reflection of X-rays by atoms in two different planes of a crystal.

This is referred to as the Bragg equation or Bragg's law. For a first-order or principal reflection, $n = 1$, so the above equation becomes $2d \sin\theta = \lambda$.

Example 12.17

The principal diffraction angle of a crystal is 28° for X-rays of wavelength 1.62 Å. What is the interplanar spacing?

Solution

To determine the interplanar distance d, we use Bragg's equation:

$$n\lambda = 2d \sin\theta$$

Given:

$$n = 1 \text{ for principal diffraction angle}$$
$$\lambda = 1.62 \text{ Å}$$
$$\theta = 28°$$

Solve for d by substituting in Bragg's equation:

$$d = \frac{n\lambda}{2\sin\theta} = \frac{1 \times 1.62 \text{ Å}}{2 \times \sin 28} = \frac{1.62 \text{ Å}}{0.9389} = 1.725\text{Å}$$

Example 12.18

What is the principal diffraction angle of a crystal if X-rays of wavelength 1.54 Å are used? The sets of parallel crystal planes are 1.90 Å apart.

Solution

To determine the diffraction angle, we use Bragg's equation:

$$n\lambda = 2d \sin\theta$$

Given:

$$n = 1 \text{ for principal diffraction angle}$$
$$\lambda = 1.54 \text{ Å}$$
$$d = 1.90 \text{ Å}$$

Solve for θ by substituting in Bragg's equation:

$$\sin\theta = \frac{n\lambda}{2d} = \frac{1 \times 1.54\ \text{Å}}{2 \times 1.90\ \text{Å}} = 0.4053$$

$$\theta = 23.9°$$

Example 12.19

The so-called Miller indexes are very helpful in describing the separation between atomic planes. For example, for a cubic lattice, the separation of the (*hkl*) planes is given by the equation:

$$d_{hkl} = \frac{a}{\left(h^2 + k^2 + l^2\right)^{1/2}}$$

where a is the unit cell length. Sodium chloride crystallizes in a face-centered cubic lattice with a unit cell edge length of 5.64 Å. Calculate the perpendicular distances (d_{hkl}) between the planes having the following Miller indices (100), (111), (210), and (221).

Solution

Simply substitute the given information in the given equation.

$$d_{hkl} = \frac{a}{\left(h^2 + k^2 + l^2\right)^{1/2}}$$

$$d_{100} = \frac{5.64\ \text{Å}}{\left(1^2 + 0^2 + 0^2\right)^{1/2}} = 5.64\ \text{Å}$$

$$d_{111} = \frac{5.64\ \text{Å}}{\left(1^2 + 1^2 + 1^2\right)^{1/2}} = 3.26\ \text{Å}$$

$$d_{210} = \frac{5.64\ \text{Å}}{\left(2^2 + 1^2 + 0^2\right)^{1/2}} = 2.52\ \text{Å}$$

$$d_{221} = \frac{5.64\ \text{Å}}{\left(2^2 + 2^2 + 1^2\right)^{1/2}} = 1.88\ \text{Å}$$

Example 12.20

Cesium metal crystallizes in a body-centered cubic lattice with a unit cell length of 6.10 Å. Calculate the glancing angle at which the reflection from the (222) planes of this crystal lattice will be observed when an X-ray beam of wavelength 1.54 Å is used.

Solution

Use the equation given below to solve for d_{222}. Then substitute this value in the Bragg equation and solve for the angle θ.

$$d_{hkl} = \frac{a}{\left(h^2 + k^2 + l^2\right)^{1/2}}$$

$$d_{222} = \frac{6.10\ \text{Å}}{\left(2^2 + 2^2 + 2^2\right)^{1/2}} = 1.76\ \text{Å}$$

$$d_{222} = \frac{\lambda}{2\sin\theta} \quad \Rightarrow \quad \sin\theta = \frac{\lambda}{2d_{222}} = \frac{1.54\ \text{Å}}{2 \times 1.76\ \text{Å}} = 0.4375$$

$$\theta = 25.9°$$

12.10. Problems

1. The bond length and dipole moment of HI are 1.61 Å and 0.45 D respectively. Calculate the magnitude of the charge, in units of electronic charge e, on the hydrogen and iodine.
2. The distance between the centers of H and F (i.e., bond length) in an HF molecule is 0.92 Å. Given that the dipole moment is 1.82 D, what is the percent ionic character of the molecule?
3. The interatomic distance between Cs and I in a CsI molecule is 332 pm. Calculate the dipole moment of CsI if the percent ionic character of CsI is believed to be 96%.
4. In a laboratory experiment, a student determined the vapor pressure of ethanol to be 350 mmHg at 60°C. If the normal boiling point of ethanol is 78.3°C, calculate the molar enthalpy of vaporization of ethanol.
5. The normal boiling point of mercury is 357°C and its molar heat of vaporization is 59 kJ/mol. At what pressure will mercury have its boiling point reduced by 37°C?
6. The molar heat of vaporization of chloroform ($CHCl_3$) is 29.4 kJ/mol. If the vapor pressure at 80°C is 1455 mmHg, what is the normal boiling point of $CHCl_3$?
7. At 30°C and 100°C the vapor pressure of carbon tetrachloride (CCl_4) are 142.5 mmHg and 1463 mmHg respectively. Calculate the heat of vaporization of CCl_4.
8. The heat of vaporization of water is 40.8 kJ/mol and its vapor pressure at 50°C is 92.7 mmHg. What will be the vapor pressure at 110°C?

9. The vapor pressure of hexafluorobenzene, C_6F_6, is 92.5 mm at 27°C and 483 mm at 67°C. Calculate (a) the molar heat of vaporization, (b) the vapor pressure at 0°C, (c) the normal boiling point of C_6F_6.
10. The vapor pressure of methyl iodide (CH_3I) at −45.5°C and 25.5°C are 9.5 mm and 400 mm respectively. Calculate (a) the molar heat of vaporization, (b) the vapor pressure at 100°C, (c) the normal boiling point of CH_3I.
11. Aluminum (Al) crystallizes in a face-centered cubic cell. Determine the number of Al atoms in each unit cell.
12. Vanadium (V) crystallizes in a body-centered cubic cell. Calculate the number of V atoms in each unit cell.
13. The element iridium (Ir) crystallizes in a face-centered cubic cell, and the length of the unit cell is 3.83 Å. Calculate the density of Ir in g/cm^3.
14. Polonium (Po) crystallizes in a simple cubic structure with a unit cell edge of 3.34 Å. Calculate (a) the number of Po atoms in each unit cell, (b) the volume of the unit cell, (c) the density of Po in g/cm^3.
15. Sodium chloride (NaCl) crystallizes in a face-centered cubic lattice. (a) Calculate the number of Na^+ ions and Cl^- ions in each NaCl unit cell. (b) If the density of NaCl is 2.15 g/cm^3, what is the unit cell length in Å?
16. An unknown metal *M* crystallizes in a face-centered cubic structure with a unit cell length of 4.08 Å. The density was determined to be 19.3 g/cm^3. What is the atomic mass of the metal?
17. Nickel metal (Ni) has a face-centered cubic unit cell with an edge length of 3.52 Å. The density of Ni is 8.90 g/cm^3. Use these data to calculate the value of Avogadro's number. (The atomic mass of Ni is 58.7 g/mol.)
18. Gold (Au) crystallizes in a face-centered cubic lattice with a unit cell length of 4.08 Å. What is the radius of a gold atom in Å?
19. Europium crystallizes in a body-centered cubic lattice. The unit cell edge length is 4.60 Å. Calculate the radius of a europium atom.
20. Aluminum crystallizes in a cubic lattice with a unit cell edge of 4.04 Å. The density is 2.70 g/cm^3. How many Al atoms are in each unit cell?
21. Zinc oxide crystallizes with a cubic crystal structure. The ionic radii are 0.74 Å for Zn^{2+} and 1.40 Å for O^{2-}. Calculate the radius ratio and suggest the possible crystal structure of ZnO.
22. Indium phosphide (InP) crystallizes with a cubic structure. The ionic radii are 0.81 Å for In^{3+} and 2.12 Å for P^{3-}. Calculate the radius ratio and describe the crystal structure of InP.
23. X-rays of wavelength 1.54 Å are diffracted from a gallium crystal at an angle of 12.7°. What is the spacing between the planes of gallium atoms responsible for this diffraction, assuming $n = 1$?
24. The distance between layers of atoms parallel to the cube face of a sodium chloride crystal is 2.80 Å. What X-ray wavelength will be diffracted at an angle of 6.25° from this crystal, assuming $n = 1$?

25. Bragg's law states that $n\lambda = 2d \sin\theta$ where $n = 1, 2, 3\ldots$ are called the first-order reflection, second-order reflection, third-order reflection, etc. If X-rays of wavelength 1.54 Å are diffracted from a NaCl crystal, calculate the glancing angles at which the first, second, and third order reflections are obtained from atomic planes separated by 2.80 Å.
26. Gold metal (Au) crystallizes in a face-centered cubic lattice with a unit cell length of 4.08 Å. Calculate the separations of the planes (100), (111), and (311).
27. Cesium bromide (CsBr) crystallizes in a body-centered cubic structure and has a density of 4.44 g/cm^3. Determine (a) the unit cell edge length, (b) the separations of the planes (002), (211), and (222).
28. Diamond crystallizes in a face-centered cubic lattice. There are 4 unshared atoms in the unit cell along with 8 atoms shared by 8 unit cells at the corners and 6 atoms shared by 2 unit cells on the faces. The reflection from the (222) planes was observed at a glancing angle of 20.2° using X-rays of wavelength 0.712 Å. Calculate (a) the unit cell edge length, (b) the density of diamond in g/cm^3.

13

Solution Chemistry

13.1. Solution and Solubility

13.1.1. Some definitions

A *solution* is a homogeneous mixture of two or more substances. It is usually made up of a solute and a solvent. Generally,

$$\text{Solute} + \text{Solvent} = \text{Solution}$$

A *solute* is any substance that is dissolved in a solvent. For example, when granulated sugar dissolves in water to give a clear sugar solution, the sugar is the solute, while water is the solvent. Relative to the solvent, a solute is usually present in small amounts.

A *solvent* is any substance in which a solute dissolves. It is usually the part of the solution that is present in the largest amount.

Two liquids are said to be *miscible* if they form a single phase (homogeneous solution) or dissolve in each other in all proportions. For example, ethanol and water are miscible. If two liquids do not form a single phase (or do not dissolve in each other) in any appreciable amount, they are said to be *immiscible*. For example, water and oil are immiscible. When mixed, they separate into two distinct layers.

Substances that are only slightly soluble in a given solvent are said to be *insoluble*.

An *aqueous* solution is one in which water is the solvent.

A *dilute* solution is one that contains a small amount of solute compared to the maximum amount the solvent can dissolve at that temperature.

A *concentrated* solution is one that contains a large amount of solute compared to the maximum amount the solvent can dissolve at that temperature.

A *saturated* solution is one that is in equilibrium with undissolved solute at a given temperature and pressure:

$$\text{Solute}_{(\text{solid})} \rightleftharpoons \text{Solute}_{(\text{dissolved})}$$

In other words, it contains the *maximum amount of solute that can be dissolved* at that particular temperature.

An *unsaturated* solution contains less solute than the maximum amount (saturated solution) possible at the same temperature.

A *supersaturated* solution is a solution that contains more solute than the saturated solution at the same temperature. This type of solution is very unstable. When it is agitated, or a speck of the solute is added to it, the excess solute will begin to crystallize out rapidly from the solution until the concentration becomes equal to that of the saturated solution.

The *solubility* of a solute in a solvent, at a given temperature, is the maximum amount of solute in grams that can dissolve in 100 g of solvent at that temperature.

13.2. Concentration of Solutions

13.2.1. Percent by mass

The *percent concentration of a solution* may be expressed as the *mass of the solute per 100 mass units of solution.*

$$\begin{aligned}\%\ \text{Solute} &= \frac{\text{Mass of solute}}{\text{Mass of solution}} \times 100\% \\ &= \frac{\text{Mass of solute}}{\text{Mass of solute} + \text{Mass of solvent}} \times 100\%\end{aligned}$$

For example, a 10% solution of NaCl contains 10 g of NaCl per 100 g of solution.

Example 13.1

Calculate the percentage of Na_2SO_4 in a solution prepared by dissolving 20 g of Na_2SO_4 in 80 g of H_2O.

Solution

$$\begin{aligned}\%\ Na_2SO_4 &= \frac{\text{Mass of } Na_2SO_4}{\text{Mass of } Na_2SO_4 + \text{Mass of } H_2O} \times 100 \\ &= \frac{20\ \text{g}}{20\ \text{g} + 80\ \text{g}} \times 100 = 20\%\end{aligned}$$

Example 13.2

How many grams of ethanol are dissolved in 60 g of water to make a 40% solution?

Solution

$$\% \text{ Ethanol} = \frac{\text{Mass of ethanol}}{\text{Mass of ethanol} + \text{Mass of } H_2O} \times 100$$

Let the mass of ethanol be y g.

$$\begin{aligned} 40\% &= \frac{y}{y + 60 \text{ g}} \times 100 \\ 0.40 &= \frac{y}{y + 60 \text{ g}} \\ 0.40y + 24 &= y \\ 24 &= 0.60y \\ y &= 40 \text{ g} \end{aligned}$$

13.2.2. Molarity

Molarity (M), or molar concentration, is defined as moles of solute per liter of solution. (Note: some textbooks may use dm^3 in place of liter.)

$$\text{Molarity} = \frac{\text{Number of moles of solute}}{\text{Number of liters of solution}}$$

This implies that:

$$\text{Number of moles of solute} = \text{Molarity} \times \text{Volume in liters}$$

If the concentration is expressed in grams per liter or dm^3, then molarity may be calculated as follows:

$$\begin{aligned} \text{Molarity} &= \frac{\text{Concentration in grams per liter}}{\text{Formula mass}} \\ &= \frac{\text{Grams of solute}}{\text{Formula mass of solute} \times \text{Volume in liters of solution}} \end{aligned}$$

Example 13.3

Calculate the molarity of a solution which contains 4.9 g of H_2SO_4 in 3.0 L of solution.

Solution

Method 1

Use the expression for mass concentration, i.e.:

$$\text{Molarity} = \frac{\text{Mass in grams of solute}}{\text{MW of solute} \times \text{Volume in liters of solution}}$$

$$\text{MW of } H_2SO_4 = 98 \text{ g/mol}$$

$$\text{Mass of } H_2SO_4 = 4.9 \text{ g}$$

$$\text{Volume of solution} = 3.00 \text{ L}$$

$$\text{Molarity} = \frac{4.9 \text{ g}}{98 \text{ g/mol} \times 3.00 \text{ L}} = 0.0167 \frac{\text{mol}}{\text{L}} \text{ (or M)}$$

Method 2

Step 1: Calculate the number of moles of H_2SO_4:

$$\text{No. of moles} = \frac{\text{Mass of } H_2SO_4}{\text{MW of } H_2SO_4} = \frac{4.9 \text{ g}}{98 \text{ g/mol}} = 0.0500 \text{ mol}$$

Step 2: Calculate the molarity of H_2SO_4 from the definition:

$$\text{Molarity} = \frac{0.050 \text{ mol}}{3.00 \text{ L}} = 0.0167 \text{ M} \cong 0.017 \text{ M}$$

Example 13.4

Calculate the molarity of each of the following solutions:

(a) 4.90 g of H_2SO_4 per liter of solution
(b) 0.40 g of NaOH in enough water to make 0.50 L of solution
(c) 6.00 g HNO_3 in enough water to make 250 cm^3 of solution

Solution

(a) Determine the number of moles of H_2SO_4:

$$\text{Molar mass of } H_2SO_4 = 2 \times 1 + 32 + 4 \times 16 = 98$$

$$(4.90 \text{ g } H_2SO_4)\left(\frac{1 \text{ mol of } H_2SO_4}{98 \text{ g of } H_2SO_4}\right) = 0.0500 \text{ mol}$$

Find the molarity from the definition:

$$\text{Molarity} = \frac{\text{Number of moles of solute}}{\text{Number of liters of solution}}$$

$$= \frac{0.0500\ \text{mol}}{1.00\ \text{L}} = 0.0500\ \text{M}$$

(b) Determine the number of moles of NaOH:

$$(0.40\ \text{g NaOH})\left(\frac{1\ \text{mol of NaOH}}{40\ \text{g of NaOH}}\right) = 0.010\ \text{mol}$$

Find the molarity using the definition:

$$\text{Molarity} = \frac{\text{Number of moles of solute}}{\text{Number of liters of solution}}$$

$$= \frac{0.010\ \text{mol}}{0.50\ \text{L}} = 0.020\ \text{M}$$

(c) Determine the number of moles of HNO_3:

$$(6.0\ \text{g HNO}_3)\left(\frac{1\ \text{mol of HNO}_3}{63.0\ \text{g of HNO}_3}\right) = 0.095\ \text{mol}$$

Find the molarity from the definition:

$$\text{Molarity} = \frac{\text{Number of moles of solute}}{\text{Number of liters of solution}}$$

$$= \frac{0.095\ \text{mol}}{0.250\ \text{L}} = 0.38\ \text{M}$$

13.2.3. Normality

At times, it may be desirable to express the concentration of a solution as "Normality". The *normality* (N) of a solution is defined as *the number of equivalent weights (or simply equivalents) of solute per liter of solution.*

$$\text{Normality (N)} = \frac{\text{Number of equivalent weights of solute}}{\text{Number of liters of solution}} \quad \text{or} \quad \left(\frac{\text{eq}}{\text{L}}\right)$$

Normality may also be expressed as:

$$\text{Normality (N)} = \frac{\text{Mass of solute in grams}}{\text{Equivalent weight of solute} \times \text{Liters of solution}}$$

The term "equivalent weight" or "equivalent" of an element is defined as the atomic weight divided by its valency or oxidation number. In acid–base reactions, this definition differs. The equivalent weight of an acid is defined as the mass of the acid that will yield 1 mol of hydrogen ions or react with 1 mol of hydroxide ions.

Table 13-1 Molar Mass and Equivalent Mass of Some Acids and Bases

Substance	*Formula*	*Molar mass*	*No. of replaceable H^- or OH^- ions*	*Equivalent mass*
Acids				
Nitric acid	HNO_3	63	1	63
Acetic acid	CH_3COOH	60	1	60
Alanine	$CH_3CH(NH_2)COOH$	89	1	89
Oxalic acid	H_2OCCO_2H	90	2	45
Sulfuric acid	H_2SO_4	98	2	49
Phosphoric acid	H_3PO_4	98	3	32.7
Bases				
Sodium hydroxide	NaOH	40	1	40
Ammonia	NH_3	17	1	17
Barium hydroxide	$Ba(OH)_2$	171.4	2	85.7
Aluminium hydroxide	$Al(OH)_3$	78	2	26

It is obtained by dividing the formula mass of the acid by the acidity (i.e., the number of ionizable hydrogen ions). Similarly, the equivalent weight of a base is defined as the mass of the base that will produce 1 mol of hydroxide ions, or react with 1 mol of hydrogen ions. It can be obtained by dividing the formula mass by the number of ionizable hydroxide ions per molecule. Table 13-1 shows the equivalent weights of some common acids and bases.

Example 13.5

What is the normality of a solution containing 9.8 g of H_2SO_4 in 2 L of solution? (MW of H_2SO_4 = 98 g/mol, equivalent weight (EW) = 49 g/equiv.)

Solution

Step 1: Calculate the number of equivalents:

$$\text{No. of equivalents} = \frac{\text{Mass in grams}}{\text{EW}} = \frac{9.8\ \text{g}}{49\ \text{g/equiv}} = 0.2\ \text{equiv}$$

Step 2: Calculate the normality:

$$\text{Normality (N)} = \frac{\text{Number of equivalent weights of solute}}{\text{Number of liters of solution}} \quad \text{or} \quad \left(\frac{\text{eq}}{\text{L}}\right)$$

$$= \frac{0.2\ \text{eq}}{2\ \text{L}} = 0.1\ \text{N}$$

Example 13.6

Calculate the normality of a solution of phosphoric acid (H_3PO_4) prepared by dissolving 49 g of H_3PO_4 in enough water to make 750 cm^3 of solution.

Solution

First change the grams of H_3PO_4 to equivalents.

$$\text{Equivalent mass of } H_3PO_4 = \frac{\text{Molar mass of } H_3PO_4}{\text{Number of replaceable } H^+ \text{ ions}}$$

$$= \frac{98 \text{ g}}{3 \text{ eq}} = 32.7 \text{ g/eq}$$

Next, determine the equivalents of H_3PO_4.

$$\text{Equivalents of } H_3PO_4 = (49.0 \text{ g})\left(\frac{1 \text{ equivalent}}{32.7 \text{ g}}\right)$$

$$= 1.50 \text{ equivalents of } H_3PO_4$$

$$\text{Normality} = \frac{\text{Number of equivalents}}{\text{Liters of solution}}$$

$$= \frac{1.50 \text{ equivalents}}{0.75 \text{ L}} = 2.0 \text{ N}$$

13.2.4. Mole fraction

Mole fraction is defined as *the ratio of the number of moles of a given component in a mixture to the total number of moles of all the components* in the mixture. For example, the mole fraction (X_A) of a component A in a mixture containing components A, B, C, and D, is given by

$$X_A = \frac{n_A}{n_{\text{Total}}} = \frac{n_A}{n_A + n_B + n_C + n_D}$$

where n_A, n_B, n_C, and n_D, are the number of moles of components A, B, C, and D.

Example 13.7

Calculate the mole fraction of NaOH in a solution made by dissolving 1.0 mol of solute in 1000 g of water.

Solution

$$\chi_{NaOH} = \frac{n_{NaOH}}{n_{NaOH} + n_{H_2O}}$$

$$n_{NaOH} = 1.0 \text{ mol}$$

$$n_{H_2O} = \frac{\text{Grams of } H_2O}{\text{MW of } H_2O} = \frac{1000 \text{ g}}{18.0 \text{ g/mol}} = 55.5 \text{ mol}$$

$$\chi_{NaOH} = \frac{1}{1 + 55.5} = 0.018$$

13.2.5. Molality

Molality (m) is defined as *the number of moles of solute per kilogram of solvent*. Thus, a 1 molal solution contains 1 mol of solute in 1 kg of solvent.

$$\text{Molality} = \frac{\text{Moles of solute}}{\text{kg of solvent}}$$

But:

$$\text{Moles of solute} = \frac{\text{Grams of solute}}{\text{MW of solute}}$$

Hence molality may also be expressed as:

$$\text{Molality} = \frac{\text{Mass of solute}}{\text{MW of solute} \times \text{kg of solvent}}$$

Example 13.8

A solution of sodium carbonate (Na_2CO_3) was prepared by dissolving 1.06 g of Na_2CO_3 in 100 g of deionized water. What is the molality of Na_2CO_3 in this solution? (Density of water = 1.00 g/mL.)

Solution

$$\text{Molality} = \frac{\text{Moles of solute}}{\text{kg of solvent}}$$

$$\text{Moles of } Na_2CO_3 = \frac{1.06 \text{ g}}{106 \text{ g/mol}} = 0.010$$

$$\text{Molality} = \frac{0.010 \text{ mol}}{0.100 \text{ kg}} = 0.10\, m$$

Example 13.9

What is the molality of a 20% aqueous solution of ethanol (C_2H_5OH)?

Solution

We can assume for convenience that we have 100 g of solution, containing 20 g of ethanol and 80 g of water.

$$\text{Mass of } C_2H_5OH = 20 \text{ g}$$

$$\text{Moles of } C_2H_5OH = \frac{\text{Moles of ethanol}}{\text{MW of ethanol}} = \frac{20 \text{ g}}{46 \text{ g/mol}} = 0.435$$

$$\text{Mass of water} = 100 \text{ g} - 20 \text{ g} = 80 \text{ g} = 0.080 \text{ kg}$$

$$\text{Molality} = \frac{\text{Moles of solute}}{\text{kg of solvent}}$$

$$\text{Molality} = \frac{0.435 \text{ mol}}{0.080 \text{ kg}} = 5.435\ m$$

13.2.6. Dilute solutions

A dilute solution is prepared by adding solvent to a concentrated solution. When a concentrated solution is diluted, the number of moles of solute remains unchanged. Only the net volume is changed, due to adding more solvent. From the molarity equation, we can write an expression for the number of moles of solute, i.e.:

$$\text{Molarity (M)} = \frac{\text{Number of moles of solute}}{\text{Number of liters of solution}} \quad \text{or} \quad \left(\frac{\text{mol}}{\text{L}}\right), \quad \text{and}$$

$$\text{Moles of solute} = \text{Molarity (M)} \times \text{Liters of solution (L)} = \text{constant}$$

Therefore:

$$M_1\left(\frac{\text{mol}}{\text{L}}\right)V_1(\text{L}) = M_2\left(\frac{\text{mol}}{\text{L}}\right)V_2(\text{L})$$

or simply:

$$M_1 \times V_1 = M_2 \times V_2$$

where M_1 and V_1 are the molarity and volume before dilution, and M_2 and V_2 are the molarity and volume after dilution.

Example 13.10

Hydrochloric acid is normally purchased at a 12.0 M concentration. Calculate the volume of this stock solution needed to prepare 500 mL of 0.500 M aqueous solution.

Solution

$$M_1 \times V_1 = M_2 \times V_2$$

$$M_1 = 12.0 \text{ M}$$

$$V_1 = ?$$

$$M_2 = 0.500 \text{ M}$$

$$V_2 = 500 \text{ mL}$$

$$V_1 = \frac{M_2 V_2}{M_1}$$

$$V_1 = \frac{0.5 \text{ M} \times 500 \text{ mL}}{12 \text{ M}} = 20.8 \text{ mL}$$

For problems involving mixing, always remember that no solute is lost or gained. The total amount of solute present before mixing is the same as the total amount of solute after mixing. That is:

Moles of solute # 1 + moles of solute # 2 = Final moles of solute

$$M_1 V_1 + M_2 V_2 = M_3 V_3$$

Here M_1, M_2, V_1, and V_2 represent the molar concentration and volume of solutions 1 and 2 before mixing. M_3 and V_3 represent the molar concentration and volume obtained after mixing. Note that $V_3 = V_1 + V_2$.

Example 13.11

50 cm^3 of 0.25 M NaOH solution was mixed with 250 cm^3 of 0.75 M NaOH solution. What is the molarity of the resulting solution?

Solution

To solve this problem, remember that the number of moles of solute (= molarity × volume) after mixing is the sum of the amounts that were mixed together. Now substitute the values for V_1, V_2, and V_3 into the equation:

$$M_1 V_1 + M_2 V_2 = M_3 V_3$$

$$M_1 = 0.25 \text{ M}, \quad V_1 = 50 \text{ cm}^3, \quad M_2 = 0.75 \text{ M}, \quad V_2 = 250 \text{ cm}^3, \quad V_3 = 300 \text{ cm}^3$$

$M_3 = ?$

$0.25\ \text{M} \times 50\ \text{cm}^3 + 0.75\ \text{M} \times 250\ \text{cm}^3 = M_3 \times 300\ \text{cm}^3$

$M_3 = 0.67\ \text{M}$

13.3. Solving Solubility Problems

13.3.1. Solubility in grams per 100 g of solvent

1. Determine the amount of solute in the solution.
2. Write an expression for concentration (C) in terms of mass ratio, i.e.:

$$\text{Concentration } (C) = \frac{\text{Mass of solute (g)}}{\text{Mass of solvent (g)}}$$

3. Using the expression for C as a conversion factor, determine the amount of solute in 100 g of H_2O.
4. Make sure solubility is in proper units.

Example 13.12

A laboratory technician added 35 g of KCl to 50 g of water and heated to 80°C until all the solids were completely dissolved. When the solution was cooled to 55°C, crystals of KCl began to form. What is the solubility of KCl at 55°C?

Solution

1. Mass of solute = 35 g; mass of solvent = 50 g
2. $\text{Concentration } (C) = \dfrac{\text{Mass of solute (g)}}{\text{Mass of solvent (g)}} = \dfrac{35\ \text{g KCl}}{50\ \text{g } H_2O}$
3. Amount of KCl in dissolved in 100 g of water at 55°C

$$= 100\ \text{g of } H_2O \times \frac{35\ \text{g KCl}}{50\ \text{g } H_2O} = 70\ \text{g KCl}$$

4. The solubility of KCl at 55°C = 70 g per 100 g of H_2O.

Example 13.13

In an effort to prepare a saturated solution of sodium carbonate, an undergraduate student added 106 g of the salt to 1000 g of water at 24°C under constant stirring until no more solute dissolved. Upon filtration, the student recovered 26 g of undissolved Na_2CO_3. Calculate the solubility of Na_2CO_3 in water at 24°C.

Solution

1. Mass of dissolved Na_2CO_3 = Initial mass − Mass of undissolved Na_2CO_3

$$= 106\ \text{g} - 26\ \text{g} = 80\ \text{g}$$

2. Mass of solvent = 1000 g
3. $\text{Concentration}(C) = \dfrac{\text{Mass of solute (g)}}{\text{Mass of solvent (g)}} = \dfrac{80\ \text{g Na}_2\text{CO}_3}{1000\ \text{g H}_2\text{O}}$
4. Amount of Na_2CO_3 dissolved in 100 g of water at 24°C

$$= 100\ \text{g of H}_2\text{O} \times \frac{80\ \text{g Na}_2\text{CO}_3}{1000\ \text{g H}_2\text{O}} = 8\ \text{g Na}_2\text{CO}_3$$

5. The solubility of Na_2CO_3 at 24°C = 8 g per 100 g of H_2O.

13.3.2. Solubility in moles per liter of solvent

Solubility can also be expressed in moles per liter of the solvent (or molarity):

$$\text{Solubility in moles per liter} = \frac{\text{Mass concentration (g/L)}}{\text{Molar mass of solute (g/mol)}}$$

Example 13.14

A saturated solution was made by dissolving 25 g of sodium nitrate, $NaNO_3$, in 75 g of deionized water. Calculate the solubility of $NaNO_3$ in moles per liter at 25°C.

Solution

1. Molar mass of $NaNO_3$ = 85 g
2. Calculate solubility in grams per 100 g of solvent:

Mass of solute = 25 g; mass of solvent = 75 g

$$\text{Solubility in grams per 100 g of solvent} = \frac{25\ \text{g}}{75\ \text{g}} \times 100\ \text{g H}_2\text{O} = 33.33\ \text{g}$$

3. Convert solubility in grams per 100 g of solvent to mass concentration in grams per 1000 g of solvent or grams per liter:

$$33.3\ \text{g per 100 g of H}_2\text{O} = 333\ \text{g per 1000 g of H}_2\text{O} \quad \text{or}$$

$$333\ \text{g per liter } (1000\ \text{g} \cong 1\ \text{L})$$

4. Calculate the solubility in moles per liter (molarity):

$$\text{Solubility in moles per liter} = \frac{\text{Mass concentration (g/L)}}{\text{Molar mass of solute (g/mol)}}$$

$$\text{Solubility in moles per liter} = \frac{333.0\ \text{g/L}}{85.0\ \text{g/mol}} = 3.92\ \text{mol/L (or M)}$$

Example 13.15

The solubility of lead (II) nitrate, $Pb(NO_3)_2$, is 2.5 M. Calculate the mass of solute in 1500 g of saturated solution.

Solution

$$\text{Molarity of saturated } Pb(NO_3)_2 = 2.5$$

$$\text{Molar mass of } Pb(NO_3)_2 = 331\ \text{g/mol}$$

$$\text{Solubility in moles per liter} = \frac{\text{Mass concentration (g/L)}}{\text{Molar mass of solute (g/mol)}}$$

$$\text{Mass concentration}\left(\frac{\text{g}}{\text{L}}\right) = \text{Solubility}\left(\frac{\text{mol}}{\text{L}}\right) \times \text{Molar mass}\left(\frac{\text{g}}{\text{mol}}\right)$$

$$= 2.5\left(\frac{\text{mol}}{\text{L}}\right) \times 331\left(\frac{\text{g}}{\text{mol}}\right) = 827.4\ \text{g/L}$$

Thus, 1000 g of water dissolve 827.5 g of $Pb(NO_3)_2$.

Recall that:

$$\text{Mass of solute} + \text{Mass of solvent} = \text{Mass of solution}$$

$$\text{Mass of solution} = 1000\ \text{g} + 827.5\ \text{g} = 1827.5\ \text{g}$$

1827.5 g of solution contain 827.5 g of $Pb(NO_3)_2$, so 1500 g of solution will contain

$$1500\ \text{g} \times \frac{827.5\ \text{g}}{1827.5\ \text{g}} = 679.2\ \text{g of } Pb(NO_3)_2$$

Example 13.16

The solubility of potassium chlorate, $KClO_3$, is 25.5 g per 100 g H_2O at 70°C, and 7.15 g per 100 g H_2O at 20°C. What mass of $KClO_3$ will crystallize out of solution if 95 g of the saturated solution at 70°C is cooled to 20°C?

Solution

$$\text{Mass of solute} + \text{Mass of solvent} = \text{Mass of solution}$$

At 70°C, we have: $25.5\ \text{g} + 100\ \text{g} = 125.5\ \text{g}$

At 20°C, we have: $7.15\ \text{g} + 100\ \text{g} = 107.5\ \text{g}$

Mass of solute precipitated on cooling from 70°C to 20°C = 25.5 − 7.15 = 18.35 g.
When cooled from 70°C to 20°C, 125.5 g of saturated solution precipitate 18.35 g.

$$95\ \text{g of saturated solution will precipitate } 95\ \text{g} \times \frac{18.35\ \text{g}}{125.5\ \text{g}} = 13.89\ \text{g of } KClO_3$$

13.4. Effect of Temperature on Solubility

In general, the solubility of a solute in a given solvent varies with temperature. However, it is very difficult to predict the temperature dependence of solubility. For most ionic and molecular solids such as potassium nitrate, calcium chloride, potassium chlorate, and sugar, solubility increases with temperature. For a few substances, such as NaCl, temperature has little effect on solubility; but others, like Na_2SO_4 and $CeSO_4$, become less soluble with increasing temperature. The solubility of gases is more predictable: in general the solubility of a gas in water decreases with increasing temperature.

If the solubility of a substance is plotted as a function of temperature, a graph known as a solubility curve is obtained.

13.5. Solubility Curves

A *solubility curve* is a graph that represents the change in concentration of a saturated solution with change in temperature. The solubility is plotted on the vertical axis (y-axis) while the temperature is plotted on the horizontal axis (x-axis). From the graph, the solubility of a solute at a particular temperature can be determined. Points on the curve represent saturated solution; points above the curve represent supersaturated solution, while those below represent unsaturated solution.

Example 13.17

Figure 13-1 shows the solubility curves of sodium nitrate and potassium chlorate at various temperatures. From the graph answer the following questions:

1. What is the solubility of sodium nitrate and potassium chlorate at 65°C?
2. What mass of sodium nitrate crystals will be deposited when the saturated solution is cooled from 60°C to 20°C? Note: solubility is in units of g/100 g water.

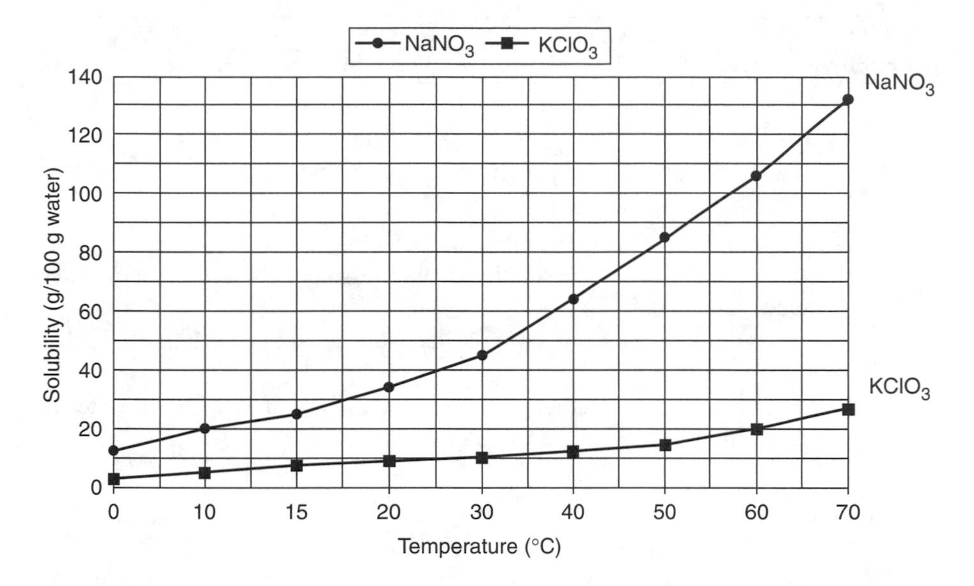

Figure 13-1 The solubility of sodium nitrate and potassium chlorate at various temperatures.

3. At what temperature will a solution containing 38 g of sodium nitrate in 100 g of water become saturated?

Solution

1. A vertical line from 65°C meets the $NaNO_3$ solubility curve at a point corresponding to about 119 g on the solubility axis. Therefore the solubility of sodium nitrate at 65°C is 119 g/100 g water. Similarly, a vertical line from 65°C meets the $KClO_3$ solubility curve at a point corresponding to about 24 g. Therefore, the solubility of $KClO_3$ at 65°C is 24 g per 100 g water.
2. Draw vertical lines from 60°C and 20°C on the x-axis to the solubility curve of sodium nitrate. Read off the corresponding solubility values from the solubility (y) axis. These are approximately 19.8 g at 60°C and 9.2 g at 20°C. So when the solution is cooled from 60°C to 20°C, the mass of sodium nitrate deposited will be 19.8 g − 9.2 g = 10.6 g per 100g of water.
3. A horizontal line from 38 g meets the solubility curve of sodium nitrate at a point, which corresponds to about 23°C. Therefore the required temperature is 23°C.

Example 13.18

The solubility of copper (II) sulfate at various temperatures is given in the table below.

Temperature (°C)	0	10	20	30	40	50	60	70	80
Solubility (g/100 g H_2O)	23.1	27.7	31.7	37.9	44.4	53.5	62.1	73.1	83.5

Plot the solubility curve for copper (II) sulfate, and determine from it:

(a) The solubility of copper (II) sulfate at 55°C.
(b) The temperature at which a solution containing 65 g of salt in 100 g of water becomes saturated.
(c) The mass of the salt that will be deposited when a saturated solution is cooled from 60°C to 10°C.
(d) Whether a solution containing 50 g of copper (II) sulfate in 100 g water at 20°C is saturated, unsaturated, or supersaturated.

Solution

Figure 13-2 shows the solubility of copper (II) sulphate at various temperatures. The solutions to the questions are determined directly from the graph.

(a) 58 g/100 g water
(b) 63°C
(c) 34.4 g
(d) Supersaturated

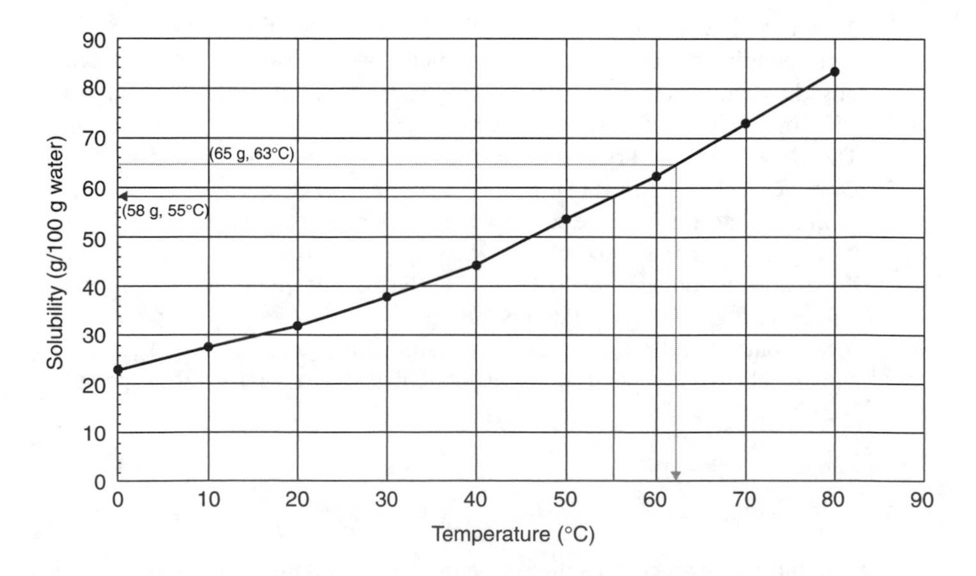

Figure 13-2 The solubility of copper (II) sulfate at various temperatures.

13.6. Effect of Pressure on Solubility

Generally, pressure change has little or no effect on the solubility of solids and liquids. However, the solubility of gases in a liquid is greatly affected by changes in pressure. Henry's law describes the effect of pressure on the solubility of a gas in quantitative terms. According to Henry's law, the solubility of a gas in a liquid at a particular temperature is directly proportional to the partial pressure of that gas above the solution:

$$\text{Solubility } (S) = k_H P$$

where k_H is the Henry's law constant for each gas at a given temperature (in moles per liter-atm), and P is the partial pressure of the gas (in atm).

Example 13.19

What is the solubility of CO_2 in a soda drink pressurized to 1.5 atm at 25°C? The Henry's law constant for CO_2 is 3.2×10^{-2} mol/L-atm at 25°C.

Solution

Substitute the given data directly into the gas solubility formula:

$$\text{Solubility } (S) = k_H P = \left(3.2 \times 10^{-2} \frac{\text{mol}}{\text{L-atm}}\right)(1.5 \text{ atm}) = 4.8 \times 10^{-2} \text{ M}$$

Example 13.20

The solubility of argon in water at 20°C is 1.5×10^{-3} M when the pressure over the solution is 1.0 atm. What is the solubility at this temperature if the argon pressure is increased to 5 atm?

Solution

Using the initial set of data, calculate the Henry's law constant. Since the temperature is kept constant, use the k_H so obtained to calculate the solubility at a partial pressure of 5 atm.

$$S = k_H P$$

$$k_H = \frac{S}{P} = \frac{(1.5 \times 10^{-3} \text{ mol/L})}{(1 \text{ atm})} = 1.5 \times 10^{-3} \text{ mol/L-atm}$$

When $P = 5$ atm, the solubility, S, will be

$$S = k_H P = \left(1.5 \times 10^{-3} \frac{\text{mol}}{\text{L-atm}}\right)(5 \text{ atm}) = 7.5 \times 10^{-3} \text{ M}$$

13.7. Problems

1. A 2.5569-g sample of sodium hydrosulfide (NaSH) is dissolved in 18.75 mL of water. Calculate the percent by mass of NaSH in this solution. Assume the density of water to be 1.00 g/mL.
2. Calculate the mass in grams of a 20.00% by mass $LiNO_3$ solution that contains 6.125 g $LiNO_3$.
3. How many grams of NaCl would be required to make 500 g of 2% NaCl solution?
4. The density of concentrated hydrochloric acid (HCl) is 1.19 g/cm^3. Calculate the mass of anhydrous HCl in 2.5 cm^3 of the concentrated acid containing 37% HCl by mass.
5. What is the molarity of a solution containing

   (a) 10.4 g of HNO_3 per liter (L) of solution?
   (b) 6.3 g of KOH per 500 cm^3 of solution?
   (c) 0.25 mol of Na_2CO_3 per 250 cm^3 of solution?
   (d) 5.00 mol of glucose ($C_6H_{12}O_6$) per 20 L of solution?

6. Calculate the molarity of calcium hydrogen carbonate solution containing 25 g of $Ca(HCO_3)_2$ in 850 cm^3 of solution.
7. The density of concentrated sulfuric acid (H_2SO_4) containing 98% by mass of H_2SO_4 is 1.84 g/cm^3. Calculate the volume of the liquid needed to prepare 500 cm^3 of 0.125 M sulfuric acid.
8. Calculate the volume in cm^3 of sulfur trioxide gas at STP that will dissolve in 500 cm^3 of water to produce 0.10 M solution of sulfuric acid.
9. Calculate the normality of a solution of 6.25 g of carbonic acid (H_2CO_3) in 500 cm^3 of solution. (Note: carbonic acid can supply two hydrogen ions per molecule.)
10. What volume of solution is required to provide 22.40 g barium hydroxide ($Ba(OH)_2$) from a 2.5 N $Ba(OH)_2$ solution in reactions that replace the two hydroxide ions?
11. How many grams of solute are required to prepare 2.00 L of 1.00 N solution of phosphoric acid (H_3PO_4)?
12. Calculate the mole fractions of solute and solvent in a solution prepared by dissolving 50 g of sodium nitrate ($NaNO_3$) in 750 g of water (H_2O).
13. Calculate the mole fraction of all the components of a solution containing 0.375 mol methanol (CH_3OH), 1.275 mol acetone (CH_3COCH_3), and 1.750 mol carbon tetrachloride (CCl_4).

14. What is the mole fraction of table salt (sodium chloride, NaCl) in an aqueous solution containing 40.0 g of the salt and 100 g of water?
15. Calculate the molality of a sulfuric acid solution containing 35 g of sulfuric acid in 250 g of water.
16. A chemist wants to prepare a 0.05 M solution of iron (II) sulfate ($FeSO_4$). Calculate the number of grams of $FeSO_4$ which she must add to 300 g of water to achieve this concentration.
17. The density of a 1.50-M aqueous solution of ethyl alcohol (C_2H_5OH) is $0.857 g/cm^3$ at 25°C. Calculate the molality of the solution.
18. The density of a 6.50-molal acetic acid (CH_3COOH) solution is 1.045 g/cm^3. What is the molarity of the solution?
19. The density of an aqueous solution of sulfuric acid containing 20% H_2SO_4 by mass is 1.175 g/cm^3. Find the molar concentration and molality of the solution.
20. What is the molarity of nitric acid solution prepared by mixing 150 cm^3 of 7.50 M HNO_3 with 2 L of water?
21. Calculate the concentration of the resulting solution if 10 mL of a 12.0 M stock solution of hydrochloric acid is diluted to 500 mL.
22. What is the molarity of the solution prepared by mixing 125 cm^3 of 2.50 M KOH solution with 250 cm^3 of 0.75 M KOH solution?
23. How would you prepare 5.00 L of 1.00 M HCl solution from a stock solution that contains 18% HCl by mass and has a density of 1.05 g/cm^3?
24. A solution is prepared by dissolving 25 g of anhydrous sodium carbonate, Na_2CO_3, in sufficient double-distilled water to produce 500.00 cm^3 of the solution. A 10.00-cm^3 portion of this solution was further diluted to 500.00 cm^3. Calculate the concentration of the final Na_2CO_3 solution in

    (a) mol/L
    (b) g/L

25. A solution was prepared by diluting 100 cm^3 of 0.75 M Li_3PO_4 with water to a final volume of 1.0 L. Determine (a) the molar concentrations of Li^+ and PO_4^{3-} in the original solution, (b) the molar concentrations of Li_3PO_4, Li^+, and PO_4^{3-} in the final solution after dilution.
26. The following data were obtained in the determination of the solubility of copper (II) sulfate at 60°C:

$$\text{Mass of dish} = 30.25 \text{ g}$$

$$\text{Mass of dish} + \text{saturated solution of } CuSO_4 \text{ at } 60°C = 62.15 \text{ g}$$

$$\text{Mass of dish} + \text{solid } CuSO_4 \text{ after heating to dryness} = 46.70 \text{ g}$$

    Calculate the solubility of $CuSO_4$ (a) in grams per 100 g of water at 60°C; (b) in moles per liter. Assume the density of water at this temperature to be 1.00 g/mL.

27. The solubility of NH_4Cl is 70 g per 100 g H_2O at 90°C, and 41.75 g per 100 g H_2O at 30°C. What mass of NH_4Cl will crystallize out if 200 g of the saturated solution is cooled from 90°C to 30°C?
28. A 20-mL saturated solution of $AgNO_3$ was prepared by adding 125.25 g $AgNO_3$ to deionized water at 25°C. If the solubility of $AgNO_3$ is 9.75 mol/L, determine the mass of the salt remaining undissolved under these conditions.
29. At a particular temperature, the solubility of anhydrous Na_2SO_4 was found to be 4.25 M. Calculate the amount of this salt that dissolves in 20 mL of water. (Density of water = 1.00 g/mL).
30. The solubility of an organic salt at 22°C is determined to be 45 g per 100 g of water. Calculate the mass of the substance present in 100 g of its saturated solution at this temperature.
31. Use figure 13-3 to answer the following questions:
    (a) How many grams of KNO_3 will dissolve in 100 g water at 64°C?
    (b) How many grams of KNO_3 will dissolve in 50 g water at 60°C?
    (c) How many grams of $NaNO_3$ will dissolve in 400 g water at 73°C?
    (d) Describe how you would prepare a saturated solution of $NaNO_3$ at 40°C in 25 g water.
    (e) Estimate the solubility of ammonia, NH_3, at 15°C.
    (f) Starting with a saturated solution of ammonia in 100 g water at 15°C, estimate the mass of ammonia that will be liberated as the temperature of the solution is raised to 55°C.
32. Use the solubility curve in figure 13-3 to answer the following questions:
    (a) At what temperatures will solutions containing
        (i) 100 g of KNO_3
        (ii) 70 g of NH_3

        in 100 g of water become saturated?

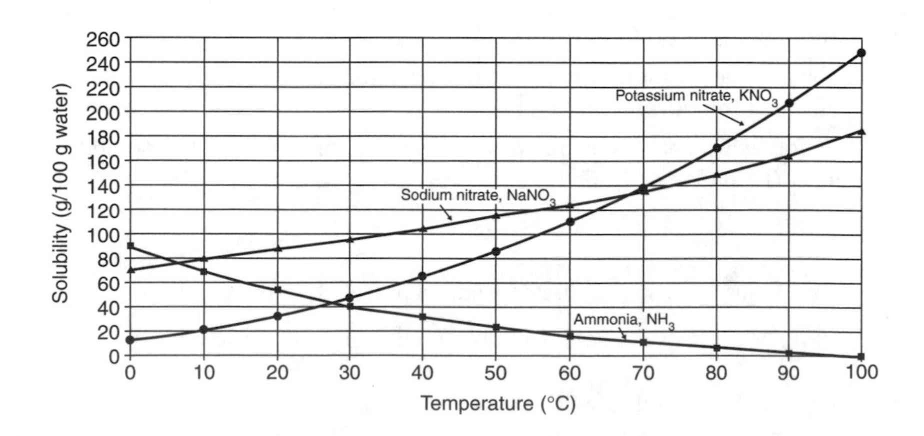

Figure 13-3 The solubility of KNO_3, $NaNO_3$ and NH_3 at various temperatures.

(b) At what temperature do $NaNO_3$ and KNO_3 have the same solubility?
(c) Calculate the mass of KNO_3 that will be deposited by cooling 300 g of the saturated solution from 80°C to 50°C.
(d) What mass of ammonia will be evolved if 200 g of the saturated solution at 10°C is gently warmed to 30°C?

33. The Henry's law constant for N_2 is 6.8×10^{-4} mol/L-atm at 25°C. What is the solubility of N_2 in water at 25°C if the partial pressure of nitrogen is 0.78 atm?
34. The solubility of pure oxygen gas in water at 25°C and 1.00 atm is 2.56×10^{-5} M. What is the concentration of oxygen in water at 0.21 atm and 25°C?
35. The Henry's law constant for CO_2 is 3.2×10^{-2} mol/L-atm at 25°C, and the partial pressure of CO_2 in air is 0.000400 atm. What is the concentration of CO_2 in:

    (a) A 2.0-L bottle of soda open to the atmosphere at 25°C?
    (b) A 2.0-L bottle of soda covered under a CO_2 pressure of 4.5 atm and at 25°C?

14

Volumetric Analysis

14.1. Introduction

Volumetric analysis is a chemical analytical procedure based on measurement of volumes of reaction in solutions. It uses *titration* to determine the concentration of a solution by carefully measuring the volume of one solution needed to react with another. In this process, a measured volume of a standard solution, the *titrant*, is added from a *burette* to the solution of unknown concentration. When the two substances are present in exact stoichiometric ratio, the reaction is said to have reached the *equivalence* or *stoichiometric point*.

In order to determine when this occurs, another substance, the *indicator*, is also added to the reaction mixture. This is an organic dye which changes color when the reaction is complete. This color change is known as the *end point*; ideally, it will coincide with the equivalence point. For various reasons, there is usually some difference between the two, though if the indicator is carefully chosen, the difference will be negligible.

A typical titration is based on a reaction of the general type

$$aA + bB \longrightarrow \text{products}$$

where A is the titrant, B the substance titrated, and a:b is the stoichiometric ratio between the two. Some indicators include Litmus, Methyl Orange, Methyl Red, Phenolphthalein, and Thymol Blue.

14.2. Applications of Titration

Titration can be applied to any of the following chemical reactions:

- Acid–base
- Complexation
- Oxidation–reduction
- Precipitation

Only acid–base and oxidation–reduction titration will be treated here, though the fundamental principles are the same in all cases.

14.3. Acid–Base Titrations

Acid–base titration involves measuring the volume of a solution of the acid (or base) that is required to completely react with a known volume of a solution of a base (or acid). The relative amounts of acid and base required to reach the equivalence point depend on their stoichiometric coefficients. It is therefore critical to have a balanced equation before attempting calculations based on acid–base reactions.

Below we define some of the common terms associated with acid–base reactions.

14.3.1. A molar solution

A *molar solution* is one that contains one mole of the substance per liter of solution. For example, a molar solution of sodium hydroxide contains 40 g (NaOH = 40 g/mol) of the solute per liter of solution. As described in chapter 13, the concentration of a solution expressed in moles per liter of solution is known as the *molarity* of the solution.

14.3.2. Standard solution

A standard solution is one of known concentration. For example, titration of an unknown acid might use a carefully prepared 0.1250 M solution of sodium carbonate.

14.3.3. Standardization

Standardization involves determining the concentration of a solution by accurately measuring the volume of that solution required to react with an exactly known amount of a primary standard. The solution thus standardized is called a secondary standard and is used to analyze unknown samples. An example of a primary standard is sodium carbonate solution. It can readily be prepared by dissolving pure anhydrous sodium carbonate in water. A standardized base can be used to determine the concentration of an unknown acid solution, and vice-versa.

14.4. Calculations Involving Acid–Base Titration

Generally, calculations concerning acid–base reactions involve the concepts of reaction stoichiometry, number of moles, mole ratio, and molarity. The key requirements here are: the volume and concentration (molarity or normality) of titrant; a balanced chemical equation between the reacting substances; and the mole ratio of the titrant to the substance being titrated.

14.4.1. Calculation involving mass and percentage of substance titrated

Consider the reaction between titrant A and titrated substance B. The balanced equation is

$$aA + bB \longrightarrow cC + dD$$

The mole ratio of B to A is given by

$$R = \frac{b}{a}$$

Note: R as used here is just a symbol, and not the gas constant. Any other letter such as K, U, or B could have been used.

To calculate the mass and percentage of substance titrated, follow these steps:

1. Write a balanced equation for the reaction between A and B.
2. Calculate the moles of titrant A.

$$\text{moles of } A = \frac{\text{mL}_A M_A}{1000 \text{ mL/L}}$$

3. Calculate the moles of titrated substance B:

$$\text{moles of } B = \text{moles of } A \times R = \frac{\text{mL}_A M_A}{1000 \text{ mL/L}} \times R$$

4. Calculate the mass of substance (B) titrated:

$$\text{Mass of } B \text{ in grams} = \text{moles of } B \times \text{ MW of } B = \frac{\text{mL}_A M_A}{1000 \text{ mL/L}} \times R \times \text{MW}_B$$

5. Calculate the percentage of substance B in the sample:

$$\%B = \frac{\text{Mass of } B \text{ in grams}}{\text{Mass of sample in grams}} \times 100 = \left(\frac{g_B}{g_{\text{sample}}}\right) \times 100$$

Example 14.1

A chemistry student was given a mini-project to analyze for ethanoic acid in a 100-g sample of a marine plant extract. The student conducted a titration and found that it took 20 mL of 0.625 M NaOH to neutralize 25 mL of a solution of the marine plant extract. Calculate:

(a) The mass of ethanoic acid in the sample
(b) The percentage of ethanoic acid in the sample

Solution

1. $NaOH+CH_3COOH \longrightarrow CH_3COONa+H_2O$

2. Moles of NaOH $= \dfrac{mL_{NaOH} \times M_{NaOH}}{1000\ mL/L} = \dfrac{20\ mL \times 0.625\ M}{1000\ mL/L} = 0.0125$

3. Moles of ethanoic acid = moles of NaOH $\times R$

$$R = 1$$

$$\text{Moles of ethanoic acid} = \frac{mL_{NaOH} \times M_{NaOH}}{1000\ mL/L} \times R = \frac{20\ mL \times 0.625\ M}{1000\ mL/L} \times 1$$
$$= 0.0125$$

4.
$$\text{Mass of ethanoic acid} = \text{moles of ethanoic acid} \times \text{MW of ethanoic acid}$$
$$= 0.0125\ mol \times 60.03\ g/mol$$
$$= 0.7504\ g$$

5. $\%$ Ethanoic acid $= \left(\dfrac{g_{\text{ethanoic acid}}}{g_{\text{sample}}}\right) \times 100 = \dfrac{0.7504}{100} \times 100 = 0.75\%$

Example 14.2

A research chemist analyzed a 25-g sample of polluted water taken from a lake suspected to contain trichloroethanoic acid (which has one acidic hydrogen per molecule). Titration of this sample with 0.25 M NaOH required 55 mL to reach the equivalence point. Calculate the mass percent of trichloroethanoic acid in the water sample.

Solution

$$Cl_3CCOOH+NaOH \longrightarrow Cl_3CCOONa+H_2O$$

$$R = 1$$

$$\text{Moles of NaOH} = \frac{mL_{NaOH} \times M_{NaOH}}{1000\ mL/L} = \frac{55\ mL \times 0.25\ M}{1000\ mL/L} = 0.0138$$

$$\text{Moles of } Cl_3CCOOH = \text{Moles of NaOH} = 0.0138 \quad \text{since } R = 1$$

$$\text{Mass of } Cl_3CCOOH = \text{moles of } Cl_3CCOOH \times \text{MW of } Cl_3CCOOH$$
$$= 0.0138\ mol \times 163.5\ g/mol$$
$$= 2.256\ g$$

$$\text{Mass percent of } Cl_3CCOOH = \frac{2.256\ g}{25\ g} \times 100 = 9.03\%$$

14.4.2. Calculations involving molarity, mass concentration, solubility, and percentage purity from a standardizing titration

In a titration experiment such as acid–base or oxidation–reduction titration, the following relationship holds between reacting solutions at the equivalence point:

$$\frac{\text{Volume of acid} \times \text{Molarity of acid}}{\text{Volume of base} \times \text{Molarity of base}} = \frac{\text{Moles of acid}}{\text{Moles of acid}}$$

This can be shortened to:

$$\frac{M_A V_A}{M_B V_B} = \frac{n_A}{n_B} = \text{mole ratio}$$

where:

M_A = molarity of acid

M_B = molarity of base

V_A = volume of acid

V_B = volume of base

n_A = moles of acid

n_B = moles of base

For example, the balanced equation for the reaction between HNO_3 and Na_2CO_3 is:

$$2HNO_3 + Na_2CO_3 \longrightarrow 2NaNO_3 + H_2O + CO_2$$

The equation indicates that 2 mol of the acid react with 1 mol of the base. The relationship at the equivalence point is given by:

$$\frac{M_A V_A}{M_B V_B} = \frac{2}{1}$$

Other formulas of interest include:

$$\text{Molarity} = \frac{\text{Concentration in g/L or dm}^{-3}}{\text{Molar mass}}$$

And:

$$\text{Molarity} = \frac{\text{Moles of solute}}{\text{Liters of solution}}$$

Example 14.3

A sample of solid $Mg(OH)_2$ was stirred in water at 50°C for several hours until the solution became saturated. Undissolved solid $Mg(OH)_2$ was filtered out and

150 mL of the resulting solution was titrated with 0.100 M HNO_3. The titration required 75.50 mL of the acid for complete neutralization. Calculate

(a) the molarity of the $Mg(OH)_2$
(b) the mass of $Mg(OH)_2$ contained in 150 mL of the saturated solution
(c) the solubility of $Mg(OH)_2$ in water, in g per 100 mL at 50°C.

Solution

This is an example of acid–base reaction. First we write the balanced equation for the reaction, then, using the formula method (or the conversion factor method) we can determine the molarity of the base, $Mg(OH)_2$.

$$Mg(OH)_2(aq) + HNO_3(aq) \longrightarrow Mg(NO_3)_2(aq) + 2H_2O$$

(a) Calculate the molarity of $Mg(OH)_2$:

$$\frac{M_A V_A}{M_B V_B} = \frac{n_A}{n_B}$$

where

$$M_A = \text{molarity of the acid} = 0.100 \text{ M}$$

$$M_B = \text{molarity of the base} = ?$$

$$V_A = \text{volume of acid} = 75.5 \text{ mL}$$

$$V_B = \text{volume of base} = 150 \text{ mL}$$

$$n_A = \text{mol of acid} = 1$$

$$n_B = \text{mol of base} = 2$$

$$M_B = \frac{M_A V_A n_B}{V_B n_A} = \frac{0.100 \text{ M} \times 75.5 \text{ mL} \times 1}{150 \text{ mL} \times 2} = 0.0252 \text{ M}$$

(b) Calculate the moles and hence the mass of $Mg(OH)_2$ contained in 150 mL of the saturated solution:

$$\text{mol of } Mg(OH)_2 = \left(0.0252 \frac{\text{mol}}{\text{L}}\right)\left(\frac{1 \text{ L}}{1000 \text{ mL}}\right)(150 \text{ mL}) = 3.8 \times 10^{-3} \text{ mol}$$

$$\text{Mass of } Mg(OH)_2 = \left(58.3 \frac{\text{g}}{\text{mol}}\right)\left(3.8 \times 10^{-3} \text{ mol}\right) = 0.22 \text{ g}$$

(c) Now calculate the solubility of $Mg(OH)_2$ in water in g per 100 mL at 50°C.

$$\text{Solubility in} \frac{\text{mol}}{\text{L}} = \text{molarity in} \frac{\text{mol}}{\text{L}}$$

Using the molarity of $Mg(OH)_2$ from part (a):

$$\text{Solubility} = \left(\frac{0.0252 \text{ mol Mg(OH)}_2}{\text{L}}\right)\left(\frac{58.3 \text{ g}}{\text{mol Mg(OH)}_2}\right)\left(\frac{1 \text{ L}}{1000 \text{ mL}}\right)$$

$$= 1.5 \times 10^{-3} \text{ g/mL} \quad \text{or} \quad 0.15 \text{ g/100 mL}$$

Example 14.4

14.30 g of hydrated sodium carbonate, $Na_2CO_3{\cdot}yH_2O$, were dissolved in water and the solution made up to 500 mL. 25 mL of this solution required 20.0 mL of 0.25 M HNO_3 for complete neutralization using methyl orange indicator. Calculate the number of moles of water of crystallization, y, in the $Na_2CO_3{\cdot}yH_2O$ sample.

Solution

We will start by determining the molarity of the sodium carbonate solution, as in previous titration problems. From this we can calculate the actual moles of sodium carbonate present, which will tell us the mass. Subtracting this from the mass of the original sample will tell us the mass of water present, which we can use to find the moles of water; then the ratio between these two numbers of moles will give us y.

1. Write a balanced equation for the reaction and calculate the molarity of Na_2CO_3.

$$Na_2CO_3 + 2\ HNO_3 \longrightarrow NaNO_3 + 2\ H_2O + CO_2$$

$$\frac{M_A V_A}{n_A} = \frac{M_B V_B}{n_B} \Rightarrow M_B = \frac{M_A V_A n_B}{V_B n_A} = \frac{0.25 \text{ M} \times 20 \text{ mL} \times 1}{25 \text{ mL} \times 2} = 0.100 \text{ M}$$

2. Calculate the number of moles and hence the mass of Na_2CO_3:

$$\text{mol Na}_2\text{CO}_3 \text{in 500 mL} = \left(0.100\frac{\text{mol}}{\text{L}}\right)\left(\frac{1 \text{ L}}{1000 \text{ mL}}\right)(500 \text{ mL})$$

$$= 0.050 \text{ M}$$

$$\text{Mass of anhydrous Na}_2\text{CO}_3 = \left(106\frac{\text{g}}{\text{mol}}\right)(0.050 \text{ mol}) = 5.30 \text{ g}$$

$$\text{Mass of water of crystallization } = 14.3 - 5.3 = 9 \text{ g}$$

3. Calculate the moles of water of crystallization:

$$\frac{\text{Mass of anhydrous } Na_2CO_3}{\text{Mass of water of crystallization}} = \frac{\text{Molar mass of anhydrous } Na_2CO_3}{\text{Molar mass of } yH_2O}$$

$$\frac{5.3}{9.0} = \frac{106}{18y}$$

$$y = \frac{106 \times 9}{18 \times 5.3} = 10$$

The formula is $Na_2CO_3 \cdot 10H_2O$.

Example 14.5

8.5 g of potassium hydroxide were dissolved in water and the solution made up to 1 L. Using methyl orange as an indicator, 25 mL of this solution neutralized 30.25 mL of a solution containing 5.75 g/L of impure HCl. Calculate:

(a) The concentration of the pure HCl in mol/L
(b) The concentration of the pure acid in g/L
(c) The percentage by mass of pure HCl
(d) The percentage of impurity in the HCl

Solution

(a) First, calculate the molarity of KOH:

$$\text{Molar mass of KOH} = 56.1 \text{ g/mol}$$

$$\text{Molarity of KOH} = \left(8.5\frac{\text{g KOH}}{\text{L}}\right)\left(\frac{1 \text{ mol}}{56.1 \text{ g}}\right) = 0.152 \text{ mol/L} = 0.152 \text{ M}$$

(b) Write a balanced equation for the reaction and calculate the molarity (concentration in mol/L) of the acid:

$$KOH(aq) + HCl(aq) \longrightarrow KCl(aq) + H_2O(l)$$

$$\frac{M_A V_A}{n_A} = \frac{M_B V_B}{n_B} \Rightarrow MA = \frac{M_B V_B n_A}{V_A n_A} = \frac{0.152 \text{ M} \times 25 \text{ mL} \times 1}{30.25 \text{ mL} \times 1} = 0.125 \text{ M}$$

(c) Calculate the concentration of the pure acid in grams per liter:

$$\text{Concentration of HCl in g/L} = \left(0.125\frac{\text{mol}}{\text{L}}\right)\left(36.5\frac{\text{g}}{\text{mol}}\right) = 4.57 \text{ g/L}$$

(d) Calculate the percentage of pure HCl:

$$\% \text{ by mass of pure HCl} = \frac{\text{Mass of pure HCl}}{\text{Mass of impure HCl}} \times 100$$

$$= \frac{4.57 \text{ g/L}}{5.75 \text{ g/L}} \times 100 = 79.49\%$$

(e) Calculate the percentage of impurity in the HCl:

$$\% \text{ impurity} = \frac{\text{Mass of impurity}}{\text{Mass of impure HCl}} = \frac{(5.75 - 4.57)}{5.75}$$

$$\times 100 = 20.51\% = 20.5\%$$

Alternatively, we can calculate the % of impurity by subtracting the % of pure HCl from 100, i.e., % impurity = (100.00 − 79.49) = 20.51% = 20.5%.

14.5. Back Titrations

The direct titration method is only applicable to substances that are soluble in water. Some compounds, such as metal carbonates or hydroxides, that are insoluble in water can be determined by using a procedure known as *back titration*. Here, an excess of a standard solution of a titrant (say acid A) is deliberately added to a given amount of the substance (say base B) to be determined. The amount of excess acid A is then determined by titrating against a second standard titrant (say base C).

$$\text{Acid } A + \text{Base } B \longrightarrow \text{Products} + \text{Excess Acid } A$$

$$\text{Excess Acid } A + \text{Base } C \longrightarrow \text{Products}$$

Thus two standard titrant solutions are employed, both of which are used in the calculations. We know the initial amount of A; the second titration lets us calculate the unreacted excess, which in turn lets us calculate the amount of A consumed in the reaction with B, which then lets us calculate the amount of B.

Example 14.6

50 mL of 0.500 M HCl were added to 1.125 g of a sample of limestone (impure $CaCO_3$). The excess acid required 25 mL of 0.500 M KOH for neutralization. Calculate the percentage purity of the $CaCO_3$.

Solution

1. Write a balanced equation for the neutralization of excess HCl and determine the number of moles of excess HCl.

$$HCl(aq) + KOH(aq) \longrightarrow KCl(aq) + H_2O(l)$$

$$\text{Moles of KOH used to neutralize excess HCl} = \left(0.500\frac{\text{mol}}{\text{L}}\right)\left(\frac{1\text{ L}}{1000\text{ mL}}\right) \times (25\text{ mL}) = 0.0125\text{ mol.}$$

From the balanced equation, 1 mol KOH = 1 mol HCl

$$\text{mol of excess HCl} = 0.0125\text{ mol}$$

2. Determine the moles of HCl that actually reacted with the limestone. This is obtained as the difference between the initial amount of HCl added to the limestone, and the excess HCl.

$$\text{mol of HCl initially added} = \left(0.500\frac{\text{mol}}{\text{L}}\right)\left(\frac{1\text{ L}}{1000\text{ mL}}\right)(50\text{ mL}) = 0.0250\text{ mol}$$

$$\text{HCl used} = \text{Initial HCl added} - \text{Excess HCl} = 0.0250\text{ mol} - 0.0125\text{ mol} = 0.0125\text{ mol}$$

3. Write a balanced equation for the initial reaction between HCl and $CaCO_3$. From the equation determine the moles of $CaCO_3$ reacted.

$$2\ HCl(aq) + CaCO_3(s) \longrightarrow CaCl_2(aq) + H_2O(l) + CO_2(g)$$

2 mol of HCl reacted with 1 mol of $CaCO_3$

0.0125 mol HCl will react with 0.0063 mol of $CaCO_3$

Thus mol of $CaCO_3$ = 0.0063 mol.

4. Calculate the percentage purity:

$$\%\ \text{Purity of } CaCO_3 = \frac{\text{Mass of pure } CaCO_3 \text{ in the limestone}}{\text{Mass of limestone (impure } CaCO_3)} \times 100$$

$$\text{Mass of pure } CaCO_3 = (0.0063\text{ mol}) \times \left(100\ \frac{\text{g}}{\text{mol}}\right) = 0.63\text{ g}$$

$$\text{Mass of limestone used} = 1.125\text{ g}$$

$$\%\ \text{purity of } CaCO_3 = \frac{0.6300}{1.125} \times 100\% = 56\%$$

5. Thus the percentage purity of $CaCO_3$ is 56%.

Example 14.7

2.21 g of a divalent metal carbonate (MCO_3) were dissolved in 50 mL of 1.0 M HNO_3. The resulting solution was made up to 250 mL. Using a suitable indicator, 25 mL of the acid solution required 20 mL of 0.10 M LiOH solution for complete neutralization. Calculate the relative molecular mass of the metal carbonate, and the relative atomic mass of M.

Solution

1. Write a balanced equation for the neutralization of excess HNO_3 and determine the number of moles of excess HNO_3:

$$HNO_3(aq) + LiOH(aq) \longrightarrow LiOH(aq) + H_2O(l)$$

Concentration of HNO_3 solution that

$$\text{reacted with LiOH} = \left(0.100\frac{\text{mol}}{\text{L}}\right)\left(\frac{20\text{ mL}}{25\text{ mL}}\right) = 0.080\frac{\text{mol}}{\text{L}}.$$

The actual volume of solution used is 250 mL (or 0.25 L). Therefore the number of moles of the acid present is 0.080/4 = 0.020 mol.
From the balanced equation, 1 mol LiOH = 1 mol HNO_3
Mole of excess HNO_3 = 0.020 mol.

2. Determine the moles of HNO_3 that actually reacted with the metal carbonate. This is obtained as the difference between the initial amount of HNO_3 added to the carbonate, and the excess HNO_3.

$$\begin{aligned}\text{mol of } HNO_3 \text{ initially added} &= \left(1.00\frac{\text{mol}}{\text{L}}\right)\left(\frac{1\text{ L}}{1000\text{ mL}}\right)(50\text{ mL})\\ &= 0.050\text{ mol}\\ HNO_3 \text{ used} &= \text{Initial } HNO_3 \text{ added} - \text{Excess } HNO_3\\ &= 0.0500\text{ mol} - 0.020\text{ mol} = 0.030\text{ mol}\end{aligned}$$

3. Write a balanced equation for the initial reaction between HNO_3 and MCO_3. From this determine the moles of MCO_3 reacted.

$$2\ HNO_3(aq) + MCO_3(s) \longrightarrow M(NO_3)_2(aq) + H_2O(l) + CO_2(g)$$

2 mol of HNO_3 reacted with 1 mol of MCO_3
0.030 mol HNO_3 will react with 0.015 mol of MCO_3
Thus mol of MCO_3 = 0.015 mol.

4. Calculate the relative molecular mass of MCO_3 and the relative atomic mass of M.

$$\text{Molar mass of } MCO_3 = (2.214\text{ g})\left(\frac{1}{0.0150\text{ mol}}\right) = 147.6\text{ g/mol}$$

The relative atomic mass is calculated as follows:

$$M + \mathrm{C} + 3 \times \mathrm{O} = 147.6$$
$$M + 12 + 3 \times 16 = 147.6$$
$$M = 87.6$$

A quick look at the periodic table suggests M = Sr. Thus the metal carbonate is $SrCO_3$.

Example 14.8

The active ingredient in a Milk of Magnesia tablet is $Mg(OH)_2$. In a titration experiment, a student dissolved one tablet weighing 500 mg in 45 mL of 1.0 M HCl. The resulting solution was made up to 500 mL with water, and 20 mL of this diluted solution needed 15.5 mL of 0.10 M NaOH solution to neutralize the excess acid. Determine the mass of $Mg(OH)_2$ in each tablet. What is the mass percent of $Mg(OH)_2$ in the tablet?

Solution

1. Write a balanced equation for the neutralization of excess HCl and determine the number of moles of excess HCl.

   $$\mathrm{HCl(aq) + NaOH(aq) \longrightarrow NaCl(aq) + H_2O(l)}$$

   Moles per liter of NaOH used to

   $$\text{neutralize excess HCl} = \left(0.100\ \frac{\text{mol}}{\text{L}}\right)\left(\frac{15.5\ \text{mL}}{20\ \text{mL}}\right) = 0.0775\frac{\text{mol}}{\text{L}}.$$

   The actual volume of solution used is 500 mL. Therefore the number of moles of the acid present is $\frac{0.0775}{2} = 0.0388$ mol.
   From the balanced equation, 1 mol NaOH = 1 mol HCl.
   mol of excess HCl = 0.0388 mol.
2. Determine the moles of HCl that actually reacted with the $Mg(OH)_2$. This is obtained as the difference between the initial amount of HCl added to the Milk of Magnesia, and the excess HCl.

$$\text{mole of HCl initially added} = \left(1.00\frac{\text{mol}}{\text{L}}\right)\left(\frac{1\ \text{L}}{1000\ \text{mL}}\right)(45\ \text{mL})$$
$$= 0.0450\ \text{mol}$$

$$\text{HCl used} = \text{Initial HCl added} - \text{Excess HCl}$$
$$= 0.0450\ \text{mol} - 0.0388\ \text{mol} = 0.0063\ \text{mol}$$

3. Write a balanced equation for the initial reaction between HCl and $Mg(OH)_2$. From this determine the moles of $Mg(OH)_2$ reacted.

$$2\ HCl(aq) + Mg(OH)_2(s) \longrightarrow MgCl_2(aq) + 2\ H_2O(l)$$

2 mol of HCl reacted with 1 mol of $Mg(OH)_2$

0.0063 mol HCl will react with 0.0031 mol of $Mg(OH)_2$

Thus mol of $Mg(OH)_2 = 0.0031$ mol.

4. The mass of $Mg(OH)_2$ per tablet is obtained from the number of moles and the molar mass.

$$\text{Mass of } Mg(OH)_2 \text{ in 1 tablet} = (0.0031\ \text{mol})\left(58.30\frac{\text{g}}{\text{moL}}\right)$$

$$= 0.1822\ \text{g} \quad \text{or} \quad 182.2\ \text{mg}$$

$$\text{Mass percent of } Mg(OH)_2 = \frac{\text{Mass of } Mg(OH)_2}{\text{Mass of 1 tablet}} \times 100\%$$

$$= \frac{182.2\ \text{mg}}{500\ \text{mg}} \times 100\%$$

$$= 36.4\%$$

The active $Mg(OH)_2$ represents 36.4% of the tablet.

14.5.1. Kjeldahl nitrogen determination

The Kjeldahl titration is a volumetric analysis method used for the determination of nitrogen and/or protein in organic samples. The analysis, which employs the principles of back-titration, consists of three parts:

- Digesting the sample in boiling H_2SO_4, which contains a catalyst to convert all the organic nitrogen into NH_4 ions.
- Making the solution basic to convert the NH_4 ions into NH_3, which is distilled into a known excess of an acid such as HCl or HNO_3.
- Back-titrating the excess acid (needed to react with NH_3) with an alkaline solution such NaOH to determine the amount of NH_3 in the sample.

The following chemical conversions are involved:

1. Digestion

$$\text{N (in protein or organic compound)} \longrightarrow NH_4^+$$

2. Distillation of NH_3

$$NH_4^+(aq) + OH^-(aq) \longrightarrow NH_3(g) + H_2O(l)$$

3. Collection of NH_3 in HCl or HNO_3

$$NH_3(g) + H^+(aq) \longrightarrow NH_4^+(aq)$$

4. Back titration of unreacted acid

$$H^+(aq) + OH^-(aq) \longrightarrow H_2O(l)$$

Example 14.9

1.2505 g of an unknown protein was analyzed by the Kjeldahl method. The sample was digested in H_2SO_4 to convert N to NH_4^+ ions. A concentrated solution of NaOH was added to convert all of the nitrogen in the protein to NH_3. The liberated ammonia was collected into a flask containing 100 mL of 0.125 M HCl. The excess acid required 9.50 mL of 0.0625 M NaOH for complete titration. Calculate the percent of nitrogen in the sample.

Solution

Step 1: Calculate the total number of moles of HCl in the distillation flask which received the NH_3.

$$\left(0.100\ \text{L} \times 0.125\ \frac{\text{mol}}{\text{L}}\right) = 0.0125\ \text{mol HCl}$$

Step 2: Calculate the total number of moles of NaOH required for complete titration of the excess acid.

$$\left(0.0095\ \text{L} \times 0.0625\frac{\text{mol}}{\text{L}}\right) = 0.000594\ \text{mol NaOH}$$

Step 3: Calculate the moles of ammonia produced. This is the difference between the moles of HCl and the moles of NaOH needed to completely neutralize the excess acid.

$$(0.0125\ \text{mol HCl}) - (0.000594\ \text{mol NaOH}) = 0.01191\ \text{mol NH}_3$$

Step 4: Calculate the mass of nitrogen in the sample. From the reaction stoichiometry, one mole of nitrogen in the protein sample produces one mole of ammonia. Therefore, the moles of nitrogen in the sample are equal to the moles of ammonia produced, which lets us determine the mass of nitrogen.

$$\text{Mass of N} = (0.01191\ \text{mol NH}_3)\left(\frac{1\ \text{mol N}}{1\ \text{mol NH}_3}\right)\left(\frac{14.01\ \text{g N}}{1\ \text{mol N}}\right)$$

$$= 0.1668\ \text{g N}$$

Step 5: Calculate the percentage by mass of N in the sample:

$$\%N = \frac{\text{Mass of nitrogen in sample}}{\text{Mass of sample analyzed}} \times 100$$

$$= \frac{0.1668}{1.2505} \times 100 = 13.3\%$$

14.6. Problems

1. A 35.25 cm^3 sample of HNO_3 required 27.50 cm^3 of 0.25 M NaOH solution for complete neutralization. What is the molarity of the acid?
2. A 50.00 cm^3 sample of $H_2C_2O_4$ required 45.00 cm^3 of 0.45 M NaOH solution for complete neutralization. What is the molarity of the acid?
3. Calculate the mass of potassium hydroxide required to completely neutralize 250 cm^3 of a solution containing 24.5 g of H_2SO_4 per liter of solution.
4. In a titration experiment, 7.50 cm^3 of 0.0075 M benzoic acid, C_6H_5COOH, completely neutralized 12.75 cm^3 of NaOH. What is the concentration of the NaOH in g/L?
5. In a standardization experiment, exactly 0.7525 g of primary standard potassium hydrogen phthalate, KHP (molar mass = 204.23), was dissolved in water. If titrating this solution required 22.30 cm^3 of a sodium hydroxide solution, what is the concentration of NaOH (a) in moles per liter, (b) in grams per liter? (Note: KHP and NaOH react in 1:1 ratio.)
6. A solution of nitric acid, HNO_3, contains 5.85 g of HNO_3 per liter of solution. 22.50 cm^3 of this solution neutralized 25.00 cm^3 of an unknown base, MOH, whose concentration is 8.57 g/L. Find (a) the molarity of the MOH, (b) the relative atomic mass of M, (c) the chemical identity of M, with the aid of the periodic table.
7. 7.125 g of soda ash (impure Na_2CO_3) are dissolved in water and the solution made up to 250 cm^3. Using methyl orange indicator, 25 cm^3 of this solution require 23.50 cm^3 of 0.50 M HNO_3 for neutralization. Find the percentage purity of the Na_2CO_3.
8. A 7.575 g sample of impure oxalic acid, $H_2C_2O_4$, was dissolved in water and the solution was made up to 500 cm^3. 25.00 cm^3 of this solution required 20.25 cm^3 of 0.400 M KOH for complete reaction. Find the percentage by mass of impurities in the oxalic acid sample.
9. Exactly 10.1225 g of washing soda crystals, ($Na_2CO_3 \cdot xH_2O$), were dissolved in water and made up to 250 cm^3 of solution. Using methyl orange indicator, 25.00 cm^3 of this solution required 14.10 cm^3 of 0.50 M HCl for complete neutralization. Calculate (a) the percentage of water in the crystal, (b) the number of molecules of water of crystallization, x, in the sample of Na_2CO_3.
10. 2.075 g of a diprotic organic acid (H_2A) were dissolved in water and made up to 250 cm^3 of solution. By titration, it was found that 20 cm^3 of this solution

required 25 cm^3 of 0.100 M NaOH for complete neutralization. Calculate the molar mass of the acid.

11. The nitrogen content in an egg sample was determined by the Kjeldahl procedure. The ammonia produced from 0.4557g of the sample was absorbed in a flask containing 50 mL of 0.1087 M HCl. If exactly 15 mL of 0.1125 M NaOH solution were required for the complete neutralization of the excess acid, what is the percentage of nitrogen in the sample?
12. A 1.975 mL aliquot of an extract from a bioactive natural product solution containing 17.25% (wt/wt) nitrogen was analyzed by the Kjeldahl method. The liberated ammonia was absorbed 10.00 mL of 0.0525 M HNO_3. What is the concentration of the natural product (in mg/mL) in the original solution if the excess acid required 5.75 mL of 0.0215 M NaOH for complete neutralization?
13. A 5.25 g sample of an ammonium salt was heated to boiling in 70.00 mL of 1.20 M NaOH solution until all ammonia was liberated. What is the mass percent of ammonia in the salt if the excess NaOH required 22.75 mL of 0.500 M HNO_3 for complete neutralization?
14. A 25.0 g sample of impure LiOH was dissolved in water and the solution made up to 1 L in a volumetric flask. 25.0 mL of this solution was titrated with 0.918 M HNO_3 and required 23.15 mL to reach the endpoint. What is the percentage purity of the LiOH sample used in this experiment?

15

Ideal Solutions and Colligative Properties

15.1. Colligative Properties

Colligative properties of solutions are those that depend only on the number of solute particles (molecules or ions) in the solution rather than on their chemical or physical properties. The colligative properties that can be measured experimentally include:

- Vapor pressure depression
- Boiling point elevation
- Freezing point depression
- Osmotic pressure

Noncolligative properties, on the other hand, depend on the identity of the dissolved species and the solvent. Examples include solubility, surface tension, and viscosity.

15.2. Vapor Pressure and Raoult's Law

15.2.1. Vapor pressure

The addition of a solute to a solvent typically causes the vapor pressure of the solvent (above the resulting solution) to be lower than the vapor pressure above the pure solvent. As the concentration of the solute in the solution changes, so does the vapor pressure of the solvent above a solution.

The vapor pressure of a solution of a nonvolatile solute is always lower than that of the pure solvent. For example, an aqueous solution of NaCl has a lower vapor pressure than pure water at the same temperature.

15.2.2. Raoult's law

The addition of solute to a pure solvent depresses the vapor pressure of the solvent. This observation, first made by Raoult, is now commonly known as Raoult's law. The law states that the lowering of vapor pressure of a solution containing nonvolatile solute is proportional to the mole fraction of the solute. If P_A^0 represents the vapor pressure of pure solvent, P_A the vapor pressure of the solution, and χ_B the mole fraction of the solute, then, according to Raoult's law:

$$\frac{P_A^0 - P_A}{P_A^0} = \chi_B$$

This equation may be rearranged as follows:

$$\frac{P_A^0}{P_A^0} - \frac{P_A}{P_A^0} = 1 - \frac{P_A}{P_A^0} = \chi_B$$

$$\frac{P_A}{P_A^0} = 1 - \chi_B \quad \text{or} \quad P_A = P_A^0(1 - \chi_B)$$

Recall that for a solute B dissolved in a solvent A:

$$\chi_A + \chi_B = 1 \quad \text{so} \quad \chi_A = 1 - \chi_B$$

Therefore:

$$P_A = P_A^0 \chi_A$$

In other words, the law may be stated as *the vapor pressure of a solution is proportional to the mole fraction of the solvent.*

The amount of vapor pressure lowering (ΔP) can be calculated by multiplying the vapor pressure of the pure solvent with the mole fraction of the solute. If we are to calculate the vapor pressure lowering, ΔP, rather than the actual vapor pressure, we can do so using the original form of Raoult's law. Here is a quick derivation.

Since:

$$\Delta P_A = P_A^0 - P_A$$

we have

$$\frac{P_A^0 - P_A}{P_A^0} = \chi_B$$

$$\frac{\Delta P_A}{P_A^0} = \chi_B$$

$$\Delta P_A = P_A^0 \chi_B$$

so that *the change in vapor pressure depends on the mole fraction of the solute.*

Example 15.1

What is the vapor pressure of a solution prepared by dissolving 100 g of glucose ($C_6H_{12}O_6$) in 150 g of water at 25°C? The vapor pressure of pure water at 25°C is 23.8 mmHg.

Solution

First, we need to determine the mole fraction of water. To do this, we will need the number of moles of water and glucose:

$$n_{H_2O} = 150 \text{ g } H_2O \times \frac{1 \text{ mol } H_2O}{18.0 \text{ g } H_2O} = 8.33$$

$$n_{C_6H_{12}O_6} = 100 \text{ g } C_6H_{12}O_6 \times \frac{1 \text{ mol } C_6H_{12}O_6}{180 \text{ g } C_6H_{12}O_6} = 0.56$$

The mole fraction of water is:

$$\chi_{H_2O} = \frac{n_{H_2O}}{n_{H_2O} + n_{C_6H_{12}O_6}} = \frac{8.33}{8.33 + 0.55} = 0.938$$

The vapor pressure of the solution is

$$P_{soln.} = P^0_{H_2O}\chi_{H_2O} = 23.8 \text{ mmHg} \times 0.938 = 22.3 \text{ mmHg}$$

Example 15.2

The vapor pressure of water at 25°C is 23.8 torr. If 2.25 g of a nonvolatile solute dissolved in 25 g of water lowers the vapor pressure by 0.22 torr, what is the molecular weight of the solute?

Solution

Calculate the mole fraction of the solute using Raoult's law:

$$\Delta P_{soln} = 0.22 \text{ torr}, \qquad P^0_{H_2O} = 23.8 \text{ torr}$$

$$\Delta P_{soln} = P^0_{solv}\chi_{solute}$$

$$\chi_{solute} = \frac{\Delta P_{soln}}{P^0_{H_2O}} = \frac{0.22 \text{ torr}}{23.8 \text{ torr}} = 0.0092$$

But:

$$\chi_{solute} = \frac{n_{solute}}{n_{solute} + n_{H_2O}}$$

$$n_{H_2O} = 25 \text{ g } H_2O \times \frac{1 \text{ mol } H_2O}{18.0 \text{ g } H_2O} = 1.389 \text{ mol}$$

$$0.0092 = \frac{n_{solute}}{n_{solute} + 1.389}$$

$$(0.0092)(n_{solute} + 1.389) = n_{solute}$$

$$(0.0092)(1.389) = n_{solute} - (0.0092)n_{solute}$$

$$0.9908n_{solute} = 0.0128$$

$$n_{solute} = 0.0129 \text{ mol}$$

$$\text{Molecular weight of solute} = \frac{\text{Mass of solute in grams}}{\text{Number of moles of solute}} = \frac{2.25 \text{ g}}{0.0129 \text{ mol}} = 174.2 \text{ g/mol}$$

Thus, *we can determine the molecular weight of a substance without knowing anything about its composition*; this is one reason colligative properties are so useful.

15.2.3. Ideal solutions with two or more volatile components

A solution may sometimes contain two or more volatile components. If the solution behaves ideally, then we can still apply Raoult's law. For example, if an ideal solution contains two components A and B, the partial pressures of A and B vapors above the solution are given by Raoult's law:

$$P_A = \chi_A P_A^0 \quad \text{and} \quad P_B = \chi_B P_B^0$$

The total vapor pressure over the solution is obtained as the sum of the partial pressures of each volatile component. Thus:

$$P_{total} = P_A + P_B = \chi_A P_A^0 + \chi_B P_B^0$$

Example 15.3

Consider an ideal solution with a mole fraction of benzene, C_6H_6, equal to 0.30, and mole fraction of toluene, C_7H_8, equal to 0.70. If the vapor pressure of pure C_6H_6 is 68 torr and that of pure C_7H_8 is 21 torr at 22°C, what is the total vapor pressure at this temperature?

Solution

Calculate the partial pressure of each component:

$$P_{C_6H_6} = P^0_{C_6H_6}\chi_{C_6H_6}, = 68 \times 0.30 = 20.4 \text{ torr}$$

$$P_{C_7H_8} = P^0_{C_7H_8}\chi_{C_7H_8}, = 21 \times 0.70 = 14.7 \text{ torr}$$

And the total pressure:

$$P_{\text{total}} = P_{C_6H_6} + P_{C_7H_8} = 20.4 + 14.7 = 35.1 \text{ torr}$$

15.3. Elevation of Boiling Point

Besides lowering the vapor pressure, adding a nonvolatile solute to a solvent raises the boiling point of the resulting solution relative to that of the pure solvent. This increase, the *boiling point elevation*, is given by the equation:

$$\Delta T_b = ik_b m$$

where k_b is the boiling point constant, which depends on the solvent, and m is the molality of solute. $\Delta T_b = T_{b(\text{soln})} - T_{b(\text{solvent})}$. The factor i, called the van't Hoff factor, is equal to the number of moles of solute ions per mole of the solute. Ionic compounds (or electrolytes) dissolve to give more than one particle per solute; in general they ionize completely, providing as many particles as there are ions in the compound. Thus, for NaCl, which ionizes to give Na^+ and Cl^-, i is 2; similarly, $Al(NO_3)_3$ ionizes in aqueous solution to give an i value of 4. For nonelectrolytes, that is, substances which do not ionize (such as glucose) i is 1. For such solutes, the boiling point elevation (ΔT_b) is given by the equation:

$$\Delta T_b = k_b m$$

Note that $\Delta T_b = T_{b(\text{soln})} - T_{b(\text{solvent})}$. For a solution containing a nonvolatile solute, ΔT_b is always positive, since the boiling points of such solutions are always higher than those of pure solvents. Values of k_b for some common solvents are given in table 15-1.

Table 15-1 Freezing Point Depression and Boiling Point Elevation Constants

Solvent	*Formula*	*Freezing point (°C)*	k_f $\left(°C \frac{kg}{mol}\right)$	*Boiling point (°C)*	k_b $\left(°C \frac{kg}{mol}\right)$
Acetic acid	CH_3COOH	16.6	3.59	118.1	3.08
Benzene	C_6H_6	5.5	5.07	80.1	2.61
Camphor	$C_{10}H_{16}O$	179.8	40	208	5.95
Carbon disulfide	CS_2	−111.5	3.83	46.2	2.34
Carbon tetrachloride	CCL_4	−22.8	30	76.6	5.03
Chloroform	$CHCl_3$	−63.5	4.7	61.2	3.63
Cyclohexane	C_6H_{12}	6.6	20	80.7	2.79
Diethyl ether	$C_4H_{10}O$	−116.2	1.79	34.5	2.02
Ethyl alcohol	C_2H_5OH	−117.3	1.99	78.4	1.22
Naphthalene	C_10H_8	80.2	6.92	218	6.34
Water	H_2O	0	1.86	100	0.51

Example 15.4

The boiling point of pure carbon tetrachloride, CCl_4 is 76.6°C. If the boiling point of a solution made by dissolving 2.036 g of naphthalene, $C_{10}H_8$, in 25 g of CCl_4 is 79.8°C, calculate the molar boiling point elevation constant. Assume that the solute is nonvolatile and a nonelectrolyte.

Solution

First compute the molality from the boiling point elevation.

$$\text{Moles of } C_{10}H_8 = \frac{\text{Grams of } C_{10}H_8}{\text{Molar mass of } C_{10}H_8} = \frac{2.036 \text{ g}}{128 \text{ g/mol}} = 0.0159 \text{ mol}$$

$$\text{Molality, } m = \frac{\text{Moles of } C_{10}H_8}{\text{kg of } CCl_4} = \frac{0.0159 \text{ mol}}{0.025 \text{ kg}} = 0.636\, m$$

Now calculate k_b:

$$\Delta T_b = k_b m$$

$$k_b = \frac{\Delta T_b}{m} = \frac{(79.8 - 76.6)^\circ\text{C}}{0.636 \text{ mol/kg}} = \frac{3.2^\circ\text{C}}{0.636 \text{ mol/kg}} = 5.03^\circ\text{C kg/mol}$$

Example 15.5

What is the boiling point of an aqueous solution containing 25 g of the non-electrolyte urea, $CO(NH_2)_2$, per 500 g of water? k_b for water is 0.51°C kg/mol.

Solution

First calculate the molality of the urea solution:

$$n_{\text{urea}} = \frac{\text{g of urea}}{\text{MW of urea}} = \frac{25 \text{ g}}{60 \text{ g/mol}} = 0.4167 \text{ mol}$$

$$m = \frac{n_{\text{urea}}}{\text{kg of urea}} = \frac{0.4167 \text{ mol}}{0.5 \text{ kg}} = 0.833\, m$$

Calculate the boiling point of the solution:

$$\Delta T_b = k_b m$$

$$\Delta T_b = k_b m = \left(0.51^\circ\text{C} \frac{\text{kg}}{\text{mol}}\right)\left(0.833 \frac{\text{mol}}{\text{kg}}\right) = 0.42^\circ\text{C}$$

But

$$\Delta T_b = T_{b(\text{soln})} - T_{b(\text{solvent})}$$

$$T_{b(\text{soln})} = T_{b(\text{solvent})} + \Delta T_b = 100^\circ\text{C} + 0.42^\circ\text{C} = 100.42^\circ\text{C}$$

Example 15.6

Calculate the boiling point of a solution prepared by dissolving 50 g of $Ca(NO_3)_2$ in 500 g H_2O.

Solution

In this case, the solute is an electrolyte and will ionize completely in water. So the following expression will be used to calculate the boiling point elevation:

$$\Delta T_b = ik_b m$$

For $Ca(NO_3)_2$, $i = 3\,(1\ Ca^{2-}, 2\ NO_3^-)$.
First calculate moles of $Ca(NO_3)_2$:

$$\text{Mol } Ca(NO_3)_2 = 50\text{ g} \times \frac{1\text{ mol } Ca(NO_3)_2}{102\text{ g}} = 0.490\text{ mol } Ca(NO_3)_2$$

Calculate molality of the solution:

$$\text{Molality, } m, = \frac{\text{Mol solute}}{\text{kg solvent}}$$

$$= \frac{0.490\text{ mol}}{0.500\text{ kg}} = 0.980\ m$$

Now calculate the ΔT_b and $T_{b(soln)}$:

$$\Delta T_b = ik_b m = 3 \times \left(0.51^{\circ}\text{C}\ \frac{\text{kg}}{\text{mol}}\right)\left(0.980\ \frac{\text{mol}}{\text{kg}}\right) = 1.5^{\circ}\text{C}$$

$$T_{b(\text{soln})} = T_{b(\text{solvent})} + \Delta T_b = 100^{\circ}\text{C} + 1.5^{\circ}\text{C} = 101.5^{\circ}\text{C}$$

15.4. Depression of Freezing Point

The addition of a nonvolatile solute to a solvent lowers the freezing point of the resulting solution below that of the pure solvent. The magnitude of the freezing-point depression (ΔT_f) is proportional to the molality of the solute (i.e., the number of moles of the solute per kg of solvent). In mathematical terms:

$$\Delta T_f = ik_f m$$

The constant k_f is the molar freezing point constant of the solvent. The factors i and m have the same meaning as in the previous section. When dealing with solutes that are nonelectrolytes, the expression is simplified to:

$$\Delta T_f = k_f m$$

Similarly, $\Delta T_f = T_{f(\text{solvent})} - T_{f(\text{soln})}$. For a solution containing a nonvolatile solute, ΔT_f is always positive since the freezing points of such solutions are always lower than those of pure solvents. Values of k_f for some common solvents are given in table 15-1.

Example 15.7

Calculate the freezing point of a solution made by dissolving 50 g of glucose in 500 mL of water. Assume that the solute is nonvolatile. (The molecular weight of glucose is 180 g/mol.)

Solution

First calculate the molality:

$$\text{Moles of solute } (C_6H_{12}O_6) = (50\text{ g } C_6H_{12}O_6)\left(\frac{1\text{ mol } C_6H_{12}O_6}{180\text{ g } C_6H_{12}O_6}\right) = 0.2778\text{ mol}$$

$$\text{Molality} = \frac{\text{Moles of solute}}{\text{kg of solvent}} = \left(\frac{0.2778\text{ mol } C_6H_{12}O_6}{0.500\text{ kg } H_2O}\right) = 0.556\ m$$

Then calculate the freezing point depression:

$$\Delta T = k_f m = \left(1.86^{\circ}\text{C}\ \frac{\text{kg}}{\text{mol}}\right)\left(0.56\ \frac{\text{mol}}{\text{kg}}\right) = 1.04^{\circ}\text{C}$$

But:

$$\Delta T_f = T_{f(\text{solvent})} - T_{b(\text{soln})}, \quad \text{i.e., } 1.04^{\circ}\text{C} = 0.00^{\circ}\text{C} - T_{f(\text{soln})}$$

$$T_{f(\text{soln})} = (0.00 - 1.04)^{\circ}\text{C} = -1.04^{\circ}\text{C}$$

Example 15.8

Ethylene glycol, $C_2H_6O_2$, is a nonvolatile nonelectrolyte commonly used in automobile antifreeze. How much ethylene glycol must be added to 1.00 L of water so that the resulting solution will not freeze at -30°C?

Solution

First calculate molality from the expression:

$$\Delta T_f = k_f m$$

$$\Delta T = 30^{\circ}\text{C}, \qquad k_f = 1.86^{\circ}\text{C}\ \frac{\text{kg}}{\text{mol}}$$

$$m = \frac{\Delta T}{k_f} = \frac{30^{\circ}\text{C}}{1.86^{\circ}\text{C kg/mol}} = 16.13\ \frac{\text{mol}}{\text{kg}}$$

Since 1 L of water has a mass of 1 kg, the actual amount of solute is:

$$1 \text{ kg } H_2O \times \frac{16.13 \text{ mol ethylene glycol}}{\text{kg water}} = 16.13 \text{ mol}$$

Now convert to mass in grams:

$$\text{MW of } C_2H_6O_2 = 62 \text{ g/mol}$$

$$\text{Mass of ethylene glycol used} = (16.13 \text{ mol})\left(\frac{62 \text{ g}}{\text{mol}}\right) = 1000 \text{ g}$$

Example 15.9

A solution made by dissolving 1.75 g of an organic compound in 100 g of nitrobenzene freezes at 4.88°C. What is the molecular weight of the compound? The freezing point of pure nitrobenzene is 5.70°C and k_f is 7.00° C/m.

Solution

First, calculate ΔT and m:

$$\Delta T = T_{f(\text{solv})} - T_{f(\text{soln})} = 5.70 - 4.88 = 0.82^{\circ}\text{C}$$

$$m = \frac{\Delta T_f}{k_f} = \frac{0.82^{\circ}\text{C}}{7.00^{\circ}\text{C/m}} = 0.12\, m$$

$$\text{or} \quad 0.12 \text{ mol solute/kg nitrobenzene}$$

Now calculate the molar mass:

$$\text{Moles of solute} = \text{molality } (m) \times \text{kg of solvent}$$
$$= (0.12 \text{ mol solute/kg nitrobenzene})$$
$$\times (0.10 \text{ kg nitrobenzene}) = 0.012 \text{ mol}$$

$$\text{Molar mass} = \frac{\text{Mass in grams}}{\text{No. of moles}} = \frac{1.75 \text{ g}}{0.012 \text{ mol}} = 145.8 \text{ g/mol}$$

15.5. Osmosis and Osmotic Pressure

It is a common observation that when a solution and its pure solvent are separated by a semi-permeable membrane, the pure solvent will diffuse across the membrane to dilute the solution. This process is referred to as osmosis. Osmosis may be defined as the movement of water (or solvent) molecules from a less concentrated to a more concentrated solution across a semi-permeable membrane so as to equalize concentrations. Like any reversible process, the diffusion of solvent molecules through the membrane occurs in both directions, but the rate of diffusion from the pure-solvent region to the solvent-plus-solute region is higher than the reverse rate. The pressure that must be applied to equalize the flow of solvent molecules in both directions across the membrane is known as osmotic pressure, $\prod$. For a dilute solution, the following relationship holds:

$$\Pi = \frac{nRT}{V}$$

where n is the number of moles of solute in V liters of the solution, R is the gas constant (0.0821 L-atm/mol-K), and T is temperature (in K). When the concentration is expressed in terms of molarity, we have

$$\Pi = \mathrm{M}RT$$

since $\mathrm{M} = n/V$.

For very dilute solution, the molarity (M) is approximately equal to molality (m). Thus,

$$\Pi = mRT$$

Also, for an electrolyte solution, the expression for osmotic pressure is modified to contain the van't Hoff factor, i:

$$\Pi = i\mathrm{M}RT$$

Example 15.10

The osmotic pressure of a solution prepared by dissolving 38 g of insulin in 600 mL of water at 25°C is 0.62 atm. What is the molar mass of insulin?

Solution

Calculate the molarity of the solution from the osmotic pressure

$$\Pi = \left(\frac{n}{V}\right)RT = \mathrm{M}RT$$

$$\mathrm{M} = \frac{\Pi}{RT} = \frac{0.62\text{ atm}}{\left(0.0821\,\frac{\text{L-atm}}{\text{mol-K}}\right)(298\text{ K})} = 0.0253\,\frac{\text{mol}}{\text{L}}$$

Now calculate molar mass from the mass and molarity of insulin solution:

$$\text{Molar mass} = \left(\frac{38\text{ g}}{600\text{ mL}}\right)\left(\frac{1000\text{ mL}}{1\text{ L}}\right)\left(\frac{1\text{ L}}{0.0253\text{ mol}}\right) = 2{,}500\text{ g/mol}$$

Example 15.11

The freezing point of human blood is -0.58°C. Calculate the osmotic pressure of blood at 37°C if k_f is 1.86°C.kg/mol. Assume the molarity and molality to be equal.

Solution

First calculate the molality

$$\Delta T = k_f m$$

$$m = \frac{\Delta T}{k_f} = \frac{0.58^{\circ}\text{C}}{\left(1.86^{\circ}\text{C}\ \dfrac{\text{kg}}{\text{mol}}\right)} = 0.312\ m \approx 0.312\ \text{M}$$

From the given assumption, the molarity = 0.312 M.

Now calculate the osmotic pressure:

$$\Pi = MRT = \left(0.312\ \frac{\text{mol}}{\text{L}}\right)\left(0.0821\ \frac{\text{L-atm}}{\text{mol-K}}\right)(310\ \text{K}) = 7.94\ \text{atm}$$

Example 15.12

What would be the osmotic pressure at 50°C of an aqueous solution containing 45 g of glucose ($C_6H_{12}O_6$) per 250 mL of solution?

Solution

$$\Pi = \left(\frac{n}{V}\right)RT = \left(\frac{m}{\text{MW}.V}\right)RT$$

MW of glucose = 180 g/mol

$V = 250$ mL $= 0.250$ L

Mass of glucose = 45 g

$T = 323$ K

$R = 0.0821$ L-atm/mol-K

$$\Pi = \left(\frac{m}{\text{MW}.V}\right)RT$$

$$= \left(\frac{45\ \text{g}}{(180\ \text{g/mol})(0.25\ \text{L})}\right)(0.0821\ \text{L-atm/mol-K})(323\ \text{K}) = 26.5\ \text{atm}$$

Example 15.13

An aqueous solution of sucrose ($C_{12}O_{22}O_{11}$) has an osmotic pressure of 2.70 atm at 298 K. (a) How many moles of sucrose were dissolved per liter of solution? (b) How many grams of sucrose were dissolved per liter of solution?

Solution

$$\Pi = MRT$$

$$\text{Molarity, M} = \frac{\Pi}{RT}$$

$$\text{Molar mass of } C_{12}H_{22}O_{11} = 342 \text{ g/mol}$$

$$\text{Osmotic pressure, } \Pi = 2.70 \text{ atm}$$

$$R = 0.0821 \text{ L-atm/mol-K}$$

$$T = 298 \text{ K}$$

(a) The number of moles of sucrose dissolved in 1 L of solution can be obtained from the molarity, M:

$$\text{M} = \frac{\Pi}{RT} = \frac{2.70 \text{ atm}}{(0.0821 \text{ L-atm/K-mol})(298 \text{ K})} = 0.110 \text{ mol/L}$$

(b) Mass in grams per liter = Molarity × Molar mass

$$= \left(0.110 \frac{\text{mol}}{\text{L}}\right)\left(342 \frac{\text{g}}{\text{mol}}\right) = 37.74 \text{ g/L}$$

15.6. Problems

1. Calculate the vapor pressure of water above a solution at 25°C prepared by dissolving 35 g of glucose ($C_6H_{12}O_6$) in 1000 g of water. The vapor pressure of pure water is 23.8 mmHg.
2. What is the vapor pressure at 25°C of a solution prepared by dissolving 115 g of sucrose (molar mass = 342.3 g/mol) in 700 g of water? The vapor pressure of pure water is 23.8 mmHg and the density is 0.997 g/mL at 25°C.
3. The vapor pressure of water at 25°C is 23.8 mmHg. How many grams of sucrose must be added to 250 g of water to lower the vapor pressure by 1.25 mmHg at 25°C? The molar mass of sucrose is 342.3g/mol.
4. At 45°C, the vapor pressure of pure water is 71.88 mmHg and that of a urea solution (NH_2CONH_2) solution is 71.20 mmHg. Calculate the molality of the solution.

5. What is the vapor pressure of a solution prepared by mixing 45.5 g of Na_3PO_4 with 255 g of water at 25°C? The vapor pressure of pure water at 25°C is 23.76 mmHg. (*Hint*: Note that Na_3PO_4 is a strong electrolyte. When 1 mol of Na_3PO_4 dissolves in water, it produces 3 mol of Na^+ ions and 1 mol of PO_4^{3-} ions. Therefore the number of solute particles present in the solution will be four times the concentration of the original solute.)
6. A solution is prepared by dissolving 35 g of $NaNO_3$ in 500 g of water. What is the vapor pressure of this solution at 25°C? The vapor pressure of pure water at 25°C is 23.8 mmHg.
7. Consider an ideal solution with a mole fraction of methanol (CH_3OH) equal to 0.23, and mole fraction of ethanol (C_2H_5OH) equal to 0.77. If the vapor pressure of pure CH_3OH is 94 mmHg and that of pure C_2H_5OH is 44 mmHg at 20°C, what is the total vapor pressure at this temperature?
8. Assuming ideal behavior, calculate the mole fraction of toluene in the vapor phase that is in equilibrium with a solution having the following composition at 20°C:

$$\chi^1_{\text{benzene}} = 0.75 \text{ mmHg}, \quad \chi^1_{\text{toluene}} = 0.25 \text{ mmHg}, \quad P_{\text{Total}} = 8.20 \text{ mmHg},$$

$$P^0_{\text{benzene}} = 9.90 \text{ mmHg}, \quad \text{and} \quad P^0_{\text{toluene}} = 2.80 \text{ mmHg}$$

9. The freezing point depression constant, k_f, for camphor is 40°C kg/mol. Calculate the freezing point of a solution made by dissolving 15.75 g of a nonvolatile solute (of a molecular weight 130) in 500 g of camphor. The normal freezing point of pure camphor is 178.4°C.
10. What is the freezing point of a solution made by dissolving 25 g of sucrose (MW = 342) in 800 g of water? ($k_f = 1.86$°Ckg/mol.)
11. The molecular weight of an unknown compound may be determined from freezing point depression data. A solution containing 30 g of a nonvolatile organic solute dissolved in 750 g of water freezes at −0.576°C. Calculate the molecular weight of the solute.
12. A solution made by dissolving 12.5 g of an unknown compound in 30 g of benzene has a freezing point of −3.375°C. Determine the molar mass of the compound.
13. The boiling point elevation constant, k_b, for nitrobenzene is 5.24°C kg/mol. The normal boiling point of nitrobenzene is 210.8°C. Calculate the boiling point (in °C) of a solution made by dissolving 5.15 g of naphthalene ($C_{10}H_8$, MW = 128.2 g/mol) in 50 g of nitrobenzene.
14. An unknown organic compound extracted from a mango was found to contain 39.6% C, 7.75% H, and 52.7% O upon analysis. A 1.005-g sample of this compound was dissolved in 100 g of water. The freezing point of the solution was −0.101°C. (a) What is the molecular mass of this compound? (b) What are its empirical and molecular formulas?
15. Estimate the freezing points of the following aqueous solutions: (a) 0.15 *m* NaCl (b) 0.25 *m* Na_2SO_4 and (c) 0.05 *m* $Al_2(SO_4)_3$.

16. A solution made by dissolving 1.500 g of an unknown compound *Z* in 5.00 g benzene had a freezing point of 2.07°C. Given that the freezing point of pure benzene is 5.46°C, determine the molecular weight of the unknown compound. The molal freezing point depression constant of benzene is 5.12°C/*m*.
17. A 20% (w/w) aqueous solution of an unknown organic compound has a freezing point of −0.07°C. If the empirical formula of the compound is $C_6H_{10}O_5$, determine the molecular formula.
18. The average osmotic pressure of 10 seawater samples taken from different oceans is approximately 30.15 atm at 25°C. What is the molar concentration of an aqueous solution of hemoglobin that is isotonic with seawater?
19. A solution is prepared by dissolving 15.25 g of a protein fraction in sufficient amount of water and making up to volume in a 1 L flask. What is the molar mass of the protein if the osmotic pressure of the solution at 25°C is 76 mmHg?
20. What is the molar concentration of solute particles in human blood if the osmotic pressure is 8.77 atm at 37°C?
21. An aqueous solution contains 4.5 g of bovine insulin per 500 mL of solution at 25°C. What is the osmotic pressure of the solution in mmHg? The molecular weight of the insulin is 5700 g/mol.
22. What osmotic pressure (in atmospheres) would you expect for each of the following aqueous solutions? (a) 10 g of LiCl in 250 mL at 25°C, (b) 5.5 g of $CaCl_2$ in 500 mL at 10°C, (c) 0.125 M $Fe(NH_4)_2(SO_4)_2$ at 0°C.

16

Chemical Kinetics

16.1. Rates of Reaction

Chemical kinetics is the aspect of chemistry that deals with the speed or rate of chemical reactions and the mechanisms by which they occur. The *rate* of a chemical reaction is a measure of how fast the reaction occurs, and it is defined as the change in the amount or concentration of a reactant or product per unit time.

The *mechanism* of a reaction is the series of steps or processes through which it occurs.

16.2. Measurement of Reaction Rates

Most experimental techniques for determining reaction rates involve measuring of the rate of disappearance of a reactant, or the rate of appearance of a product. For a reaction in which the reactant Y is converted to some products:

$$\text{Rate} = \frac{\text{Concentration of } Y \text{ at time } t_2 - \text{Concentration of } Y \text{ at time } t_1}{t_2 - t_1}$$

$$\text{Rate} = \frac{\Delta[Y]}{\Delta t}$$

where $[Y]$ indicates the molar concentration of the reactant of interest, and Δ refers to a change in the given amount. Rate for a reactant, by this definition, is a negative number. For a product, it is positive.

Example 16.1

For the reaction

$$C_2H_6(g) \rightarrow C_2H_4(g) + H_2(g)$$

the initial concentration of C_2H_6 gas is 2.500 mol/L and its concentration after 45 s is 1.175 mol/L. What is the average rate of this reaction?

Solution

Assume that the reaction started at time $t = 0$ and use the expression for average rate:

$$\text{Average rate} = -\frac{\Delta(\text{mol/L of } C_2H_6)}{\Delta t}$$

$$\text{Average rate} = -\frac{(\text{mol/L of } C_2H_6 \text{ at } t = 45 \text{ s}) - (\text{mol of } C_2H_6 \text{ at } t = 0 \text{ s})}{45 \text{ s} - 0 \text{ s}}$$

$$= -\frac{(1.175 - 2.500) \text{ mol/L}}{(45 - 0) \text{ s}}$$

$$= 0.0294 \text{ mol/s}$$

The rate of reaction always changes with time. For example, consider a reaction in which reactant A is converted to product B. The average rate may be derived as:

$$\text{Average rate} = \frac{\text{Change in the concentration of } B}{\text{Change in time}}$$

$$\text{Average rate} = \frac{\Delta(\text{moles of } B)}{\Delta t}$$

Suppose 1.0 mol/L of A undergoes thermal decomposition to form B. No B is initially present. If after 30 min, 0.65 mol/L of B is produced, the average rate of reaction will be:

$$\text{Average rate} = \frac{\Delta(\text{mol/L of } B)}{\Delta t}$$

$$\text{Average rate} = \frac{(\text{mol/L of } B \text{ at } t = 30 \text{ min}) - (\text{mol/L of } B \text{ at } t = 0 \text{ min})}{30 \text{ min} - 0 \text{ min}}$$

$$= \frac{0.65 \text{ mol/L} - 0 \text{ mol/L}}{30 \text{ min} - 0 \text{ min}}$$

$$= 0.0217 \text{ mol/min}$$

We can also calculate the average rate of reaction with respect to A if we know how much of A is left after a given interval. The rate expression here will have a negative sign in front of it since the concentration of A will decrease with time. However, the value of the rate must be a positive quantity.

$$\text{Average rate} = -\frac{\Delta(\text{moles of } A)}{\Delta t}$$

Example 16.2

The following data were obtained for the conversion (thermal electrocyclization) of 1,3,5-hexatriene to 1,3-cyclohexadiene (see figure 16-1).

Using the data in table 16-1, determine the average rate for this reaction and tabulate your results.

Solution

Calculate the average rate by taking two adjacent time intervals and their corresponding concentrations. For example:

For the time interval 0 to 0.56 h;

$$\text{Average rate} = -\frac{\Delta\,[1,3,5\text{-hexatriene}]}{\Delta t}$$

$$\text{Average rate} = -\frac{(1.150 - 1.700)\ \text{mol}}{(0.56 - 0)\ \text{h}} = 0.982\ \text{mol/h}$$

For the time interval 0.56 to 1.39 h;

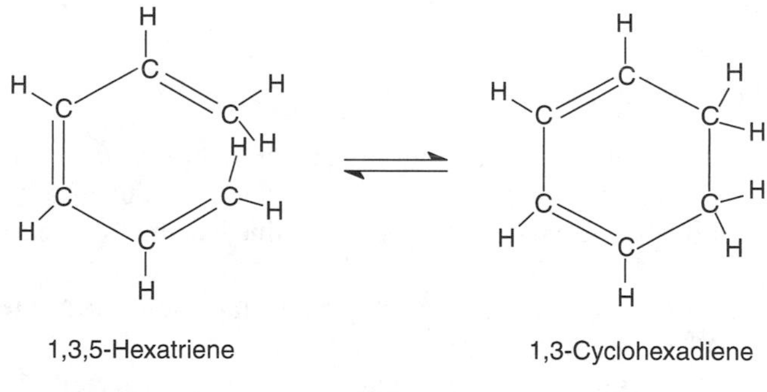

1,3,5-Hexatriene 1,3-Cyclohexadiene

Figure 16-1 The conversion (thermal electrocyclization) of 1,3,5-hexatriene to 1,3-cyclohexadiene.

Table 16-1 Kinetic Data for the Conversion of 1,3,5-Hexatriene to 1,3-Cyclohexadiene

Time (h)	*Concentration of 1,3,5-hexatriene (mol)*
0	1.700
0.56	1.150
1.39	0.596
2.22	0.319
3.33	0.142
4.17	0.079

Table 16-2 Rate Data for the Conversion of 1,3,5-Hexatriene to 1,3-Cyclohexadiene

Time (h)	*Concentration of 1,3,5-hexatriene (mol)*	*Average rate (mol/h)*
0	1.700	—
0.56	1.150	0.982
1.39	0.596	0.668
2.22	0.319	0.334
3.33	0.142	0.160
4.17	0.079	0.075

$$\text{Average rate} = -\frac{(0.596 - 1.150)\ \text{mol}}{(1.39\ -\ 0.596)\ \text{h}} = 0.668\ \text{mol/h}$$

For the time interval 1.39 to 2.22 h;

$$\text{Average rate} = -\frac{(0.319 - 0.596)\ \text{mol}}{(2.22\ -\ 1.39)\ \text{h}} = 0.334\ \text{mol/h}$$

Complete the remaining calculations. The results are tabulated in table 16-2.

16.2.1. Instantaneous rate

The value of the rate at a particular time is known as the *instantaneous rate* and will be different from the average rate. Its value can be obtained from the plot of concentration (mol/L) vs. time (s) as the slope of a line tangent to the curve at a given point.

Consider the following kinetic data for the decomposition of N_2O_5 to gaseous NO_2 and O_2 at 40°C (see table 16-3).

A plot of $[N_2O_5]$ vs. time is shown in figure 16-2. From this curve, the instantaneous rate of reaction at any time t can be obtained from the slope of the tangent to the curve. This corresponds to the value of $\Delta[N_2O_5]/\Delta t$ for the tangent at a given instant. The instantaneous rate at the beginning of the reaction ($t = 0$) is known as the *initial rate*. For this reaction the initial rate is calculated as:

$$\begin{aligned}\text{Initial rate} &= -\text{slope of tangent line}\\ &= -\frac{\Delta[N_2O_5]}{\Delta t} = -\frac{(0\ -\ 1.420)\,\text{M}}{(200\ -\ 0)\ \text{min}} = 7.10\times10^{-3}\ \text{M}\end{aligned}$$

The instantaneous rate at 420 min is calculated in a similar way:

$$\begin{aligned}\text{Instantaneous rate} &= -\text{slope of tangent line}\\ &= -\frac{\Delta[N_2O_5]}{\Delta t} = -\frac{(0.100\ -\ 0.260)\,\text{M}}{(490\ -\ 360)\ \text{min}} = 1.23\times10^{-3}\ \text{M}\end{aligned}$$

Table 16-3 Kinetic Data for the Decomposition of N_2O_5 to Gaseous NO_2 and O_2 at 40°C

Time	*Concentration of N_2O_5 (mol/L)*
0	1.42
60	1.02
120	0.78
180	0.61
240	0.45
300	0.33
360	0.25
420	0.18
480	0.14
540	0.10
600	0.08
660	0.06
720	0.05

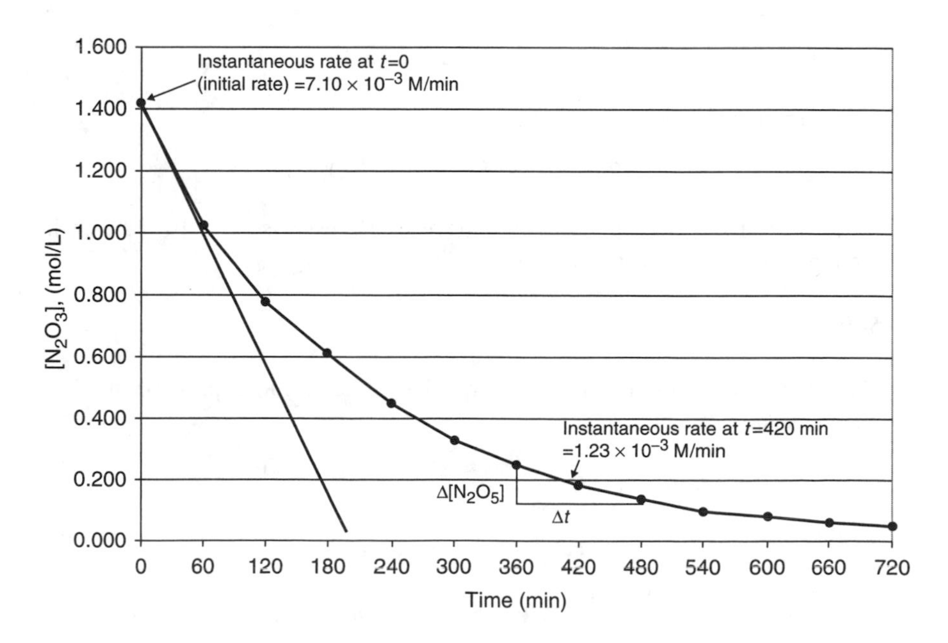

Figure 16-2 Concentration as a function of time for the decomposition of N_2O_5 at 40°C.

16.3. Reaction Rates and Stoichiometry

For the reaction

$$2\,N_2O_5(g) \longrightarrow 4\,NO(g) + O_2(g)$$

we can relate the rate of decomposition of N_2O_5 to the rates of formation of NO and O_2. The stoichiometry of the equation indicates that 2 mol N_2O_5 produce 4 mol NO and 1 mol O_2. Therefore, the rate of decomposition of N_2O_5 is half the rate of formation of NO and twice the rate of formation of O_2.

$$\frac{\Delta\,[N_2O_5]}{\Delta t} = -\frac{1}{2}\frac{\Delta\,[NO]}{\Delta t} \quad \text{and} \quad \frac{\Delta\,[N_2O_5]}{\Delta t} = -2\,\frac{\Delta\,[O_2]}{\Delta t}$$

Thus, if the rate of decomposition of N_2O_5 is known, the rates of formation NO and O_2 can be deduced. Consider the general reaction:

$$aA + bB \longrightarrow cC + dD$$

Using the coefficients in the balanced equation, the rate is given by:

$$\text{Rate} = -\frac{1}{a}\frac{\Delta\,[A]}{\Delta t} = -\frac{1}{b}\frac{\Delta\,[B]}{\Delta t} = \frac{1}{c}\frac{\Delta\,[C]}{\Delta t} = \frac{1}{d}\frac{\Delta\,[D]}{\Delta t}$$

and:

$$\frac{\Delta\,[A]}{\Delta t} = \frac{a}{b}\frac{\Delta\,[B]}{\Delta t} = -\frac{a}{c}\frac{\Delta\,[C]}{\Delta t} = -\frac{a}{d}\frac{\Delta\,[D]}{\Delta t}$$

Example 16.3

Consider the reaction

$$2\,NO(g) + O_2(g) \longrightarrow NO_2(g)$$

If the rate of formation of NO_2 within the reaction vessel is 2.5×10^{-5} M/s, what are the rates of disappearance of NO and O_2?

Solution

Using the coefficients in the balanced equation, write the rate expression involving reactants and product:

$$\text{Rate} = -\frac{1}{2}\frac{\Delta\,[NO]}{\Delta t} = -\frac{\Delta\,[O_2]}{\Delta t} = \frac{\Delta\,[NO_2]}{\Delta t}$$

Then use this expression to calculate the other rates.

$$-\frac{\Delta\,[NO]}{\Delta t} = 2\frac{\Delta\,[NO_2]}{\Delta t} = 2\left(2.5 \times 10^{-5}\ \text{M/s}\right) = 5.0 \times 10^{-5}\ \text{M/s}$$

$$-\frac{\Delta\,[O_2]}{\Delta t} = \frac{\Delta\,[NO_2]}{\Delta t} = 2.5 \times 10^{-5}\ \text{M/s}$$

16.4. Collision Theory of Reaction Rates

An essential requirement for a chemical reaction to occur is collision between particles of reactants. Collision theory stipulates the conditions necessary for a reaction to occur:

1. Reactant atoms, ions, or molecules must collide with each other.
2. For the collision to result in a reaction, the colliding atoms, ions, or molecules must generate a minimum amount of total kinetic energy. The minimum energy necessary to cause colliding atoms, ions, or molecules of reactants to undergo specific reaction is known as the *energy of activation*, E_a.
3. The reactant atoms, ions, or molecules must have appropriate orientation upon collision for the reaction to occur.

Any factor that will increase the fraction of successful collisions will increase the rate of reaction.

16.4.1. Factors affecting reaction rates

A number of factors influence the rates of chemical reactions:

1. *The nature of the reactants*: The type of reaction and the nature of bonds to be broken between reactants can affect the rate of reaction. Generally, ionic reactions are faster because they simply involve the formation of new chemical bonds by the union of oppositely charged ions. Reactions involving covalent bonds are slower since they involve the breaking of covalent bonds in reacting molecules. Also, gases react faster than liquids and liquids faster than solids. This follows the order of their molecular velocities.
2. *Concentration of the reactants*: An increase in the concentration of reactants will increase the rate of collision due to effective increase in the number of particles of reactants per unit volume. This results in an increase in the rate of reaction.
3. *Temperature*: An increase in temperature causes more colliding molecules to have a higher velocity, and hence kinetic energy greater than or equal to the energy of activation. This increases the rate of reaction.
4. *Light*: The rates of many reactions are greatly increased by the presence of light of suitable wavelengths. Examples include photosynthesis in plants, certain halogenation reactions, and other organic photochemical reactions.
5. *Pressure*: A change in pressure is only applicable to reactions involving gases. It does not affect reactions involving only solids or liquids due to their low compressibility. An increase in total pressure of gaseous reactants will increase the rate of collision, and hence the rate of reaction.
6. *Surface area*: The rate of a reaction is affected by the nature of the reactants. A solid will react faster when in powdered form than when in large pieces or lumps. Grinding the solid lumps into smaller particles will increase the

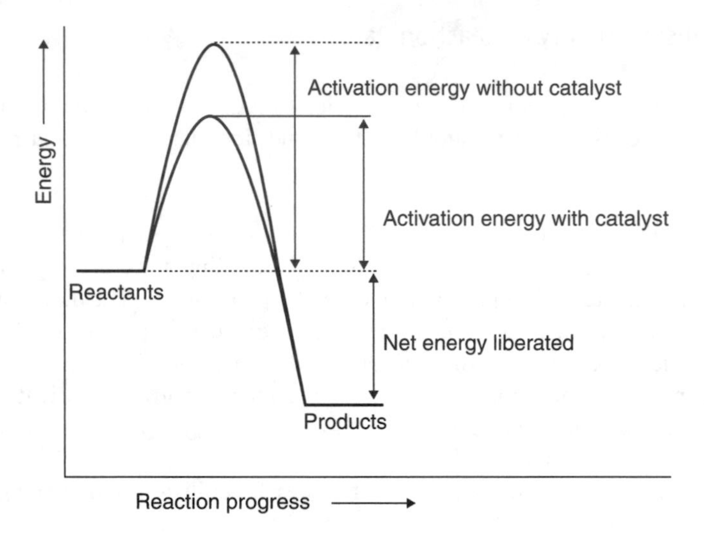

Figure 16-3 Potential energy profile showing the activation energies for the catalyzed and uncatalyzed reactions.

surface area, thereby increasing the collision frequency, and thus the rate of reaction.

7. *Catalysts*: A catalyst is a substance that can alter the rate of a reaction, but which remains chemically unchanged after the reaction is completed. Catalysts increase reaction rates by providing alternative reaction pathways that have a lower energy of activation relative to the uncatalyzed pathway (see figure 16-3).

16.5. Rate Laws and the Order of Reactions

The rates of many chemical reactions can be related to the concentrations of the reactants. For example, in the case of a one-step reaction of the form $aA + bB \rightarrow$ Products, the rate of reaction can be written as

$$\text{Rate} \propto [A]^x[B]^y$$

The rate equation may therefore be written as:

$$\text{Rate} = k[A]^x[B]^y$$

The proportionality constant k is called the rate constant (or specific rate constant) for the reaction; the [ ] represent the molar concentration of reactants A and B. The numerical values of the exponents x and y must be determined experimentally and are not necessarily related to the coefficients a and b.

The *order of the reaction* is governed by the exponents (x and y) to which the concentrations of the reactants are raised in the rate equation. Reaction order can be expressed either as order with respect to an individual reactant, or as the overall order of the reaction $(x + y)$. If x is 1, the reaction is first-order with respect to A. If y is 1, the reaction is first-order with respect to B. If x is 1, and y is 2, the reaction is first-order with respect to A, second-order with respect to B, and third-order overall.

16.6. Experimental Determination of Rate Law Using Initial Rates

The values of the exponents in a rate law can be determined by a series of experiments in which the initial rate of reaction is measured as a function of different initial concentrations. For a reaction that is zero-order with respect to a particular reactant, changing its concentration will not affect the rate as long as the concentration of the reactant is not zero. This is because a zero-order reaction is concentration-independent. If the reaction is first-order in a reactant, doubling the concentration of the reactant will double the reaction rate. Similarly, if a reaction is second-order in a given reactant, doubling the concentration causes the rate to increase by a factor of 4 (2^2). Let's illustrate this with an example.

Consider the reaction

$$2\,NO(g) + O_2(g) \longrightarrow NO_2(g)$$

The initial concentration and rate data are given table 16-4.

The general rate law for the reaction is:

$$\text{Rate} = k\,[NO]^m\,[O_2]^n$$

Let's see how we can calculate the values of m and n. To determine the rate law, we will need to evaluate the value of m at two different rates when $[O_2]$ is constant and then determine the value of n when [NO] is constant. From the data in the table, we observe that as the concentration of NO is doubled from 0.03 M to 0.06 M (comparing experiments 1 and 2 where $[O_2]$ is constant), the rate is increased 4 times (2^2) from 0.096 M/s to 0.384 M/s. This suggests the reaction rate is second-order with respect to NO. On the other hand, we also observe that as the concentration of O_2 is doubled from 0.03 M to 0.06 M (comparing experiments 1 and 3 where [NO] is constant),

Table 16-4 Initial-Rate Data for the Formation of NO_2

Experiment	*Initial [NO], M*	*Initial $[O_2]$, M*	*Initial rate of formation of NO_2, M/s*
1	0.03	0.03	0.096
2	0.06	0.03	0.384
3	0.03	0.06	0.192
4	0.06	0.06	0.768

the rate doubles, suggesting a first-order rate with respect to O_2. The rate equation is therefore:

$$\text{Rate} = k\,[\text{NO}]^2\,[\text{O}_2]^1$$

We can put the above calculation in more formal algebraic terms. To calculate the value of m, we take the ratio of the initial rates for two different initial concentrations of NO that have the same initial concentrations of O_2.

$$\frac{\text{Rate}_2}{\text{Rate}_1} = \frac{k\,[\text{NO}]^m_{\text{exp2}}\,[\text{O}_2]^n_{\text{exp2}}}{k\,[\text{NO}]^m_{\text{exp1}}\,[\text{O}_2]^n_{\text{exp1}}}$$

$$\frac{0.384}{0.096} = \frac{k\,(0.06)^m\,\cancel{(0.03)^n}}{k\,(0.03)^m\,\cancel{(0.03)^n}}$$

$$4 = 2^m \quad \Rightarrow \quad 2^2 = 2^m \quad \text{and} \quad m = 2$$

We can find the value of n in a similar way:

$$\frac{\text{Rate}_3}{\text{Rate}_1} = \frac{k\,[\text{NO}]^m_{\text{exp3}}\,[\text{O}_2]^n_{\text{exp3}}}{k\,[\text{NO}]^m_{\text{exp1}}\,[\text{O}_2]^n_{\text{exp1}}}$$

$$\frac{0.196}{0.098} = \frac{\cancel{k\,(0.03)^m}\,(0.06)^n}{\cancel{k\,(0.03)^m}\,(0.03)^n}$$

$$2 = 2^n \quad \Rightarrow \quad 2^1 = 2^n \quad \text{and} \quad n = 1$$

Hence the rate for this reaction is proportional to the product of $[\text{NO}]^2$ and $[\text{O}_2]$ or

$$\text{Rate} = k\,[\text{NO}]^2\,[\text{O}_2]^1$$

Example 16.4

Under certain experimental conditions, the rate equation for the formation of nitrosyl bromide:

$$2\,\text{NO} + \text{Br}_2 \rightarrow 2\,\text{NOBr}$$

is $R = k\,[\text{NO}]^2\,[\text{Br}_2]$.

(a) What is the order of the reaction?
(b) Given that the rate of formation of NOBr is 4.24×10^{-5} M/s, calculate the rate constant for the reaction if [NO] = 0.035 M and $[\text{Br}_2]$ = 0.070 M.

Solution

(a) Use the rate equation given. The overall order is the sum of the orders for each reactant. For this reactant $m + n = 3$. Therefore the reaction is third-order: second-order in NO and first-order in Br_2.

(b) The value of k can be calculated by substituting the given data into the rate equation:

$$k = \frac{\text{rate}}{[\text{NO}]^2\,[\text{Br}_2]} = \frac{4.24 \times 10^{-5}\ \text{mol/L-s}}{(0.035\ \text{mol/L})(0.070\ \text{mol/L})} = 0.0173\ \frac{\text{L}}{\text{mol-s}}$$

Example 16.5

The rate law for the reduction of bromate ions (BrO_3^-) by bromide ions in acidic solution:

$$\text{BrO}_3^-(\text{aq}) + 5\ \text{Br}^-(\text{aq}) + 6\ \text{H}^+(\text{aq}) \longrightarrow 3\ \text{Br}_2(\text{aq}) + 3\ \text{H}_2\text{O(l)}$$

has the general form:

$$\text{Rate} = k\left[\text{BrO}_3^-\right]^x\left[\text{Br}^-\right]^y\left[\text{H}^+\right]^z$$

Table 16-5 gives the results of experiments conducted at four different initial concentrations. From these data, determine the

(a) reaction order with respect to each reactant
(b) overall reaction order
(c) rate constant

Solution

The rate expression is

$$\text{Rate} = k\left[\text{BrO}_3^-\right]^x\left[\text{Br}^-\right]^y\left[\text{H}^+\right]^z$$

We can determine the values of x, y, and z by comparing the relevant experimental data. From the data for experiments 1 and 2, we see that the rate doubles as the concentration of BrO_3^- is increased from 0.10 M to 0.20 M while the other concentrations are held constant. Thus, the reaction is first-order with respect to BrO_3^- ($x = 1$). To determine the order with respect to Br^-, consider experiments 2

Table 16-5 Initial-Rate Data for the Reduction of Bromate Ions (BrO_3^-) by Bromide Ions in Acidic Solution

Experiment	*Initial* $[BrO_3^-]_o$ *(M)*	*Initial* $[Br^-]_o$ *(M)*	*Initial* $[H^+]_o$ *(M)*	*Initial rate, mol (L-s)*
1	0.1	0.1	0.1	0.00095
2	0.2	0.1	0.1	0.0019
3	0.2	0.2	0.1	0.0038
4	0.1	0.1	0.2	0.0038

and 3. When the concentration of Br^- is doubled from 0.10 M to 0.20 M, the reaction rate also doubles, indicating the reaction is again first-order with respect to Br^- ($y = 1$). Finally, when the concentration of H^+ is increased from 0.10 M to 0.20 M (experiments 1 and 4), the reaction rate quadruples ($\times\ 2^2$). Therefore the reaction is second-order in H^+ ($z = 2$). The complete rate equation is

$$\text{Rate} = k\left[BrO_3^-\right]\left[Br^-\right]\left[H^+\right]^2$$

(a) The reaction is first-order in BrO_3^-, first-order in Br^-, and second-order in H^+
(b) The reaction is fourth-order overall, i.e., $x + y + z = 4$
(c) Having established the rate law, the value of k can be obtained from the results of any of the four experiments. Let's use the data for experiment 1.

$$\text{Rate} = k\left[BrO_3^-\right]\left[Br^-\right]\left[H^+\right]^2$$

$$k = \frac{\text{Rate}_{\text{exp1}}}{\left[BrO_3^-\right]_{\text{exp 1}}\left[Br^-\right]_{\text{exp 1}}\left[H^+\right]^2_{\text{exp 1}}}$$

$$= \frac{(9.5 \times 10^{-4}\ \text{mol/(L-s)})}{(0.10\ \text{mol/L})(0.10\ \text{mol/L})(0.10\ \text{mol/L})^2} = 9.5\frac{L^3}{\text{mol}^3\text{-s}}$$

16.6.1. Algebraic method

We can solve for the values of x, y, and z algebraically by comparing the rates from the four experiments. Substitute the appropriate initial concentrations and rates into the general rate equation.

16.6.2. Determining the value of x

Use the results from experiments 1 and 2 in which $[Br^-]$ and $[H^+]$ are constant and only $[BrO_3^-]$ changes. Substitute the values of the initial concentrations and rates into the general rate equation and compare the results.

$$\frac{\text{rate}_2}{\text{rate}_1} = \frac{k\left[BrO_3^-\right]_2^x\left[Br^-\right]_2^y\left[H^+\right]_2^z}{k\left[BrO_3^-\right]_1^x\left[Br^-\right]_1^y\left[H^+\right]_1^z}$$

$$\frac{1.9 \times 10^{-3}}{9.8 \times 10^{-4}} = \frac{k\,(0.20)^x\,(0.10)^y\,(0.10)^z}{k\,(0.10)^x\,(0.10)^y\,(0.10)^z}$$

$$2^1 = 2^x, \quad \Rightarrow x = 1$$

16.6.3. Determining the value of y

Use the results from experiments 2 and 3 in which $[BrO_3^-]$ and $[H^+]$ are constant and only $[Br^-]$ changes. Substitute the values of the initial concentrations and rates into the general rate equation and compare the results:

$$\frac{\text{rate}_3}{\text{rate}_2} = \frac{k\left[\mathrm{BrO_3^-}\right]_3^x \left[\mathrm{Br^-}\right]_3^y \left[\mathrm{H^+}\right]_3^z}{k\left[\mathrm{BrO_3^-}\right]_2^x \left[\mathrm{Br^-}\right]_2^y \left[\mathrm{H^+}\right]_2^z}$$

$$\frac{3.8 \times 10^{-3}}{1.9 \times 10^{-3}} = \frac{k\,\cancel{(0.20)^x}\,(0.20)^y\,\cancel{(0.10)^z}}{k\,\cancel{(0.20)^x}\,(0.10)^y\,\cancel{(0.10)^z}}$$

$$2^1 = 2^y, \qquad \Rightarrow y = 1$$

16.6.4. Determining the value of z

Use the results from experiment 1 and 4 in which $[\mathrm{BrO_3^-}]$ and $[\mathrm{Br^-}]$ are constant and only $[\mathrm{H^+}]$ changes. Substitute the values of the initial concentrations and rates into the general rate equation and compare the results.

$$\frac{\text{rate}_1}{\text{rate}_1} = \frac{k\left[\mathrm{BrO_3^-}\right]_4^x \left[\mathrm{Br^-}\right]_4^y \left[\mathrm{H^+}\right]_4^z}{k\left[\mathrm{BrO_3^-}\right]_1^x \left[\mathrm{Br^-}\right]_1^y \left[\mathrm{H^+}\right]_1^z}$$

$$\frac{3.8 \times 10^{-3}}{9.5 \times 10^{-4}} = \frac{\cancel{k}\,\cancel{(0.10)^x}\,\cancel{(0.10)^y}\,(0.20)^z}{\cancel{k}\,\cancel{(0.10)^x}\,\cancel{(0.10)^y}\,(0.10)^z}$$

$$4 = 2^2 = 2^y, \qquad \Rightarrow y = 2$$

Thus, $x = 1$, $y = 1$, and $z = 2$.

(a) The reaction is first-order in $[\mathrm{BrO_3^-}]$ and $[\mathrm{Br^-}]$ and second-order in $[\mathrm{H^+}]$.
(b) Overall, the reaction is fourth-order ($x + y + z = 4$).
(c) Here let's substitute the results of the second experiment into the rate law:

$$\text{Rate} = k\left[\mathrm{BrO_3^-}\right]\left[\mathrm{Br^-}\right]\left[\mathrm{H^+}\right]^2$$

$$k = \frac{\text{Rate}_{\text{exp2}}}{\left[\mathrm{BrO_3^-}\right]_{\text{exp}2}\left[\mathrm{Br^-}\right]_{\text{exp}2}\left[\mathrm{H^+}\right]_{\text{exp}2}^2}$$

$$= \frac{\left(1.9 \times 10^{-3}\ \text{mol/(L-s)}\right)}{(0.20\ \text{mol/L})(0.10\ \text{mol/L})(0.10\ \text{mol/L})^2} = 9.5\frac{\mathrm{L^3}}{\mathrm{mol^3\text{-}s}}$$

16.7. The Integrated Rate Equation

16.7.1. First-order reactions

Consider a simple first-order reaction of the general type:

$$aA \longrightarrow \text{Products}$$

The rate law is:

$$\text{Rate} = -\frac{\Delta[A]}{\Delta t} = k[A]$$

Using calculus (we won't go into the details here), we can obtain the integrated rate law (equation):

$$\ln\frac{[A]_t}{[A]_o} = -kt \quad \text{or} \quad \log\frac{[A]_t}{[A]_o} = \frac{-kt}{2.303}$$

Here, $[A]_t$ represents the concentration of the reactant A at time t, and $[A]_o$ the initial concentration. The ratio $[A]_t/[A]_o$ is the fraction of reactant A that remains at time t.

Example 16.6

The thermal isomerization of cyclopropane to 1-propene follows first-order kinetics with a rate constant of 0.036 min^{-1} at 500°C. Calculate the concentration of cyclopropane after 6 h if the initial concentration is 0.075 mol/L.

Solution

For a first-order reaction, the integrated law is:

$$\log\frac{[A]_t}{[A]_o} = \frac{-kt}{2.303}$$

$$\log\frac{[A]_t}{[A]_o} = \frac{-\left(0.036\frac{1}{\text{min}}\right)\left(6\ \text{h} \times 60\frac{\text{min}}{\text{h}}\right)}{(2.303)} = -5.267$$

$$\frac{[A]_t}{[A]_o} = 10^{-5.267}$$

But $[\text{A}]_o = 0.075$ mol/L

$$[\text{A}]_{t=6\text{h}} = 10^{-5.267}[A]_o = 10^{-5.267} \times 0.075\ \text{mol/L} = 4.0 \times 10^{-7}\ \text{mol/L}$$

Example 16.7

The rate constant for the first-order decomposition of azomethane, $(CH_3)_2N_2$, is $1.5 \times 10^{-3}\ \text{s}^{-1}$. If the initial concentration is 0.25 M, calculate the time for it to decrease to 0.05 M.

Solution

$$\log \frac{[A]_t}{[A]_o} = \frac{-kt}{2.303}$$

$$[A]_o = 0.25 \text{ M}$$

$$[A]_t = 0.05 \text{ M}$$

$$k = 1.5 \times 10^{-3} s^{-1}$$

$$\log \frac{0.05}{0.25} = \frac{-1.5 \times 10^{-3}\text{s}^{-1}\, t}{2.303}$$

$$-0.6990 = -0.0007 \text{ s}^{-1}\, t$$

$$t = 998.6 \text{ s} \quad \text{or} \quad 16.6 \text{ min}$$

16.7.2. Graphing first-order data

By using the property of logarithms that $\log (A/B) = \log A - \log B$, we can rewrite the above integrated rate equation as follows:

$$\log [A]_t = \left(\frac{-k}{2.303}\right) t + \log [A]_o$$

This conforms to the equation of a straight line, $y = mx + b$:

$$\underset{y}{\underset{\uparrow}{\log [A]_t}} = \underset{m}{\underset{\uparrow}{\left(\frac{-k}{2.303}\right)}} \underset{x}{\underset{\uparrow}{t}} + \underset{b}{\underset{\uparrow}{\log[A]_o}}$$

Therefore, a plot of $\log [A]_t$ on the vertical axis vs. time on the horizontal axis should yield a straight line with slope $m = -k/2.303$ and intercept $b = \log [A]_o$. In terms of natural logarithms, the simplified expression would be:

$$\underset{y}{\underset{\uparrow}{\ln [A]_t}} = \underset{m}{\underset{\uparrow}{(-k)}} \underset{x}{\underset{\uparrow}{t}} + \underset{b}{\underset{\uparrow}{\ln [A]_o}}$$

The slope m is equal to $(-k)$, and the intercept b is equal to $\ln [A]_o$.

Example 16.8

Studies of the decomposition of N_2O_5 at 70°C yield the data in table 16-6. From this information, show that the reaction is first-order, and calculate the rate constant k.

Table 16-6 Rate Data for the Decomposition of N_2O_5 at 70°C

Time, min	*0*	*1*	*2*	*3*	*4*	*5*	*6*	*7*
P N_2O_5 torr	798	565	400	280	198	140	96	66

Solution

The integrated rate equation for a first-order reaction is:

$$\ln \frac{[A]_t}{[A]_o} = -kt$$

The ideal gas equation states that P is directly proportional to the number of moles of gas at constant temperature and volume:

$$PV = nRT$$

Since n/V equals the molar concentration of N_2O_5, we can substitute the partial pressure of N_2O_5 for molarity. Thus the rate equation becomes:

$$\ln \frac{P_t}{P_0} = -kt$$

where P_0 is the partial pressure at time $t = 0$ and P_t is the partial pressure after time t. When simplified, the equation becomes:

$$\ln P_t = (-k)t + \ln P_0$$

Plot ln P_t against time t using the data in table 16-6. If the graph yields a straight line, the reaction obeys first-order kinetics. Table 16-7 gives data for the decomposition of N_2O_5 at 70°C :

The plot of ln P vs. t yields a straight line (see figure 16-4). Therefore, the decomposition of N_2O_5 is a first-order process with a slope of $-0.3548\ (\text{min})^{-1}$.

Table 16-7 Kinetic Data for the Decomposition of N_2O_5 at 70°C

Time, s	$P_{(N_2O_5)}$, *torr*	*ln P*
0	798	6.682
1	565	6.337
2	400	5.991
3	280	5.635
4	198	5.288
5	140	4.942
6	96	4.564
7	66	4.190

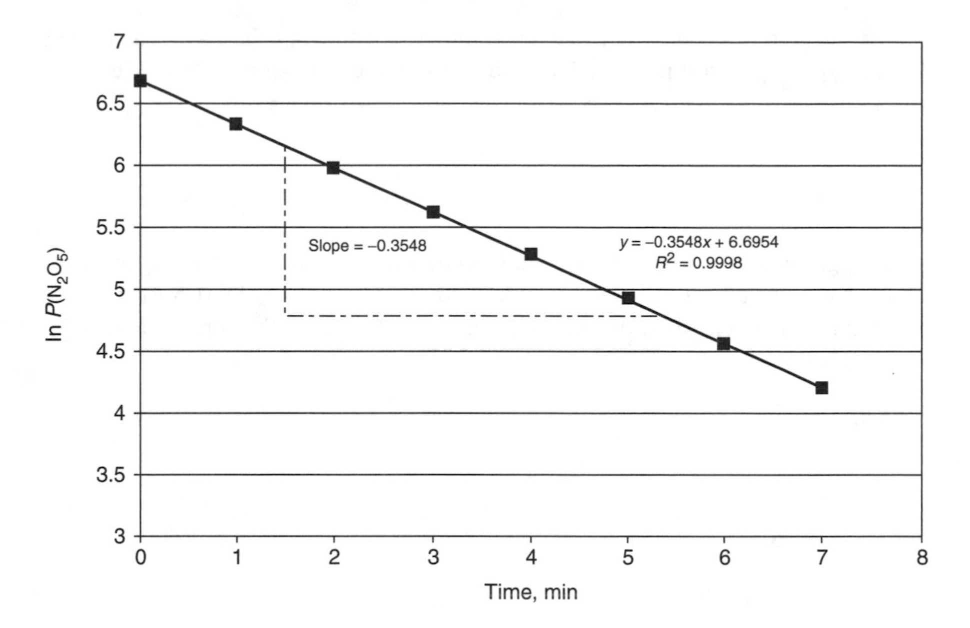

Figure 16-4 The plot of ln [N_2O_5] as a function of time for the decomposition of N_2O_5 at 70°C.

The rate constant can be obtained from the slope as follows:

$$\text{Slope} = (-k)$$

$$-0.3548 \text{ min}^{-1} = -k$$

$$k = 0.3548 \text{ min}^{-1}$$

16.7.3. Second-order reactions

Consider a reaction of the general type:

$$aA \longrightarrow \text{Products}$$

The second-order rate law is:

$$\text{Rate} = -\frac{\Delta[A]}{\Delta t} = k[A]^2$$

Using calculus (we won't go into the details here), one can obtain the integrated rate law (equation):

$$\frac{1}{[A]_t} = kt + \frac{1}{[A]_o}$$

With this equation, we can calculate the concentration of A at any time t if we know the initial concentration $[A]_o$.

The integrated second-order rate equation has the form $y = mx + b$. Therefore a plot of $1/[A]_t$ against time should yield a straight line with slope $= k$ and intercept $= 1/[A]_o$.

Example 16.9

The gas-phase decomposition of ammonia into hydrogen and nitrogen in the presence of Pt catalyst is second-order with a rate constant of 0.085 M^{-1} s^{-1}. If the initial concentration of ammonia was 0.15 M, what is the concentration after 30 min?

Solution

$$\frac{1}{[A]_t} = kt + \frac{1}{[A]_o}$$

$$[A]_o = 0.15 \text{ M}$$

$$k = 0.085 \text{ M}^{-1}\text{s}^{-1}$$

$$t = 30 \text{ min} \quad \text{or} \quad 180 \text{ s.}$$

$$[A]t = ?$$

$$\frac{1}{[A]_t} = 0.085 \text{ M}^{-1} \text{ s}^{-1} \times 180 \text{ s} + \frac{1}{0.15 \text{ M}} = 21.9667 \text{ M}^{-1}$$

$$[A]_t = 0.046 \text{ M}$$

Example 16.10

The following data were obtained for the alkaline hydrolysis of ethyl ethanoate:

$$OH^- + CH_3COOC_2H_5 \longrightarrow CH_3COO^- + CH_3CH_2OH$$

Confirm that the reaction is second-order and calculate the value of the rate constant k.

Solution

To confirm that the rate law is second-order, construct a plot of $1/[CH_3COOC_2H_5]$ vs. time using the information in table 16-8.

The plot of $1/[CH_3COOC_2H_5]$ vs. time yields a straight line with a slope of 5.5477 M^{-1} min^{-1} (see figure 16-5). The reaction is thus second-order.

Table 16-8 Kinetic Data for the Alkaline Hydrolysis of Ethyl Ethanoate

Time (s)	*Concentration of* $CH_3COOC_2H_5$ *(M)*	$1/[CH_3COOC_2H_5]$ $(M)^{-1}$
0	0.02000	50.00
5	0.01300	76.92
10	0.01010	99.01
15	0.00770	129.87
20	0.00620	161.29
25	0.00550	181.82
35	0.00430	232.56
45	0.00330	303.03
55	0.00285	350.88
85	0.00195	512.82
100	0.00165	606.06
125	0.00135	740.74

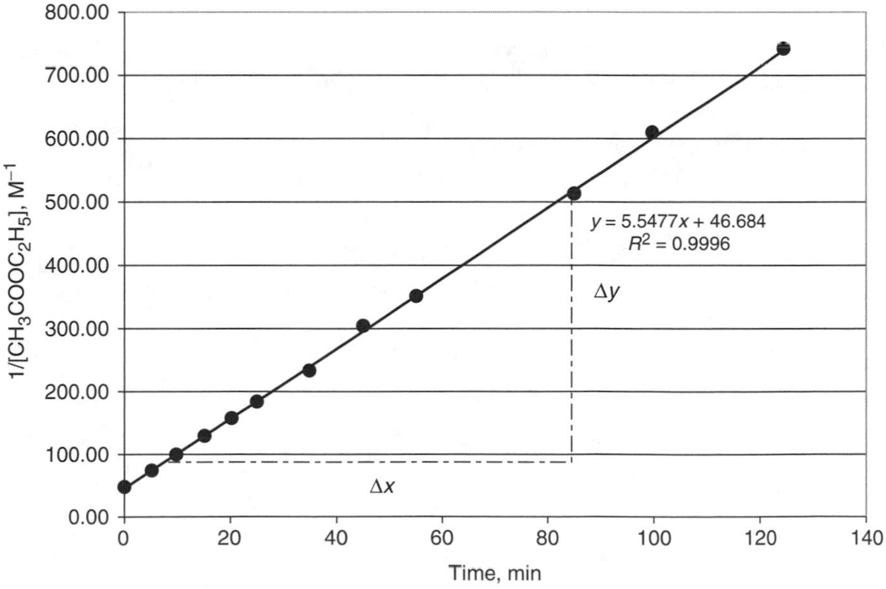

Figure 16-5 The plot of $1/[CH_3COOC_2H_5]$ versus time.

The rate constant, k, is equal to the slope of the straight line. To calculate the slope, use two widely separated points on the graph.

$$\text{Rate constant } k = \text{slope} = \frac{\Delta y}{\Delta x} = \frac{(512.82 - 76.92)\ \text{M}^{-1}}{(85.0 - 5.0)\ \text{s}} = \frac{435.9}{80.0\ \text{s}} = 5.449\ \text{M}^{-1}\ \text{s}^{-1}$$

16.8. Half-life of a Reaction

The half-life, $t_{1/2}$, of a reaction is the time needed for the reactant concentration to decrease to one-half of its original value.

Recall that the integrated rate equation for a first-order reaction is:

$$\ln \frac{[A]_t}{[A]_o} = -kt$$

At time $t = t_{1/2}$, the fraction $[A]_t/[A]_o$ becomes $\frac{1}{2}$. Substituting into the equation yields:

$$\ln \frac{1}{2} = -kt_{1/2} \quad \text{or} \quad t_{1/2} = \frac{-\ln 2}{k}$$

$$t_{1/2} = \frac{0.693}{k}$$

Thus the half-life of a first-order reaction is independent of the initial concentration of the reactant. This implies that the half-life is the same at any time during the reaction.

The half-life expression for a second-order reaction can be obtained in a similar manner. Substituting $t = t_{1/2}$ and $[A]_t = [A]_o/2$ in the integrated rate equation:

$$\frac{1}{[A]_t} = kt + \frac{1}{[A]_o}$$

yields:

$$\frac{1}{[A]_o/2} = kt_{1/2} + \frac{1}{[A]_o}$$

$$\frac{2}{[A]_o} = kt_{1/2} + \frac{1}{[A]_o}$$

$$t_{1/2} = \frac{1}{k[A]_o}$$

Thus, unlike the half-life of first-order reactions, the half-life of a second-order reaction depends on the starting concentration of the reactant.

Example 16.11

For the first-order gas phase reaction

$$2\,NOCl(g) \longrightarrow 2\,NO(g) + Cl_2(g)$$

the specific rate constant is 9.5×10^{-6}/s at 77°C. What is the half-life of NOCl if 1.25 mol NOCl were placed in a 3-L reactor?

Solution

The half-life of a first-order reaction is given by the equation:

$$t_{1/2} = \frac{0.693}{k} = \frac{0.693}{9.5 \times 10^{-6}/\text{s}} = 7.30 \times 10^4\ \text{s}$$

Note that $t_{1/2}$ does not depend on the initial concentration.

Example 16.12

The rate constant for the second-order hydrolysis of ethyl nitrobenzene at 25°C is $0.0795\ M^{-1}\ min^{-1}$. Calculate the half-life of ethyl nitrobenzene in an experiment with an initial concentration of 0.05 M.

Solution

The half-life of a second-order process is given by the equation:

$$t_{1/2} = \frac{1}{k[A]_o}$$

where A represents ethyl nitrobenzene:

$$t_{1/2} = \frac{1}{k[A]_o} = \frac{1}{\left(0.0795\ \text{M}^{-1}\ \text{min}^{-1}\right)(0.05\ \text{M})} = 251.6\ \text{min}$$

16.9. Reaction Rates and Temperature: The Arrhenius Equation

The rates of chemical reactions increase with increasing temperature. At higher temperatures, the fraction of molecules possessing the necessary energy of activation is significantly increased, causing the reaction to proceed at a faster rate. The quantitative relationship between reaction rate, temperature, and energy of activation is given by the Arrhenius equation:

$$k = A\,e^{-E_a/RT}$$

where k is the rate constant, A is a constant known as the pre-exponential factor, E_a is the energy of activation and R, the universal gas constant, is 8.314 J/mol-K. Taking logarithms, the Arrhenius equation becomes:

$$\ln k = \frac{-E_a}{RT} + \ln A \quad \text{or} \quad \log k = \log A - \frac{E_a}{2.303\ RT}$$

The Arrhenius equation at two different temperatures can be written as:

$$\log \frac{k_2}{k_1} = \frac{E_a}{2.303R}\left(\frac{1}{T_1} - \frac{1}{T_2}\right) \quad \text{or} \quad \log \frac{k_2}{k_1} = \frac{E_a}{2.303R}\left(\frac{T_2 - T_1}{T_1 T_2}\right)$$

Thus, measuring the rate constants at two different temperatures allows us to calculate E_a, which is one of the main uses of the Arrhenius equation.

Example 16.13

The kinetic data for temperature dependence of the rate constant for the reaction

$$CH_3COOC_2H_5(aq) + OH^-(aq) \longrightarrow CH_3COO^-(aq) + C_2H_5OH(aq)$$

are tabulated table 16-9.
Calculate the energy of activation, E_a, and the pre-exponential factor, A.

Solution

To determine E_a, plot ln k vs. $1/T$ (see figure 16-6) using the data in table 16-10. This should be a straight line with a slope $-E_a/R$.

$$\text{Slope} = -\frac{E_a}{R} = -5345.3$$

$$E_a = -(-5345.5) \times 8.314 \text{ J/mol} = 4.44 \times 10^4 \text{ J/mol}$$

Table 16-9 Kinetic Data for Temperature Dependence of the Rate Constant for the Hydrolysis of Ethyl Ethanoate

Temperature (°C)	*k (M/s)*
15	0.0511
25	0.104
35	0.187
45	0.332
55	0.533
65	0.783

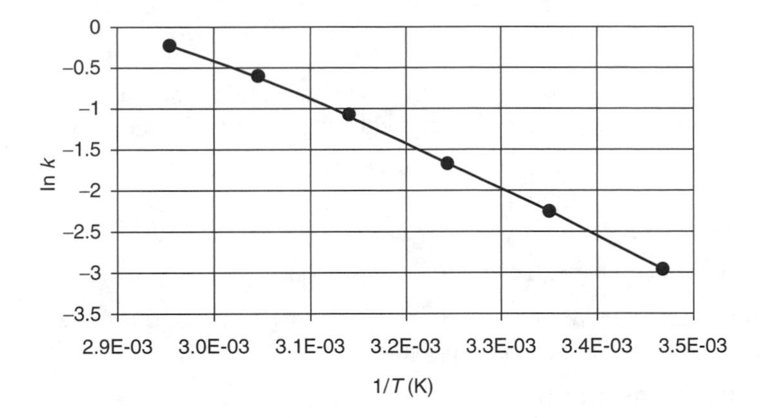

Figure 16-6 An Arrhenius plot for the alkaline hydrolysis of ethylethanoate.

Table 16-10 Determination of the Energy of Activation for the Alkaline Hydrolysis of Ethyl Ethanoate

T (K)	*k (M/s)*	*ln k*	*1/T* (K^{-1})
288.2	0.0511	−2.974	0.00347
298.2	0.104	−2.264	0.00335
308.2	0.187	−1.677	0.00324
318.2	0.332	−1.103	0.00314
328.2	0.533	−0.629	0.00305
338.2	0.783	−0.245	0.00296

Example 16.14

At 600 K, the rate constant for the first-order decomposition of nitrogen pentoxide was found to be $3.0 \times 10^{-3}\ s^{-1}$. Determine the value of the rate constant at 750 K, given that $E_a = 1.5 \times 10^2$ kJ/mol and $R = 8.314$ J/mol-K.

Solution

$$\ln \frac{k_2}{k_1} = \left(\frac{E_a}{R}\right)\left(\frac{1}{t_1} - \frac{1}{t_2}\right)$$

$$\ln \frac{k_2}{3.0 \times 10^{-3}} = \left(\frac{1.5 \times 10^2 \text{ kJ/mol} \times 1000 \text{ J/kJ}}{8.314 \text{ J/mol-K}}\right)\left(\frac{1}{600 \text{ K}} - \frac{1}{750 \text{ K}}\right)$$

$$\ln \frac{k_2}{3.0 \times 10^{-3}} = (18041.85)(0.0004) = 7.2167$$

$$\frac{k_2}{3.0 \times 10^{-3}} = e^{7.2167} = 1362.045$$

$$k_2 = 4.086 \text{ s}^{-1}$$

16.10. Problems

1. Write the rate expression in terms of the disappearance of the reactants and the appearance of the products for the following decomposition reactions:

   (a) $2\,H_2O_2(g) \rightarrow 2\,H_2O(l) + O_2(g)$
   (b) $2\,N_2O_5(g) \rightarrow 4\,NO_2(g) + O_2(g)$
   (c) $SO_2Cl_2(g) \rightarrow SO_2(g) + Cl_2(g)$
   (d) $4\,PH_3(g) \rightarrow P_4(g) + 6\,H_2(g)$

2. For the following reactions, write the rate expression in terms of the disappearance of the reactants and the appearance of the products:

   (a) $2\,NH_3(g) + 5\,O_2(g) \rightarrow 4\,NO(g) + H_2O(g)$
   (b) $2\,NO(g) + 2\,H_2(g) \rightarrow N_2(g) + H_2O(g)$
   (c) $CH_3CHO(g) \rightarrow CH_4(g) + CO(g)$
   (d) $2\,SO_2(g) + O_2(g) \rightarrow 2\,SO_3(g)$

3. The formation of nitrogen from the high-temperature reduction of nitric oxide in a hydrogen gas atmosphere is given by the equation:

   $$2\,NO(g) + H_2(g) \rightarrow N_2(g) + 2\,H_2O(g)$$

   If NO had an initial concentration of 0.025 mol per liter, what is the initial rate of production of N_2 gas?

4. The formation of hydrogen iodide from hydrogen and iodine gas is given by the equation:

   $$H_2(g) + I_2(g) \rightarrow 2\,HI(g)$$

   If iodine gas is being consumed at a rate of 0.012 M/s, what is the rate of formation of hydrogen iodide?

5. Ozone decomposes according to the reaction:

   $$2\,O_3(g) \rightarrow 3\,O_2(g)$$

   The initial concentration of oxygen is 0.050 M, and 0.130 M after 2 h of reaction. Calculate the average rate of this reaction.

6. Initial rate data (see table 16-11) were collected for the reaction $X + Y \rightarrow Z$.

   (a) What is the rate law for this reaction?
   (b) Determine the rate constant.

7. Initial rate data for the reaction $NH_4^+(aq) + CNO^-(aq) \rightarrow CO(NH_2)_2(s)$ at 80°C are given in table 16-12.

   (a) Determine the rate law.
   (b) Calculate the value of the rate constant.

Table 16-11 Data for Problem 6

Experiment	*[X] (M)*	*[Y] (M)*	*Initial rate (M/min)*
1	0.02	0.025	0.0036
2	0.01	0.025	0.0018
3	0.02	0.0125	0.0009

Table 16-12 Data for Problem 7

Experiment	*$[NH_4]$ (M)*	*$[CNO^-]$ (M)*	*Initial rate (M/s)*
1	0.014	0.02	0.002
2	0.028	0.02	0.008
3	0.014	0.01	0.001

Table 16-13 Data for Problem 8

Experiment	*$[I^-]$ (M)*	*$[OCl^-]$ (M)*	*$[OH^-]$ (M)*	*Initial rate (M/min)*
1	0.005	0.005	0.5	0.000275
2	0.0025	0.005	0.5	0.000138
3	0.0025	0.0025	0.5	0.000007
4	0.0025	0.0025	0.25	0.00014

Table 16-14 Data for Problem 9

Experiment	*$[I_2]_o$ (M)*	*$[S_2O_3^-]$ (M)*	*Initial rate (M/s)*
1	0.01	0.01	0.0004
2	0.01	0.02	0.0004
3	0.02	0.01	0.0008

8. The reaction:

$$I^-(aq) + OCl^-(aq) \rightarrow IO^-(aq) + Cl^-(aq)$$

was investigated in basic solution. Table 16-13 shows the results. Determine the

(a) Rate law and
(b) Order of the reaction.

9. Initial rate data for the reaction:

$$2\ S_2O_3^{2-}(aq) + I_2(aq) \rightarrow S_4O_6^{2-}(aq) + 2I^-(aq)$$

were collected at 25°C. The results are summarized in table 16-14. Determine (a) the order with respect to each reactant, (b) the overall rate law, and (c) the value of the rate constant.

Table 16-15 Data for Problem 10

Experiment	*$[H_2O_2]_o$ (M)*	*$[I^-]_o$ (M)*	*$[H^+]_o$ (M)*	*Initial rate (M/s)*
1	0.15	0.15	0.05	0.00012
2	0.15	0.3	0.05	0.00024
3	0.3	0.15	0.05	0.00024
4	0.15	0.15	0.1	0.00048

10. The initial rates given in table 16-15 were determined at 25°C for the reaction:

$$H_2O_2(aq) + 3\,I^-(aq) + 2\,H^+(aq) \rightarrow I_3^-(aq) + 2\,H_2O(l)$$

(a) What is the order of the reaction with respect to each reactant?
(b) Determine the overall rate law.
(c) Calculate the value of the rate constant.
(d) What is the rate of reaction when all reactant concentrations are 0.025 M?

11. The decomposition of dinitrogen pentoxide according to the equation:

$$2\,N_2O_5(g) \xrightarrow{50^\circ C} 4\,NO_2(g) + O_2(g)$$

follows first-order kinetics with a rate constant of 0.0065 s^{-1}. If the initial concentration of N_2O_5 is 0.275 M, determine:

(a) The concentration of N_2O_5 after 3 min.
(b) How long it will take for 98% of the N_2O_5 to decompose.

12. At 1200°C, CS_2 decomposes according to

$$CS_2(g) \rightarrow CS(g) + S(g)$$

The rate law is Rate $= k[CS_2]$ where $k = 1.6 \times 10^{-6}\ s^{-1}$.

(a) What is the initial rate of decomposition, given that the concentration of CS_2 is 0.25 M?
(b) If the initial concentration of CS_2 is 0.25 M, calculate the residual concentration after a reaction time of 5 h.

13. The first-order rate constant for the reaction $H_2O_2(aq) \rightarrow 2\,H_2O(l) + O_2(g)$ is $2.6 \times 10^{-4}\ s^{-1}$ at 45°C.

(a) Calculate the half-life for the reaction.
(b) Determine the concentration of H_2O_2 after four half-lives, given that the initial concentration is 0.005 M.

14. At 450°C, cyclopropane (C_3H_6) gas isomerizes to propene with first-order kinetics. The half-life for the reaction is 13 h.

(a) What is the rate constant for this reaction?

(b) Calculate the concentration of C_3H_6 remaining after 1 h if the initial concentration is 0.015 M.
(c) What fraction of C_3H_6 remains after 24 h?
(d) How many hours does it take for 80% of the C_3H_6 to react?

15. The thermal decomposition of N_2O_5, $2\,N_2O_5(g) \rightarrow 4\,NO_2(g) + O_2(g)$, is a first-order reaction. From the following kinetic data (see table 16-16), determine the value of the rate constant.

16. The following data were collected for the gas-phase isomerization of methyl isonitrile (CH_3NC) to acetonitrile (CH_3CN) at 230°C (see table 16-17).

$$CH_3NC \xrightarrow{230^\circ C} CH_3CN$$

(a) Use the kinetic data above to determine if the isomerization reaction is first- or second-order.
(b) What is the value of the rate constant?

17. The thermal decomposition of NO_2 to yield NO and O_2 is a second-order reaction with a rate constant of $0.55\ M^{-1}\ s^{-1}$ at 300°C.

$$2\,NO_2(g) \rightarrow 2\,NO(g) + O_2(g)$$

If the initial concentration of NO_2 gas is 0.0500 M:

(a) What is the concentration of NO_2 after 5 h?
(b) How long does it take for the NO_2 concentration to decrease to 0.0050 M?
(c) What is the half-life of the reaction?

18. The following kinetic data were obtained for the decomposition of HI at 400°C (see table 16-18):

$$2\,HI(g) \xrightarrow{400^\circ C} H_2(g) + I_2(g)$$

(a) Use these data to determine if the reaction is first-order or second-order.
(b) Calculate the value of the rate constant.
(c) What is the half-life of the reaction if the initial concentration of HI is 0.15 M?

Table 16-16 Data for Problem 15

Time (s)	*0*	*50*	*100*	*150*	*200*	*250*	*300*	*350*	*400*
$[N_2O_5]$ (M)	0.075	0.072	0.070	0.067	0.065	0.064	0.062	0.061	0.059

Table 16-17 Data for Problem 16

Time (s)	*0*	*300*	*600*	*900*	*1200*	*1500*	*1800*
$[CH_3NC]$ (M)	0.085	0.067	0.053	0.041	0.031	0.023	0.018

Table 16-18 Data for Problem 18

Time	*0*	*20*	*40*	*60*	*80*	*100*	*120*
[HI] (M)	0.07	0.062	0.056	0.052	0.0475	0.0446	0.042

19. The rate constant for the first-order decomposition of gaseous N_2O_5 at 55°C is $1.73 \times 10^{-4}\ s^{-1}$. If the activation energy is 121 kJ/mol, determine the rate constant at 150°C.
20. The rate constant for the second-order decomposition of gaseous NO_2 at 380°C is $10.50\ M^{-1}\ s^{-1}$. If the activation energy is 233 kJ/mol, calculate the temperature at which the rate constant is $1.050 \times 10^3\ M^{-1}\ s^{-1}$.
21. Rate constants for the decomposition of an organic peroxide, ROOR′ are $2.5 \times 10^{-7}\ s^{-1}$ at 300 K and $1.25 \times 10^{-4}\ s^{-1}$ at 425 K respectively. Calculate:

    (a) The activation energy for the reaction in kJ/mol.
    (b) The rate constant at 355 K.

22. Rate constants for the decomposition of azomethane, $(CH_3)_2N_2$, are $2.5 \times 10^{-3}\ s^{-1}$ at 300°C and $3.25 \times 10^{-2}\ s^{-1}$ at 333°C respectively. Calculate:

    (a) The activation energy for the reaction in kJ/mol.
    (b) The rate constant at 25°C.

23. The hydrolysis of a certain sugar in 0.05 M HCl follows a first-order rate equation. Variation of the rate constant with temperature is given in table 16-19. Determine graphically the activation energy and the pre-exponential factor, *A*, for the reaction.
24. The temperature dependence of the rate constant for the following reaction is given in table 16-20:

$$CH_3Br(aq) + OH^-(aq) \rightarrow CH_3OH(aq) + Br^-(aq)$$

From an appropriate graph of the data, calculate the activation energy, E_a, (in kJ/mol), and the pre-exponential or frequency factor, *A*.

Table 16-19 Data for Problem 23

Temperature (K)	*273*	*298*	*308*	*318*
$k\ (s^{-1})$	2.1×10^{-5}	6.38×10^{-4}	1.94×10^{-3}	5.84×10^{-3}

Table 16-20 Data for Problem 24

Temperature (°C)	*15*	*25*	*35*	*45*
$k\ (M^{-1}s^{-1})$	0.104	0.202	0.368	0.664

17

Chemical Equilibrium

17.1. Reversible and Irreversible Reactions

Many chemical reactions go to completion; i.e., all the reactants are converted to products. A good example is the reaction of calcium with cold water.

$$Ca(s) + 2\,H_2O(l) \longrightarrow Ca(OH)_2(s) + H_2(g)$$

There is no evidence that the reverse reaction occurs. Such reactions are said to be *irreversible*.

On the other hand, many reactions are *reversible*: the process can be made to go in the opposite direction. This means that both the reactants and products will be present at any given time. *A reversible reaction is defined as one in which the products formed can react to give the original reactants.* A double arrow is used to indicate that the reaction is reversible, as illustrated by the general equation:

$$aA + bB \rightleftharpoons cC + dD$$

At the start of the reaction, the reactants convert more quickly to products than products turn back to reactants because the reactants are present in much greater amount. Eventually the concentration of products is sufficient for the reverse reaction to become significant. The reaction is said to reach *equilibrium* when the net change in the products and reactants is zero, i.e., the rate of forward reaction equals the rate of reverse reaction. Chemical equilibria are *dynamic equilibria* because, although nothing appears to be happening, opposing reactions are occurring at the same rate.

Figure 17-1 illustrates that for a reaction in chemical equilibrium, the rate of forward reaction equals the rate of reverse reaction.

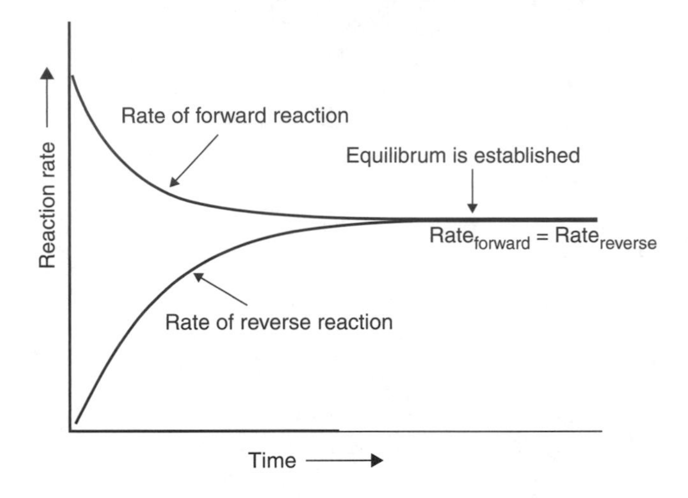

Figure 17-1 Rates of forward and reverse reactions for chemical reactions at equilibrium.

17.2. The Equilibrium Constant

When a chemical reaction is at equilibrium, the concentrations of reactants and products are constant. The relationship between the concentrations of reactants and products is given by the equilibrium expression, also known as the law of mass action. For the general reaction:

$$aA + bB \rightleftharpoons cC + dD$$

at a constant temperature, the equilibrium constant expression is written as:

$$K_c = \frac{[C]^c[D]^d}{[A]^a[B]^b}$$

where $[A]$, $[B]$, $[C]$, and $[D]$ are the molar concentrations or partial pressures of A, B, C, and D at equilibrium. The exponents a, b, c, and d in the equilibrium expression are the coefficients in the balanced equation; K_c is the equilibrium constant and is not given units. The subscript c shows that K is in terms of concentration. The numerical value for K_c is usually determined experimentally. Note that the equilibrium constant expression depends only on the reaction stoichiometry, and not on the mechanism.

Thus, *the equilibrium constant, K, is defined as the product of the molar concentrations (or partial pressures, if gaseous) of the products of a chemical reaction, each raised to the power of its coefficient in the balanced equation, divided by the product of the molar concentrations (or partial pressures, if gaseous) of the reactants, each raised to the power of its coefficient in the balanced equation.*

Example 17.1

Write the equilibrium constant expressions for the following reactions:

(a) $N_2(g) + 3\,H_2(g) \rightleftharpoons 2\,NH_3(g)$

(b) $CO(g) + Cl_2(g) \rightleftharpoons COCl_2(g)$

(c) $H_2(g) + I_2(g) \rightleftharpoons 2\,HI(g)$

Solution

(a) $K_c = \dfrac{[NH_3]^2}{[N_2][H_2]^3}$

(b) $K_c = \dfrac{[COCl_2]}{[CO][Cl_2]}$

(c) $K_c = \dfrac{[HI]^2}{[H_2][I_2]}$

Example 17.2

Phosgene, $COCl_2$, is a very toxic gas, which was used extensively during World War I. It decomposes to carbon dioxide and chlorine on heating, according to the equation:

$$COCl_2(g) \rightleftharpoons CO(g) + Cl_2(g)$$

0.75 mol $COCl_2$ was heated in a sealed 1-L vessel at 373 K. The equilibrium concentration of Cl_2 was found to be 0.25 M. Calculate the equilibrium constant for the reaction. (Note: 0.75 mol in 1 L is 0.75 M.)

Solution

At the start of the reaction:

$$[COCl_2] = 0.75\ M$$

$$[CO] = 0$$

$$[Cl_2] = 0$$

At equilibrium:

$$[COCl_2] = 0.75 - 0.25 = 0.5\ M$$

$$[CO] = 0.25\ M$$

$$[Cl_2] = 0.25\ M$$

$$K_c = \frac{[Cl_2][CO]}{[COCl]}$$

$$K_c = \frac{[0.25][0.25]}{[0.5]} = 0.125$$

17.2.1. Equilibrium constant in terms of pressure

For reactions where the reactants and products are gases, the equilibrium constant expression can be written in terms of partial pressure. For the general equation:

$$aA(g) + bB(g) \rightleftharpoons cC(g) + dD(g)$$

the equilibrium constant expression K_p can be written as:

$$K_p = \frac{(P_C)^c\,(P_D)^d}{(P_A)^a\,(P_B)^b}$$

where P_A, P_B, P_C, and P_D are the equilibrium partial pressures. The subscript p is used here because K is determined from partial pressures rather than concentrations.

Example 17.3

Write the equilibrium constant expression, K_p, in terms of partial pressure, for the decomposition of sulfur trioxide.

$$2\ SO_3(g) \rightleftharpoons 2\ SO_2(g) + O_2(g)$$

Solution

$$K_p = \frac{P^2_{SO_2} P_{O_2}}{P^2_{SO_3}}$$

Example 17.4

Write balanced chemical equations for the completely gaseous equilibrium reactions corresponding to the following expressions for the equilibrium constant.

1. $\frac{[N_2][O_2]}{[NO]^2}$

2. $\frac{[NO]^2[Cl_2]}{[NOCl]^2}$

3. $\dfrac{[CH_4][H_2S]^2}{[CS_2][H_2]^4}$
4. $\dfrac{[N_2]^2[H_2O]^6}{[NH_3]^4[O_2]^3}$

Solution

1. $2\ NO(g) \rightleftharpoons N_2(g) + O_2(g)$
2. $2\ NOCl(g) \rightleftharpoons 2\ NO(g) + Cl_2(g)$
3. $CS_2(g) + 4\ H_2(g) \rightleftharpoons CH_4(g) + 2\ H_2S(g)$
4. $4\ NH_3(g) + 3\ O_2(g) \rightleftharpoons 2\ N_2(g) + 6\ H_2O(g)$

17.2.2. Relationship between K_p and K_c

K_p and K_c are related by the expression

$$K_p = K_c(RT)^{\Delta n_{gas}}$$

where Δn_{gas} is the sum of the coefficients of the gaseous products, minus the sum of the coefficients of the gaseous reactants, in the balanced equation.

$$\Delta n = (\text{number of moles of gaseous products} - \text{number of moles of gaseous reactants})$$

Example 17.5

Calculate the value of K_c at 1000 K for the following reaction:

$$CH_4(g) + 2\ H_2S(g) \rightleftharpoons CS_2(g) + 4\ H_2(g) \qquad K_p = 4.2 \times 10^{-3} \text{ at } 1000\ \text{K}$$

Solution

1. Calculate the change in the number of moles of gases.

$$\Delta n = 5 \text{ mol product} - 3 \text{ mol reactants} = 2$$

2. Substitute Δn, R, and T into the equation $K_p = K_c(RT)^{\Delta n_{gas}}$ and solve for K_c. Use the value of R consistent with the units given in the problem. Here we use $R = 0.08206\,(\text{L-atm/K-mol})$

$$K_p = K_c(RT)^{\Delta n_{gas}}$$

$$4.2 \times 10^{-3} = K_c(0.08206 \times 1000)^2$$

$$K_c = \frac{4.2 \times 10^{-3}}{(0.08206 \times 1000)^2} = 6.24 \times 10^{-7}$$

17.3. The Reaction Quotient

For the general reaction:

$$aA + bB \rightleftharpoons cC + dD$$

the mass action expression, or equilibrium quotient K, is given as:

$$K = \frac{[C]^c[D]^d}{[A]^a[B]^b}$$

If the reaction is not at equilibrium, the calculated ratio will be different from the equilibrium constant expression. In this case, the ratio is called the *reaction quotient* Q:

$$Q = \frac{[C]^c[D]^d}{[A]^a[B]^b}$$

The concentrations $[A]$, $[B]$, $[C]$, and $[D]$ are not necessarily equilibrium concentrations; they are the actual concentrations at any point in time. When $Q = K$, the reaction is at equilibrium. (Note that we use K here without the subscript c or p for simplicity.)

17.4. Predicting the Direction of Reaction

The direction of a reaction can be predicted by comparing the magnitude of Q with the value of K_c.

1. If $Q < K_c$, the reaction will continue to move from left to right (forward) towards the product until equilibrium is achieved.
2. If $Q > K_c$, the reaction moves from right to left (reverse) towards the reactants until equilibrium is achieved.
3. If $Q = K_c$, the reaction is at equilibrium.

17.5. Position of Equilibrium

The position of equilibrium, that is, the magnitude of K, indicates in qualitative terms the extent to which a reaction at equilibrium has moved towards completion, that is, whether products or reactants are present in greater amount, as described in table 17-1.

Table 17-1 Relationship Between the Magnitude of K, Equilibrium Position, and the Concentrations of Reactants and Products

Value of K_{eq}	*Range of K_{eq}*	*Equilibrium position*	*Relative concentration of products and reactants*
Very large	10^{20}–10^{30}	Far to the right. Reaction proceeds almost 100% to completion	Mainly products
Large	$>10^3$	To the right	Products > Reactants
Unity	$10^{-3} - 10^3$	Neither to the left nor to the right	Almost the same
Small	$<10^{-3}$	To the left	Reactants > Products
Very small	10^{-20}–10^{-30}	Far to the left	Mainly reactants

Example 17.6

For the following reactions, give a qualitative description of the position of equilibrium. Also, predict the relative concentration of the products and reactants at 298 K.

1. $N_2(g) + O_2(g) \rightleftharpoons 2\,NO(g) \qquad K_{eq} = 2.45 \times 10^{45}$
2. $2\,NOCl(g) \rightleftharpoons 2\,NO(g) + Cl_2(g) \qquad K_{eq} = 1.125 \times 10^{-33}$
3. $CS_2(g) + 4\,H_2(g) \rightleftharpoons CH_4(g) + 2\,H_2S(g) \qquad K_{eq} = 9.95 \times 10^{-1}$
4. $4\,NH_3(g) + 3\,O_2(g) \rightleftharpoons 2\,N_2(g) + 6\,H_2O(g) \qquad K_{eq} = 12$

Solution

1. Equilibrium position is far to the right. The reaction has gone to completion. Mainly NO exists.
2. Equilibrium position is far to the left. The reaction will not proceed to any significant extent. Very little NO and Cl_2 exist in the gas phase.
3. The equilibrium is neither to the left nor to the right. Significant concentrations of reactants (CS_2 and H_2) and products (CH_4 and H_2S) exist.
4. The equilibrium is neither to the left nor to the right. Significant concentrations of reactants (NH_3 and O_2) and products (N_2 and H_2O) exist.

17.6. Homogeneous vs. Heterogeneous Equilibria

Homogeneous equilibria involve reactions in which all the products and reactants are in the same phase. For example, in the Haber process for ammonia synthesis:

$$N_2(g) + 3\,H_2(g) \rightleftharpoons 2\,NH_3(g)$$

the reactants and product are all gaseous.

Heterogeneous equilibria, on the other hand, involve reactions in which substances are in different phases. Consider the decomposition of solid phosphorus pentachloride to liquid phosphorus trichloride and chlorine gas:

$$PCl_5(s) \rightleftharpoons PCl_3(l) + Cl_2(g)$$

The system involves solid PCl_5 in equilibrium with liquid PCl_3 and Cl_2 gas. The equilibrium constant expression is:

$$K_c = \frac{[PCl_3][Cl_2]}{[PCl_5]}$$

The concentrations (or activities) of pure solids and liquids do not change much with temperature or pressure, so they are taken to be 1 in these calculations. Consequently, for heterogeneous equilibria, terms for liquids and solids are omitted in the K expression. Thus, the K expression for the decomposition of PCl_5 becomes:

$$K_{eq} = [Cl_2] \quad \text{or} \quad P_{Cl_2}$$

Example 17.7

Write the equilibrium constant expressions for the following reversible reactions:

1. $NH_4Cl(s) \rightleftharpoons NH_3(g) + HCl(g)$
2. $CO_2(g) + NaOH(s) \rightleftharpoons NaHCO_3(s)$
3. $NH_3(g) + H_2SO_4(l) \rightleftharpoons (NH_4)_2SO_4(s)$
4. $2\ Pb(NO_3)_2(s) \rightleftharpoons 2\ PbO(s) + 4\ NO_2(g) + O_2(g)$

Solution

Note that the concentrations of pure liquids and solids do not appear in the K expression.

1. $K_c = [NH_3][HCl]$
2. $K_c = \dfrac{1}{[CO_2]}$
3. $K_c = \dfrac{1}{[NH_3]}$
4. $K_c = [NO_2]^4[O_2]$

17.7. Calculating Equilibrium Constants

The equilibrium constant K_c can be calculated from a set of equilibrium concentrations. Once K_c is known, its value can be used to calculate equilibrium concentrations, since the initial concentrations are always known.

In general, we need to know two basic things in order to calculate the value of the equilibrium constant. They are

1. A balanced equation for the reaction system. Be sure to include the physical states of each species so that the appropriate equilibrium expression for calculating K_c or K_p can be derived.
2. The equilibrium concentrations or pressures of each reactant or product included in the equilibrium expression. Substituting these values into the equilibrium expression allows the value of the equilibrium constant to be calculated.

Calculations of K_p/K_c often fall into one of two general types, depending on the data given.

17.7.1. Calculating K_c or K_p from known equilibrium amounts

1. Write a balanced chemical equation for the reaction at equilibrium, if not already given.
2. Write the equilibrium expression for the reaction.
3. Determine the molar concentrations or partial pressures of each species involved.
4. Substitute the value of the molar concentrations or partial pressures of each species into the equilibrium expression and solve for K.

Example 17.8

The equilibrium concentrations of H_2, I_2, and HI at 700 K are 0.0058 M, 0.105 M, and 0.190 M respectively. Calculate the value of the equilibrium constant, K_c, for the reaction:

$$H_2(g) + I_2(g) \rightleftharpoons 2\,HI(g)$$

at 700 K.

Solution

1. We have a balanced equation.
2. Write the equilibrium expression for the reaction:

$$K_c = \frac{[HI]^2}{[H_2][I_2]}$$

3. Skip this step since concentrations are already known and expressed in moles per liter (M).

4. Substitute the value of the molar concentrations or partial pressures of each species into the equilibrium expression and solve for K_c.

$$K_c = \frac{[HI]^2}{[H_2][I_2]} = \frac{(0.190)^2}{(0.0058)(0.105)} = 59.3$$

17.7.2. Calculating K from initial concentrations and one equilibrium concentration

1. Write a balanced chemical equation for the reaction at equilibrium, if not already given.
2. Write the equilibrium constant expression for the reaction.
3. Determine the initial molar concentrations or partial pressures of each species involved.
4. Determine changes from initial concentrations or partial pressures, based on stoichiometric ratios. Assign x to unknown changes.
5. Use (possibly unknown) changes to calculate final concentrations.
6. Substitute the value of the molar concentrations or partial pressures of each species into the equilibrium expression and solve for K.

Example 17.9

Consider the following:

$$CH_4(g) + H_2O(g) \rightleftharpoons CO(g) + 3\ H_2(g)$$

The reaction starts with a mixture of 0.15 M CH_4, 0.30 M H_2O, 0.40 M CO, and 0.60 M H_2 and is allowed to reach equilibrium. At equilibrium the concentration of CH_4 is found to be 0.05 M. Calculate the equilibrium constant, K_c, for the reaction.

Solution

1. Write the equilibrium constant expression for the reaction:

$$K_c = \frac{[CO]\,[H_2]^3}{[CH_4][H_2O]}$$

2. Determine the initial molar concentrations of each species involved. Initial concentrations:

$$[CH_4] = 0.15\text{ M};\quad [H_2O] = 0.30\text{ M};\quad [CO] = 0.40\text{ M};\quad [H_2] = 0.60\text{ M}.$$

3. Determine changes from initial concentrations based on stoichiometric ratios, assigning x to unknown changes. Use table 17-2(a) to simplify your

Table 17-2 Data for Example 17.9

(a)	$[CH_4]$	$[H_2O]$	$[CO]$	$[H_2]$
Initial concentration (M)	0.15 M	0.30 M	0.40 M	0.60 M
Change (M)	$(-)X$	$(-)X$	$(+)X$	$(+)X$
Equilibrium concentration (M)	$(0.15 - X) = 0.05$	$(0.30 - X)$	$(X + 0.4)$	$(X + 0.6)$

(b)	$[CH_4]$	$[H_2O]$	$[CO]$	$[H_2]$
Initial concentration (M)	0.15 M	0.30 M	0.40 M	0.60 M
Change (M)	$(-)0.10$	$(-)0.10$	0.10	0.30
Equilibrium concentration (M)	0.05	0.20	0.50	0.90

calculations. We know the equilibrium concentration of CH_4 to be 0.05 M. Thus we can determine the changes for the other species.

Use the changes to calculate final concentrations as shown in table 17-2(b).

4. Substitute the value of the molar concentrations or partial pressures of each species into the equilibrium expression and solve for K:

$$K_c = \frac{[CO][H_2]^3}{[CH_4][H_2O]} = \frac{(0.50)(0.90)^3}{(0.05)(0.20)} = 36.45$$

17.8. Calculating Equilibrium Concentrations from K

If we know the equilibrium constant for a given reaction, we can calculate the concentrations in the equilibrium mixture provided we know the initial concentrations.

1. Write a balanced chemical equation for the reaction at equilibrium, if not already given.
2. Write the equilibrium constant expression for the reaction.
3. Determine the initial molar concentrations or partial pressures of each species involved.
4. Determine changes from initial concentrations or partial pressures, based on stoichiometric ratios. Assign x to unknown changes.
5. Substitute concentrations in the K expression and solve for the unknown.
6. Use the value of the unknown to determine the desired concentrations.
7. Check values as needed.

The following examples illustrate the steps involved in these calculations.

Example 17.10

One mole of PCl_5 was decomposed to gaseous PCl_3 and Cl_2 in a 1-L sealed vessel. If K_c at the decomposition temperature is 1×10^{-3}, what are the equilibrium concentrations of the reactant and products?

Solution

Step 1

$$PCl_5(g) \rightleftharpoons PCl_3(g) + Cl_2(g)$$

Step 2

$$K_c = \frac{[PCl_3][Cl_2]}{[PCl_5]} = 1.0 \times 10^{-3}$$

Step 3

$$[PCl_5]_0 = \frac{1.0 \text{ mol}}{1.0 \text{ L}} = 1.0 \text{ M}$$
$$[PCl_3]_0 = 0$$
$$[Cl_2]_0 = 0$$

Step 4 Let x equal the amount of PCl_5 that reacted to reach equilibrium. Table 17-3 shows the equilibrium concentrations of all the species.

Steps 5 and 6

$$K_c = \frac{[PCl_3][Cl_2]}{[PCl_5]} = 1.0 \times 10^{-3}$$
$$1.0 \times 10^{-3} = \frac{(x)(x)}{(1-x)} = \frac{x^2}{1-x}$$
$$x^2 + 1.0 \times 10^{-3}x + 1.0 \times 10^{-3} = 0$$

This yields a quadratic equation of the general form $ax^2 + bx + c = 0$. The roots can be obtained by using "the (almighty) quadratic formula":

$$x = \frac{-b \pm \sqrt{b^2 - 4ac}}{2a}$$

Comparing our equation with the general form shows that $a = 1.0$, $b = 1.0 \times 10^{-3}$, and $c = 1.0 \times 10^{-3}$. The value for x is obtained by substituting these values into the quadratic formula.

Table 17-3 Data for Example 17.10

	$[PCL_5]$	$[PCL_3]$	$[Cl_2]$
Initial concentration (M)	1	0	0
Change (M)	$(-)X$	$(+)X$	$(+)X$
Equilibrium concentration (M)	$1-X$	X	X

This gives $x = 0.031$ M or $x = -0.032$ M. The acceptable value of x is taken to be 0.031 M since a negative concentration is meaningless.

Therefore

$$[PCl_3]_{eq} = 0.031\text{ M}$$

$$[Cl_2]_{eq} = 0.031\text{ M}$$

$$[PCl_5]_{eq} = 1.0 - 0.031 = 0.968\text{ M}$$

Step 7

$$K_c = \frac{[PCl_3][Cl_2]}{[PCl_5]} = \frac{(0.031)(0.031)}{0.968} = 1.0 \times 10^{-3}$$

This compares favorably with the given K_c of 1.0×10^{-3}.

17.8.1. A faster method for solving equilibrium problems

Note that we could have simplified the calculation from step 7 by assuming x to be very small compared to the initial concentration of PCl_5.

In other words, x is very small compared to 1, or $1 - x$ is approximately equal to 1, so that the equation:

$$1.0 \times 10^{-3} = \frac{x^2}{1 - x}$$

becomes:

$$x^2 = 1.0 \times 10^{-3}$$

$$x = \pm\sqrt{1.0 \times 10^{-3}}$$

and:

$$x = +0.0315 \quad \text{or} \quad -0.0315$$

17.8.2. When to use the approximation method

If the answer from the approximation method is less than 10% of the original concentration, then it is reasonably valid. However, if the approximation method gives an answer outside this range, then you will have to use the quadratic equation. For example, x above was found to be 0.0315 M, compared to the initial concentration of 1.0 M; that is, x is 3.2% of 1.0 M, so this answer is good.

Example 17.11

For the equilibrium

$$H_2(g) + I_2(g) \rightleftharpoons 2\ HI(g)$$

at 763 K, K_c is 46.5. Calculate the composition of the equilibrium mixture obtained if 1.0 mol each of H_2 and I_2 were equilibrated in a closed 1.0-L vessel at 763 K.

Solution

1. $H_2(g) + I_2(g) \rightleftharpoons 2\ HI(g)$
2. $K_c = \dfrac{[HI]^2}{[H_2][I_2]}$
3. $[H_2]_0 = 1.0\ M$

   $[I_2]_0 = 1.0\ M$

   $[HI]_0 = 0\ M$
4. Assuming x equals the amount of H_2 and I_2 that reacts, then the equilibrium concentrations are as shown in table 17-4.

   Note that the change in [HI] is $2x$, which reflects the reaction stoichiometry.
5. $K_c = \dfrac{[HI]^2}{[H_2][I_2]} = 46.5$
6. $46.5 = \dfrac{(2x)^2}{(1-x)(1-x)}$

   The right side of this equation is a perfect square, so we can simply take the square root of both sides. (We could also use the quadratic equation, but that would be more work.) Taking square roots, we have:

$$6.82 = \frac{(2x)}{(1-x)}$$

$$2x + 6.82x = 6.82$$

$$8.82x = 6.82 \quad \text{or} \quad x = 0.773\ M$$

Table 17-4 Data for Example 17.11

	$[H_2]$	$[I_2]$	$[H_2]$
Initial concentration (M)	1	1	0
Change (M)	$(-)X$	$(+)X$	$(+)2X$
Equilibrium concentration (M)	$1-X$	$1-X$	$2X$

Table 17-5 Data for Example 17.12

(a)	$[NH_4CO_2NH_2]$	$[NH_3]$	$[CO_2]$
Initial concentration (M)	51.33	0	0
Change (M)	$(-)X$	$(+)2X$	$(+)X$
Equilibrium concentration (M)	$(51.33-X)$	$2X$	X
(b)			
Initial concentration (M)	51.33	0	0
Change (M)	$(-)0.714$	$(+)1.428$	$(+)0.714$
Equilibrium concentration (M)	50.62	1.428	0.714

Thus,

$$[HI]_{eq} = 1.55 \text{ M}; \quad [H_2]_{eq} = 0.227 \text{ M}; \quad [I_2]_{eq} = 0.227 \text{ M}$$

7. Check for K_c:

$$K_c = \frac{[HI]^2}{[H_2][I_2]} = \frac{(1.55)^2}{(0.227)(0.227)} = 46.62$$

Example 17.12

The equilibrium constant K_p for the reaction:

$$NH_4CO_2NH_2(g) \rightleftharpoons 2\,NH_3(g) + CO_2(g)$$

is 7.1×10^{-3} at 40°C. Calculate the partial pressures of all species present after 5.0 mol of ammonium carbamate, $NH_4CO_2NH_2$, decompose in an evacuated 2.5-L vessel and equilibrium is reached at 40°C.

Solution

1. The equation given is balanced.
2. $K_p = \dfrac{P_{NH_3}P_{CO_2}}{P_{NH_4CO_2NH_2}}$
3. Calculate the initial partial pressure of $NH_4CO_2NH_2$, using the ideal gas equation.

$$P_{NH_4CO_2NH_2} = \frac{nRT}{V} = \frac{(5 \text{ mol})\left(0.082\,\dfrac{\text{L-atm}}{\text{mol-K}}\right)(313 \text{ K})}{2.5 \text{ L}} = 51.33 \text{ atm}$$

4. Determine the equilibrium partial pressure for each species present (see table 17-5a).
5. Calculate the partial pressures of all species present at equilibrium:

$$K_p = \frac{P^2_{NH_3} \cdot P_{CO_2}}{P_{NH_4CO_2NH_2}} = \frac{(2x)^2 \cdot x}{(51.33 - x)} = \frac{4x^3}{(51.33 - x)}$$

$$7.1 \times 10^{-3} = \frac{4x^3}{(51.33 - x)}$$

6. This equation would be difficult to solve, so we will try the approximation method first:

$$7.1 \times 10^{-3} = \frac{4x^3}{51.33}$$

$$4x^3 = \left(7.1 \times 10^{-3}\right)(51.33) = 0.36444$$

$$x = \sqrt[3]{0.36444} = 0.0.714$$

The equilibrium concentrations are summarized in table 17-5b.

$$P_{NH_4CO_2NH_2} = 50.62 \text{ atm}; \quad P_{NH_3} = 1.428 \text{ atm};$$

$$P_{CO_2} = 0.714 \text{ atm}$$

7. Our assumption is valid since 0.714 is much less than 51.55.

17.9. Qualitative Treatment of Equilibrium: Le Chatelier's Principle

In 1888, Henri-Louis Le Chatelier (1850–1936), a French industrial chemist, proposed a qualitative principle for predicting the behavior of a chemical system in equilibrium. Le Chatelier's principle can be stated as:

When a chemical system in equilibrium is disturbed by imposing a stress (e.g., a change in temperature, pressure, or concentration of reactants or products) on it, the position of the equilibrium will shift so as to counteract the effect of the disturbance.

17.9.1. Factors affecting a chemical reaction at equilibrium

There are three ways we can alter a chemical system in equilibrium. We can change:

1. The concentration of one of the reactants or products.
2. The pressure or volume.
3. The reaction temperature.

When any of the above factors in an equilibrium system is altered, Le Chatelier's principle can be used to predict how the system can regain equilibrium.

Changes in concentration

If a system is in equilibrium and we increase the concentration of one of the reactants or products, the reaction will shift in the forward or reverse direction to consume the added reactant or product and restore the equilibrium. Consider the Haber process for ammonia synthesis:

$$N_2(g) + 3\,H_2(g) \rightleftharpoons 2\,NH_3(g)$$

An increase in the concentration of nitrogen or hydrogen will cause the formation of more ammonia. This will undo the added stress of increased hydrogen or nitrogen. Therefore, the equilibrium will shift forward, towards thc products. Similarly, a decrease in the concentration of ammonia (i.e., removing ammonia as soon as it is formed), will shift the reaction to the right so that more ammonia will be produced. Again the equilibrium will favor more product formation. On the other hand, if we remove some of the hydrogen gas at equilibrium, the reverse reaction, which favors the decomposition of ammonia to increase the concentration of nitrogen and hydrogen, will be favored.

The reaction quotient Q, is very helpful when analyzing the effect of an imposed stress on a system in equilibrium. When a reaction is at equilibrium, $Q = K_c$. For the Haber process:

$$K_c = \frac{[NH_3]^2}{[N_2][H_2]^3} = Q$$

If we disturb the equilibrium by adding more nitrogen gas to the system, the denominator will increase relative to the numerator. Once this happens, Q becomes less than K_c. To restore the system to equilibrium (make $Q = K_c$), the concentration of ammonia in the numerator will have to increase. For this to happen, the overall reaction must proceed from left to right to form more ammonia. On the other hand, if we disturb the equilibrium by adding more ammonia to the system, Q will become larger than K_{eq} because the concentration of ammonia in the numerator has become larger. To restore equilibrium, the concentration of hydrogen, nitrogen, or both will have to increase. This means the reverse reaction, favoring the decomposition of ammonia, will be favored.

Thus, without any memorization, we can use Q to predict how a reaction in equilibrium will respond to added stress and arrive at the same conclusion that Le Chatelier's principle will give us.

Changes in pressure

Changes in pressure or volume are only important for equilibrium reactions in which one or more substances are gaseous. There are three ways to change the pressure of such systems:

- Adding or removing a gaseous reactant or product.
- Changing the volume of the container.
- Adding an inert gas to the system

An increase in pressure does not affect the value of K_c, but the value of Q may be altered. The same principles outlined above for changes in concentration can be used to predict how the system will deal with changes in pressure. You may recall that at constant temperature and pressure, the volume of a given mass of gas is directly proportional to the number of moles of gas present. Generally, an increase in pressure (or a decrease in volume) will cause the equilibrium condition to shift in the direction where there are fewer numbers of moles of gas. A decrease in pressure (or an increase in volume) of an equilibrium mixture of gases will shift the equilibrium in the direction where there are more moles of gas.

Consider the following system at equilibrium at constant temperature.

$$2\,SO_3(g) \rightleftharpoons 2\,SO_2(g) + O_2(g)$$

2 molecules 2 molecules 1 molecule

2 volumes 2 volumes 1 volume (by Avogadro's law)

The total moles of reactants and products are 2 and 3, respectively. Therefore, an increase in pressure will favor the reverse reaction, since the equilibrium will shift in the direction where there are fewer moles of gas (3 mol products ($2\text{ mol } SO_2 + 1\text{ mol } O_2$) $\rightarrow$ 2 mol reactants (SO_3)).

In cases where the total number of gas molecules remains constant, pressure change will have no effect on the equilibrium. Consider the reaction in which 1 mol hydrogen reacts with 1 mole of iodine to form 2 mol of hydrogen iodide:

$$H_2(g) + I_2(g) \rightleftharpoons 2\,HI(g)$$

1 mol 1 mol 2 mol

1 volume 1 volume 2 volumes (by Avogadro's law)

The number of moles on both sides of the equation is the same. A change in P does not favor either side, so will have no effect on the position of equilibrium.

The addition of an inert gas that is not involved in the equilibrium reaction has no effect on the equilibrium position. True, the total pressure is increased when an inert gas is added, but the concentrations or partial pressures of the reactants and products remain unchanged. Hence the original equilibrium position is maintained.

Changes in temperature

As long as the temperature remains constant, imposing a stress (change in concentration, pressure, or volume) on a reaction at equilibrium affects only the composition of the equilibrium mixture and hence the reaction quotient Q. The value of the equilibrium constant K_c is not affected.

On the other hand, a change in temperature will alter the value of the equilibrium constant. There is a general relationship between temperature and K_{eq}. For an exothermic reaction (negative ΔH^0), K_c decreases as the temperature increases. For an endothermic reaction (positive ΔH^0), K_c increases as the temperature decreases. To use Le Chatelier's principle to describe the effect of temperature changes on a system at equilibrium, treat the heat of reaction as a product in an exothermic reaction, or as a reactant in an endothermic reaction.

The Haber synthesis of ammonia is an exothermic reaction, liberating about 92.2 kJ of heat energy:

$$N_2(g) + 3\,H_2(g) \rightleftharpoons 2\,NH_3(g) + 92.2\text{ kJ} \qquad \Delta H^0 = -92.2\text{ kJ}$$

If the temperature is increased, the equilibrium will shift to the left in favor of reactants to absorb the heat and thus annul the stress. If the temperature is decreased, the equilibrium will be displaced to the right, favoring the formation of more ammonia.

Let's consider another example. The transformation of graphite to diamond is an endothermic process:

$$\text{C (graphite)} + 2\text{ kJ} \rightleftharpoons \text{C (diamond)} \qquad \Delta H^0 = +2\text{ kJ}$$

If the temperature is increased, the equilibrium shifts in the direction that absorbs heat. This is an endothermic reaction; raising the temperature will shift the equilibrium to the right in favor of the formation of more diamond.

17.9.2. Addition of a catalyst

The condition for a chemical system to be at equilibrium is that the rates of the forward and reverse reactions be equal. The addition of a catalyst affects the rates of both the forward and reverses reactions equally. Thus, while a catalyst increases the rate at which equilibrium is reached, it affects neither the equilibrium concentrations of reactants and products nor the equilibrium constant for the reaction.

Example 17.13

The formation of sulfur trioxide is a key step in the “contact process” for manufacturing sulfuric acid. The equilibrium reaction is described by the equation:

$$2\ SO_2(g) + O_2(g) \rightleftharpoons 2\ SO_3(g)\ \Delta H = -196\ \text{kJ/mol}$$

What will be the effect on the equilibrium if:

(a) The temperature is increased?
(b) The pressure is increased?
(c) The pressure is decreased?
(d) SO_3 is removed as soon as it is formed?
(e) O_2 is removed from the reaction mixture?

Solution

(a) Since the reaction is exothermic, an increase in temperature favors the formation of reactants. This will cause the decomposition of SO_3 to yield more SO_2 and O_2, thus shifting the equilibrium towards the reactants.
(b) An increase in pressure will force the equilibrium to shift in the direction where there are fewer moles of gas. There are 3 mol of reactants and 2 mol of products. Therefore, the equilibrium will shift from left to right, in the direction of SO_3, as the pressure is increased.
(c) A decrease in pressure will force the equilibrium to shift in the direction where there are more gaseous molecules. There are 3 mol of reactant molecules and 2 mol of products. Therefore, the equilibrium will shift from right (product) to left, in the direction of reactants, as the pressure is decreased.
(d) The removal of products as soon as they are formed results in the formation of more products. Therefore, the removal SO_3 causes more SO_2 and O_2 to react to form additional SO_3, thus shifting the equilibrium from left to right.
(e) The removal of one or more reactants causes one or more products to decompose to form additional reactants. Thus removal of O_2 will shift the equilibrium from right (products) to left (reactants).

Example 17.14

Nitric oxide, an important air pollutant, is produced in automobile engines at elevated temperatures according to the reaction:

$$N_2(g) + O_2(g) \rightleftharpoons 2\ NO(g) \qquad \Delta H^0 = +180.5\ \text{kJ}$$

What is the effect of increasing the pressure on this equilibrium, assuming no change in temperature?

Solution

An increase in pressure will shift the equilibrium in the direction where there are fewer gaseous molecules. Since the reaction contains an equal number of moles of gas on both sides of the equation, changes in pressure will have no effect on the position of equilibrium.

Example 17.15

Consider the decomposition of limestone, $CaCO_3$, represented by the equation:

$$CaCO_3(s) \rightleftharpoons CaO(s) + CO_2(g) \quad \Delta H = +178.5 \text{ kJ}$$

What is the effect on the equilibrium position of

(a) An increase in pressure?
(b) Adding a catalyst?

Solution

(a) Since one of the products is a gas, an increase in pressure is equivalent to an increase in the concentration of the gas involved. Increasing the pressure will shift the position of equilibrium to the left in favor of the reactants.
(b) A catalyst does not alter the equilibrium position of a reaction. Therefore the addition of a catalyst to this reaction will have no effect.

17.10. Problems

1. Write the equilibrium constant expression (K_c) for the following reactions:

   (a) $2\ SO_2(g) + O_2(g) \rightleftharpoons 2\ SO_3(g)$

   (b) $CO(g) + H_2O(g) \rightleftharpoons CO_2(g) + H_2(g)$

   (c) $CH_4(g) + 2H_2S(g) \rightleftharpoons CS_2(g) + 4\ H_2(g)$

   (d) $N_2O_4(g) \rightleftharpoons 2\ NO_2(g)$

2. Write the equilibrium expression for K_p for the reactions in problem 1.
3. Write the equilibrium constant expression (K_c) for the following reactions:

   (a) $CO(g) + 2\ H_2(g) \xrightleftharpoons{\text{Catalyst}} CH_3OH(g)$

   (b) $CH_3Cl(aq) + OH^-(aq) \rightleftharpoons CH_3OH(aq) + Cl^-(aq)$

   (c) $Ag^+(aq) + 2\ NH_3(aq) \rightleftharpoons Ag(NH_3)_2^+(aq)$

   (d) $C_6H_5SO_2NHCH_3(aq) + NaOH(aq) \rightleftharpoons \left[C_6H_5SO_2\overline{N}CH_3\right]\ Na^+(aq)$

4. Write the equilibrium constant (K_p) expression for the following reactions:

   (a) $PCl_3(g) + Cl_2(g) \rightleftharpoons PCl_5(g)$

   (b) $2\ PCl_3(g) + O_2(g) \rightleftharpoons 2\ POCl_3(g)$

   (c) $N_2H_4(g) + 6\ H_2O_2(g) \rightleftharpoons 2\ NO_2(g) + 8\ H_2O(g)$

   (d) $NO_2(g) + SO_2(g) \rightleftharpoons NO(g) + SO_3(g)$

5. The following are the equilibrium constant expressions (K_c) for totally gaseous equilibrium reactions. Write a balanced chemical equation for each one of them.

   (a) $\dfrac{[H_2][F_2]}{[HF]^2}$ (b) $\dfrac{[CO_2]^4[H_2O]^2}{[C_2H_2]^2[O_2]^5}$

   (c) $\dfrac{[CO_2][H_2]}{[CO][H_2O]}$ (d) $\dfrac{[HCl]^4[O_2]}{[Cl_2]^2[H_2O]^2}$

6. The following are the equilibrium constant expressions (K_c) for totally gaseous equilibrium reactions. Write a balanced chemical equation for each one.

   (a) $\dfrac{[H_2NCH_2CH_2N^+H_3][OH^-]}{[H_2NCH_2CH_2NH_2][H_2O]^2}$ (b) $\dfrac{[CS_2][H_2]^4}{[CH_4][H_2S]^2}$

   (c) $\dfrac{[H_2][CO]}{[CH_2O]}$ (d) $\dfrac{[NO]^2}{[N_2][O_2]}$

7. Re-write the following K_c expressions in terms of K_p:

   (a) $\dfrac{[N_2][H_2]^3}{[NH_3]^2}$ (b) $\dfrac{[CS_2][H_2]^4}{[CH_4][H_2S]^2}$

   (c) $\dfrac{[H_2][CO]}{[CH_2O]}$ (d) $\dfrac{[NO]^2}{[N_2][O_2]}$

8. For the following reactions, give a qualitative description of the position of equilibrium. Also, predict the relative concentrations of the products and reactants at 298 K.

   (a) $2\ Cl_2(g) + 2\ H_2O(g) \rightleftharpoons 4\ HCl(g) + O_2(g)$ $K_p = 3.2 \times 10^{-14}$

   (b) $H_2(g) + I_2(g) \rightleftharpoons 2\ HI(g)$ $K_{eq} = 51$

   (c) $CH_3Cl(aq) + OH^-(aq) \rightleftharpoons CH_3OH(aq) + Cl^-(aq)$ $K_c = 1 \times 10^{16}$

   (d) $2\ O_3(g) \rightleftharpoons 3\ O_2(g)$ $K_p = 1.3 \times 10^{57}$

9. For the reaction $2\ SO_2(g) + O_2(g) \rightleftharpoons 2\ SO_3(g)$, the equilibrium constant, K_c, is 3.5×10^6 at 400°C.

   (a) At this temperature, does the equilibrium favor the product, SO_3, or the reactants, SO_2 and O_2?
   (b) Calculate the value of K_p for this reaction.
   (c) Calculate the K_c for the reaction $2\ SO_3(g) \rightleftharpoons 2\ SO_2(g) + O_2(g)$.

10. A 1.0-L reaction vessel containing SO_2 gas and O_2 gas was equilibrated at 300°c. The equilibrium concentrations of SO_2, O_2, and SO_3 in the gaseous mixture were 0.229 M, 0.115 M, and 0.0373 M respectively. Determine the K_c for the reaction $2\ SO_2(g) + O_2(g) \rightleftharpoons 2\ SO_3(g)$.

11. The following reaction was carried out in a 7.14-L vessel at 690°C:

$$2\ H_2(g) + S_2(g) \rightleftharpoons 2\ H_2S(g)$$

Upon analysis the equilibrium mixture consisted of $[H_2] = 1.5$ mol; $[S_2] = 7.8 \times 10^{-6}$ mol; and $[H_2S] = 5.35$ mol. Calculate the equilibrium constant, K_c.

12. For the reaction:

$$CH_3COOH(aq) + C_2H_5OH(aq) \rightleftharpoons CH_3COOC_2H_5(aq) + H_2O(l) \qquad K_c = 4$$

calculate the equilibrium composition of reactants and products obtained by reacting 1 mol of ethanoic acid (CH_3COOH) with 1 mol of ethanol (C_2H_5OH) in a 1.0-L flask.

13. The equilibrium constant, K_c, for the reaction:

$$H_2(g) + I_2(g) \rightleftharpoons 2\ HI(g)$$

is 60 at 425°C. What is the composition of the equilibrium mixture if 8 g of hydrogen react with 508 g of iodine vapor in a sealed vessel?

14. A sample of phosgene, $COCl_2$, was allowed to decompose by heating it in a sealed 1 dm^3 reactor. The reaction is shown as:

$$COCl_2(g) \rightleftharpoons CO(g) + Cl_2(g) \qquad K_c = 2.2 \times 10^{-10}$$

Determine the equilibrium concentrations of the products and residual reactant, assuming an initial $COCl_2$ concentration of 1.0 M.

15. Write the K_p expression for the following equilibrium reactions:

(a) $SrC_2O_4 \cdot 2\ H_2O(s) \rightleftharpoons SrC_2O_4(s) + 2\ H_2O(g)$

(b) $CaCl_2 \cdot 6\ H_2O(s) \rightleftharpoons CaCl_2(s) + 6\ H_2O(g)$

(c) $Hg_2O \cdot H_2O(s) \rightleftharpoons Hg_2O(s) + H_2O(g)$

(d) $Na_2SO_4 \cdot 10\ H_2O(s) \rightleftharpoons Na_2SO_4(s) + 10\ H_2O(g)$

16. At 25°C, the values of K_p for problem 15 (b) and (d) are 5.1×10^{-44} and 1.0×10^{-16}, respectively. Calculate the water vapor pressure above each of these hydrated salts.

17. 0.1125 mol Br_2 gas is heated in a 1.50-L flask at 1800 K. At equilibrium the bromine is 5.0% dissociated. Determine the equilibrium constant, K_c.

18. NH_4HS decomposes into ammonia and hydrogen sulfide at low temperatures:

$$NH_4HS(s) \rightleftharpoons NH_3(g) + H_2S(g)$$

(a) In a 1.0-L vessel, a 0.05 mol sample of NH_4 HS was heated at 80°C. Calculate K_c for the reaction if the vessel contained 0.03 mol H_2S at equilibrium.
(b) Using the ideal gas equation calculate the pressure in the vessel at equilibrium.
(c) What is the partial pressure of each gas in the reaction vessel?
(d) Calculate K_p for the reaction.

19. With reference to the following reaction:

$$2\ POCl_3(g) \rightleftharpoons 2\ PCl_3(g) + O_2(g) \qquad \Delta H = +572\ \text{kJ}$$

(a) What is the effect of increasing the temperature? Assume no change in pressure.
(b) What is the effect of increasing the pressure? Assume no change in temperature.
(c) Describe the effect of removing PCl_3.
(d) How does the equilibrium amount of O_2 vary if a platinum catalyst is added?

20. The initial step of the Ostwald process for the industrial production of nitric acid is the oxidation of ammonia to nitric oxide:

$$4\ NH_3(g) + 5\ O_2(g) \rightleftharpoons 4\ NO(g) + 6\ H_2O(g) \qquad \Delta H^\circ = -905.6\ \text{kJ}$$

What happens to the amount of NO when:

(a) The temperature is increased?
(b) A platinum catalyst is added?
(c) The pressure is increased by adding helium gas?
(d) Some ammonia gas is pumped out of the reactor?

18

Ionic Equilibria and pH

18.1. The Ionization of Water

Water is a weak acid. At 25°C, pure water ionizes to form a hydrogen ion and a hydroxide ion:

$$H_2O \rightleftharpoons H^+ + OH^-$$

Hydration of the proton (hydrogen ion) to form hydroxonium ion is ignored here for simplicity. This equilibrium lies mainly to the left; that is, the ionization happens only to a slight extent. We know that 1 L of pure water contains 55.6 mol. Of this, only 10^{-7} mol actually ionizes into equal amounts of $[H^+]$ and $[OH^-]$, i.e.,

$$[H^+] = [OH^-] = 10^{-7}M$$

Because these concentrations are equal, pure water is neither acidic nor basic.

18.2. Definition of Acidity and Basicity

A solution is acidic if it contains more hydrogen ions than hydroxide ions. Similarly, a solution is basic if it contains more hydroxide ions than hydrogen ions.

Acidity is defined as the concentration of hydrated protons (hydrogen ions); basicity is the concentration of hydroxide ions.

18.2.1. Ionic product of water

Pure water ionizes at 25°C to produce 10^{-7} M of $[H^+]$ and 10^{-7} M of $[OH^-]$. The product

$$K_w = [H^+] \times [OH^-] = 10^{-7}\ M \times 10^{-7}\ M = 10^{-14}\ M$$

is known as the ionic product of water. Note that this is simply the equilibrium expression for the dissociation of water. This equation holds for any dilute aqueous solution of acid, base, and salt.

Example 18.1

The hydrogen ion concentration of a solution is 10^{-5} M. What is the hydroxide ion concentration? Is the solution acidic or basic?

Solution

$[H^+] = 10^{-5}M$

$[OH^-] = ?$

$[H^+][OH^-] = 10^{-14} \qquad [OH^-] = 10^{-14}$

$$[H^+] = \frac{10^{-14}}{10^{-5}} = 10^{-9}\text{ M}$$

The solution is acidic, because $[H^+] = 10^{-5}$ M > $[OH^-] = 10^{-9}$ M

18.3. The pH of a Solution

The pH of a solution is defined as the negative logarithm of the molar concentration of hydrogen ions. The lower the pH, the greater the acidity of the solution. Mathematically:

$$\text{pH} = -\log_{10}\left[H^+\right] \quad \text{or} \quad -\log_{10}\left[H_3O^+\right]$$

This can also be written as:

$$\text{pH} = \log_{10}\frac{1}{[H^+]} \quad \text{or} \quad \log_{10}\frac{1}{[H_3O^+]}$$

Taking the antilogarithm of both sides and rearranging gives:

$$\left[H^+\right] = 10^{-\text{pH}}$$

This equation can be used to calculate the hydrogen ion concentration when the pH of the solution is known.

Example 18.2

Calculate the pH of 0.0010 M nitric acid.

Solution

$$[H^+] = 1.0 \times 10^{-3}\ M$$

$$\begin{aligned} pH &= -\log_{10}\left[H^+\right] \\ &= -\log_{10}\left[1.0 \times 10^{-3}\right] \\ &= 3 \end{aligned}$$

Example 18.3

Calculate the $[H^+]$ of a solution having a pH of 5.

Solution

$$\begin{aligned} [H^+] &= 10^{-pH} \\ &= 1.0 \times 10^{-5}\ M \end{aligned}$$

Example 18.4

What is the $\left[H^+\right]$ in a cup of coffee with a pH of 5.05?

Solution

$$\begin{aligned} \left[H^+\right] &= 10^{-pH} \\ &= 10^{-5.05} = 8.9 \times 10^{-6}\ M \end{aligned}$$

18.4. The pOH of a Solution

The pOH of a solution is defined as the negative logarithm of the molar concentration of hydroxide ions. The lower the pOH, the greater the basicity of the solution. Mathematically:

$$pOH = -\log_{10}[OH^-]$$

The relationship between pH and pOH is:

$$pH + pOH = 14$$

If the pH of a solution is known, the pOH can be calculated using the above equation.

Example 18.5

Tomato juice has a pH of 4.2. What is the pOH of the solution?

Solution

$$\text{pH} = 4.2$$
$$\text{pOH} = ?$$
$$\text{pH} + \text{pOH} = 14$$
$$4.2 + \text{pOH} = 14$$
$$\text{pOH} = 9.8$$

Example 18.6

Lemon juice has a pOH of 11.7. Calculate the pH of the lemon juice.

Solution

$$\text{pOH} = 11.7$$
$$\text{pH} = ?$$
$$\text{pH} + \text{pOH} = 14$$
$$\text{pH} + 11.7 = 14$$
$$\text{pOH} = 14 - 11.7 = 2.3$$

Example 18.7

Calculate the pH of each of the following:

(a) A 0.25 M solution of KOH
(b) A 0.0025 M solution of $Mg(OH)_2$

Solution

(a) KOH is a strong base and so it is 100% ionized in aqueous solution. Therefore a KOH will provide one OH^- per molecule: $[OH^-] = 0.25$ M.

$$[H^+][OH^-] = 1.0 \times 10^{-14}$$
$$[H^+] = \frac{1.0 \times 10^{-14}}{[OH^-]} = \frac{1.0 \times 10^{-14}}{0.25\ \text{M}} = 4 \times 10^{-14}\ \text{M}$$
$$\text{pH} = -\log_{10}[H^+] = -\log\left(4.0 \times 10^{-14}\right) = 13.4$$

$Mg(OH)_2$ is a strong base and so it is 100% ionized in aqueous solution. Therefore it will provide two OH^- per molecule. $[OH^-] = 2 \times 0.0025\ M = 0.005\ M$.

$$[H^+][OH^-] = 1.0 \times 10^{-14}$$

$$[H^+] = \frac{1.0 \times 10^{-14}}{[OH^-]} = \frac{1.0 \times 10^{-14}}{0.005\ M} = 2 \times 10^{-12}\ M$$

$$pH = -\log_{10}[H^+] = -\log\left(2.0 \times 10^{-12}\right) = 11.7$$

18.5. The Acid Ionization Constant, K_a

Protonic acids undergo ionization in aqueous solution. The ionization process may be represented by the reaction:

$$HA(aq) + H_2O(l) \rightleftharpoons H_3O^- + A^-$$

or more simply:

$$HA(aq) \rightleftharpoons H^+ + A^-$$

The equilibrium expression is:

$$K_a = \frac{[H^+][A^-]}{[HA]}$$

K_a is called the *ionization constant* and is a measure of the strength of an acid. The larger the K_a, the stronger is the acid. Strong acids undergo complete or 100% ionization and have K_a values greater than 1. Weak acids, on the other hand, are only partially ionized, with K_a values less than 1. For example, HCl is a strong acid, which ionizes in water according to the following equation:

$$HCl(aq) \longrightarrow H^+(aq) + Cl^-(aq)$$

The equilibrium expression for the reaction is:

$$K_a = \frac{[H^+][Cl^-]}{[HCl]} = 2 \times 10^6$$

Since HCl dissociates fully, [HCl] will be nearly zero, making K_a very large. Acetic acid (CH_3COOH), on the other hand, is a weak acid which dissociates according to the following equation:

$$CH_3COOH(aq) \rightleftharpoons H^+(aq) + CH_3COO^-(aq)$$

The equilibrium constant measured at 25°C is 1.8×10^{-5}.

18.5.1. Definition of pK_a

Like pH, pK_a is defined as the negative logarithm of the acid ionization constant, K_a. That is:

$$pK_a = -\log K_a$$

A large K_a value will have a corresponding small pK_a and vice-versa.

18.6. Calculating pH and Equilibrium Concentrations in Solutions of Weak Acids

Acid–base equilibrium problems are solved using the same methods as any other equilibrium problems.

Example 18.8

The pH of 0.10 M HCN is 3.5. Calculate the value of K_a for hydrocyanic acid.

Solution

1. Write the balanced equation for the dissociation reaction.

$$HCN(aq) \rightleftharpoons H^+(aq) + CN^-(aq)$$

2. Write the expression for K_a.

$$K_a = \frac{[H^+][CN^-]}{[HCN]}$$

3. Calculate the hydrogen ion concentration $[H^+]$ from the given pH.

$$pH = -\log[H^+]$$
$$[H^+] = 10^{-pH} = 10^{-3.5} = 3.0 \times 10^{-4}\ M$$
$$\text{And } [CN^-] = [H^+] = 3.0 \times 10^{-4}\ M$$

4. Calculate the concentrations in the equilibrium mixture, i.e., [HCN], $[H^+]$, and $[CN^-]$ (see table 18-1).
5. Calculate K_a by substituting the various equilibrium concentrations in the expression for K_a.

$$K_a = \frac{[H^+][CN^-]}{[HCN]} = \frac{(0.0003)(0.0003)}{(0.0997)} = 9.03 \times 10^{-7}$$

Table 18-1 Data for Example 18.8

	[HCN]	$[H^+]$	$[CN^-]$
Initial concentration (M)	0.1	0	0
Change (M)	(−)0.0003	(+)0.0003	(+)0.0003
Equilibrium concentration (M)	0.0997	0.0003	0.0003

Table 18-2 Data for Example 18.9

	[HF]	$[H^+]$	[F]
Initial concentration (M)	0.1	0	0
Change (M)	$(-)x$	$(+)x$	$(+)x$
Equilibrium concentration (M)	$0.1-x$	x	x

Example 18.9

Calculate the equilibrium concentrations of the various species present in a 0.1 M solution of HF ($K_a = 6.5 \times 10^{-4}$). What is the pH of the solution?

Solution

1. Write the balanced equation for the equilibrium dissociation reaction.

$$HF(aq) \rightleftharpoons H^+(aq) + F^-(aq)$$

2. Write an expression for K_a.

$$K_a = \frac{[H^+][F^-]}{[HF]}$$

3. Set up a table listing the initial and equilibrium concentrations of the species present (see table 18-2).
4. Substitute the equilibrium concentrations into the K_a expression and then solve for the unknown, x.

$$K_a = \frac{(x).(x)}{0.1-x} = \frac{x^2}{0.1-x} = 6.5 \times 10^{-4}$$

$$x^2 + 6.5 \times 10^{-4} \times -6.5 \times 10^{-5} = 0$$

This is a quadratic equation. Solve for x using the quadratic formula:

$$x = \frac{-b \pm \sqrt{b^2 - 4ac}}{2a}, \quad \text{where } a = 1, b = 6.5 \times 10^{-4}, \text{ and } c = -6.5 \times 10^{-5}$$

$$x = \frac{-\left(6.5 \times 10^{-4}\right) \pm \sqrt{\left(6.5 \times 10^{-4}\right)^2 - 4(1)\left(-6.5 \times 10^{-5}\right)}}{2(1)}$$

$$x = \frac{-\left(6.5 \times 10^{-4}\right) \pm 0.0173}{2}$$

$$x = 0.0083 \quad \text{or} \quad -0.0090$$

Since x is the concentration of H^+, it must have a positive value. Thus $[H^+] =$ 0.0083 M.

We can also solve this problem by using the approximation method. Here we assume that $x << 0.1$:

$$0.1 - x \approx 0.1 \quad \text{and hence} \quad \frac{x^2}{0.1 - x} = \frac{x^2}{0.1} = 6.5 \times 10^{-4}$$

On solving, $x^2 = (6.5 \times 10^{-4})(0.1)$ or $x = 0.0081$ (see below for a discussion of how to decide if this answer is valid).

Therefore at equilibrium $[H^+] = [F^-] = 0.0081$ M and $[HF] = 0.092$ M.

5. Now calculate the pH of the solution.

$$\begin{aligned} pH &= -\log\left[H^+\right] \\ &= -\log(0.008) \\ &= 2.1 \end{aligned}$$

18.6.1. When to use the approximation method

The approximation method works best when x is less than 10% of the original concentration as explained in chapter 17. In the preceding example x was found to be 0.008 compared to the initial concentration of 0.10 M: that is, x is 8% of 0.10. If this criterion is not met, use the quadratic equation to solve for x.

18.7. Percent Dissociation of Weak Acids

Since weak acids are only partially dissociated in aqueous solution, the percent dissociation or ionization can be used in conjunction with K_a to measure the strength of an acid. It is calculated from the initial concentration of the acid and the concentration of hydrogen ion at equilibrium. The percent dissociation is defined as follows:

$$\text{Percent dissociation} = \frac{\text{Amount of acid dissociated (mol/L)}}{\text{Initial concentration (mol/L)}}$$

For a monoprotic acid HA:

$$\%\ \text{Dissociation} = \frac{[\text{H}^+]_{\text{eq}}}{[\text{HA}]_{\text{initial}}} \times 100\ \%$$

Example 18.10

An etching solution is prepared by dissolving 1.0 mol of HF in 1 L of water. If this solution contains 0.025 M of hydrogen ions, calculate the percent dissociation of HF.

Solution

$$\%\ \text{Dissociation} = \frac{[\text{H}^+]_{\text{eq}}}{[\text{HF}]_{\text{initial}}} \times 100\ \%$$

$$= \frac{0.025\ \text{M}}{1.0\ \text{M}} \times 100 = 2.5\%$$

Example 18.11

Calculate the pH and the degree of dissociation of a 0.10 M benzoic acid ($HC_7H_5O_2$) solution. K_a for benzoic acid is 6.6×10^{-5}.

Solution

1. Write an equation for the reaction.

$$HC_7H_5O_2(aq) \rightleftharpoons H^+(aq) + C_7H_5O_2^-(aq)$$

2. Write an expression for the dissociation constant.

$$K_a = \frac{[H^+][C_7H_5O_2^-]}{[HC_7H_5O_2]}$$

3. Calculate the equilibrium concentrations of all species (see table 18-3).
4. Substitute in the expression for K_a. Determine x and hence $[H^+]$.

$$K_a = \frac{(x)(x)}{0.10 - x} = \frac{x^2}{0.10 - x} = 6.6 \times 10^{-5}$$

Table 18-3 Data for Example 18.11

	$[HC_7H_5O_2]$	$[H^+]$	$[C_7H_5O_2]$
Initial concentration (M)	0.1	0	0
Change (M)	$(-)x$	$(+)x$	$(+)x$
Equilibrium concentration (M)	$0.10 - x$	x	x

Assume $x << 0.10$

$$6.6 \times 10^{-5} = \frac{x^2}{0.10}$$

$$x = 2.6 \times 10^{-3} = [H^+] \text{ (assumption valid)}$$

$$pH = -\log(2.6 \times 10^{-3}) = 2.59$$

5. Calculate the percent dissociation:

$$\% \text{ Dissociation} = \frac{[H^+]}{[HA]} \times 100 = \frac{2.6 \times 10^{-3}}{0.10} \times 100 = 2.6\%$$

Example 18.12

The accumulation of lactic acid ($HC_3H_5O_3$) in muscle tissue during exertion or "workout" often results in pain and tiredness. Calculate pK_a for lactic acid, if a 0.0500 M solution is 4% ionized.

Solution

1. Write an equation for the reaction.

$$HC_3H_5O_3(aq) \rightleftharpoons H^+(aq) + C_3H_5O_3^-(aq)$$

2. Write an expression for the dissociation constant.

$$K_a = \frac{[H^+][C_3H_5O_3^-]}{[HC_3H_5O_3]}$$

3. From the percent dissociation given, calculate the equilibrium concentration of hydrogen ions.

$$\% \text{ Dissociation} = \frac{[H^+]_{eq}}{[HC_3H_5O_3]_0} \times 100 = 4$$

$$[H^+]_{eq} = \frac{4 \times [HC_3H_5O_3]_0}{100}$$

But

$$[HC_3H_5O_3]_0 = 0.0500\ M$$

Hence

$$[H^+]_{eq} = 0.0020\ M$$

4. Find K_a by substituting the equilibrium concentrations of the species in solution. At equilibrium:

$$[H^+]_{eq} = [C_3H_5O_3^-]_{eq} = 0.0020\ M$$

$$[HC_3H_5O_3]_{eq} = [HC_3H_5O_3]_0 - [H^+]_{eq} = 0.050 - 0.0020 = 0.0480\ M$$

$$K_a = \frac{[H^+][C_3H_5O_3^-]}{[HC_3H_5O_3]} = \frac{(0.0020)(0.0020)}{0.0480} = 8.3 \times 10^{-5}$$

5. Calculate pK_a.

$$\begin{aligned} pK_a &= -\log K_a \\ &= -\log(8.3 \times 10^{-5}) = 4.1 \end{aligned}$$

18.8. The Base Dissociation Constant, K_b

Generally, a weak base (B) accepts a proton from water to form the conjugate acid of the base (BH^+) and OH^- ions.

$$B(aq) + H_2O(l) \rightleftharpoons BH^+(aq) + OH^-(aq)$$

The base dissociation constant for this equilibrium reaction is:

$$K_b = \frac{[BH^+][OH^-]}{[B]}$$

The weak base B may be a neutral molecule such as ammonia or an organic amine, or an anion such as fluoride (F^-) or cyanide (CN^-).

Problems involving equilibria in solutions of weak bases are solved in the same way as used for weak acid problems.

Example 18.13

Aniline ($C_6H_5NH_2$) is an organic base used in the synthesis of the electronic conductor poly(aniline). Calculate the pH and the concentrations of all species present in a 0.25 M solution of aniline ($K_b = 4.3 \times 10^{-10}$).

Table 18-4 Data for Example 18.13

	$[C_6H_5NH_2]$	$[C_6H_5NH_3^+]$	$[OH^-]$
Initial concentration (M)	0.25	0	0
Change (M)	$(-)x$	$(+)x$	$(+)x$
Equilibrium concentration (M)	$0.25-x$	x	x

Solution

1. Write an equation for the reaction and an expression for K_b.

$$C_6H_5NH_2(aq) + H_2O(l) \rightleftharpoons C_6H_5NH_3^+(aq) + OH^{-(aq)}$$

$$K_b = \frac{[C_6H_5NH_3^+][OH^-]}{[C_6H_5NH_2]}$$

2. Set up a table showing the species and their concentrations at equilibrium (see table 18-4).
3. Solve for the value of x from the equilibrium equation to find the concentrations of all species at equilibrium.

$$K_b = 4.3\times10^{-10} = \frac{[C_6H_5NH_3^+][OH^-]}{[C_6H_5NH_2]} = \frac{(x)(x)}{(0.25-x)}$$

If we assume that $x << 0.25$,

$$x^2 = (0.25)\left(4.3\times10^{-10}\right) = 1.075\times10^{-10}$$

$$x = 1.037\times10^{-5}$$

So our assumption was justified. Thus, $[C_6H_5NH_3^+] = [OH^-] = 1.037\times10^{-5}$ M and

$$[C_6H_5NH_2]_{eq} = 0.25 - 0.00001037 \quad \text{or} \quad 0.24999\text{ M} \approx 0.25\text{ M}$$

4. Now calculate the hydrogen ion concentration and the pH. Note that the small $[H^+]$ concentration is produced from the subsidiary equilibrium dissociation of water:

$$[H^+][OH^-] = 10^{-14}$$

$$[H^+] = \frac{10^{-14}}{[OH^-]} = \frac{10^{-14}}{1.0\times10^{-5}} = 1.0\times10^{-9}\text{ M}$$

$$\text{pH} = -\log[H^+] = -\log\left(1.0\times10^{-9}\right) = 9$$

So the solution is mildly basic as expected.

18.9. Relationship Between K_a and K_b

The relationship between K_a for a weak acid and K_b for its conjugate base can be deduced from the equilibrium conditions of the acidic ionization of an acid HA and the basic ionization of its conjugate base A^- in aqueous solution. The derivation is shown below:

1. $\mathrm{HA} \rightleftharpoons \mathrm{H^+} + \mathrm{A^-}$

   $$K_a = \frac{[\mathrm{H^+}][\mathrm{A^-}]}{[\mathrm{HA}]}$$

2. $\mathrm{A^-} + \mathrm{H_2O} \rightleftharpoons \mathrm{HA} + \mathrm{OH^-}$

   $$K_b = \frac{[\mathrm{HA}][\mathrm{OH^-}]}{[\mathrm{A^-}]}$$

3. Multiply the expressions for K_a and K_b and eliminate common terms:

   $$K_a \times K_b = \frac{[\mathrm{H^+}]\cancel{[\mathrm{A^-}]}}{\cancel{[\mathrm{HA}]}} \times \frac{\cancel{[\mathrm{HA}]}[\mathrm{OH^-}]}{\cancel{[\mathrm{A^-}]}} = [\mathrm{H^+}][\mathrm{OH^-}]$$

4. Recall that $[\mathrm{H^-}][\mathrm{OH^-}] = K_w$, so that $K_a K_b = K_w$

   Of course this also holds between K_b for a weak base and K_a for its conjugate acid.

Example 18.14

Calculate the pH of a 0.10 M KF solution. The K_a for HF is 6.4×10^{-4} ($K_w = 1.0 \times 10^{-14}$).

Solution

1. Write an equation for the reaction of KF in water

   $$\mathrm{F^-(aq)} + \mathrm{H_2O(l)} \rightleftharpoons \mathrm{HF(aq)} + \mathrm{OH^-(aq)}$$

2. Write an expression for K_b for the above reaction.

   $$K_b = \frac{K_w}{K_a} = \frac{1 \times 10^{-14}}{6.4 \times 10^{-4}} = 1.56 \times 10^{-11} = \frac{[\mathrm{HF}][\mathrm{OH^-}]}{[\mathrm{F^-}]}$$

3. Solve for $[\mathrm{OH^-}]$, and hence for $[\mathrm{H^+}]$.

   $$K_b = 1.56 \times 10^{-11} = \frac{[\mathrm{HF}][\mathrm{OH^-}]}{[\mathrm{F^-}]}$$

Let $[HF] = [OH^-] = x$ at equilibrium. Initial $[F^-] = 0.10$ M. Equilibrium value $[F^-]_{eq} = 0.10 - x$. Substitute these in the K_b expression:

$$1.56 \times 10^{-11} = \frac{(x)(x)}{0.10 - x} = \frac{x^2}{0.10 - x}$$

Assume that $x << 0.10$ and solve the resulting equation:

$$x^2 = 1.56 \times 10^{-12} \quad \text{or} \quad x = 1.25 \times 10^{-6}$$

$$[OH^-] = 1.25 \times 10^{-6} \text{ M}$$

But

$$[H^+] = \frac{K_w}{[OH^-]} = \frac{1.0 \times 10^{-14}}{1.25 \times 10^{-6}} = 8.0 \times 10^{-9} \text{ M}$$

4. Now solve for the pH of the solution

$$\text{pH} = -\log[H^+] = -\log(8.0 \times 10^{-9}) = 8.10$$

Example 18.15

The dissociation constant of benzoic acid, C_6H_5COOH, is 6.46×10^{-5}. Calculate the K_b and pK_b for the benzoate ion, $C_6H_5COO^-$.

Solution

$$K_a K_b = K_w$$

$$K_b = \frac{K_w}{K_a} = \frac{1.0 \times 10^{-14}}{6.46 \times 10^{-5}} = 1.55 \times 10^{-10}$$

$$pK_b = -\log K_b = -\log(1.55 \times 10^{-10}) = 9.8$$

18.10. Salt Hydrolysis: Acid–Base Properties of Salts

In a neutralization reaction, an acid reacts with a base to form water and a salt, a substance composed of an anion and a cation. Characteristically, salts are strong electrolytes that completely dissociate into anions and cations in water. Salt hydrolysis refers to the reaction of the anionic or cationic component of a salt (or both) with water. It is essentially a proton-transfer reaction between salt and water. Salt hydrolysis is the same process as the ionization of a weak acid or base. Therefore the mathematical treatment of the equilibrium process is the same.

There are three different kinds of salts:

- Acidic salts which ionize in water to produce acidic solutions. These are typically salts of a weak base and a strong acid. Examples include NH_4Cl, $AlCl_3$, and $Fe(NO_3)_3$.
- Basic salts which ionize in water to produce basic solutions. These are typically salts of a strong base and a weak acid. Examples include KNO_2, CH_3 COONa, and Na_2S.
- Neutral salts which ionize in water to produce a neutral (or nearly neutral) solution. These are typically salts of a strong acid and a strong base. Examples include $NaNO_3$, KI, and $BaCl_2$.

18.10.1. The hydrolysis constant, K_h

Consider the hydrolysis reaction of ammonium chloride:

$$NH_4Cl(s) \longrightarrow NH_4^+(aq) + Cl^-(aq) \quad (1)$$

$$NH_4^+(aq) + H_2O(l) \rightleftharpoons NH_3(aq) + H_3O^+(aq) \quad (2)$$

We can write the equilibrium expression for the hydrolysis reaction (Equation 2):

$$K_h = \frac{[NH_3][H_3O^+]}{[NH_4^+]} \quad (3)$$

The equilibrium constant K_h is known as the hydrolysis constant.

18.10.2. Relationship between K_h and K_w

For the weak acid HA:

$$HA + H_2O \rightleftharpoons H_3O^+ + A^- \quad \text{and} \quad K_a = \frac{[H_3O^+][A^-]}{[HA]}$$

For the hydrolysis of the conjugate base A^-:

$$A^- + H_2O \rightleftharpoons HA + OH^- \quad \text{and} \quad K_h = \frac{[HA][OH^-]}{[A^-]}$$

Now multiply the expressions for K_a and K_h and eliminate common terms:

$$K_aK_h = \frac{[H_3O^+]\cancel{[A^-]}}{\cancel{[HA]}} \times \frac{\cancel{[HA]}[OH^-]}{\cancel{[A^-]}} = [H_3O^+][OH^-] \quad (4)$$

We know that:

$$K_w = [H_3O^+][OH^-]$$

Table 18-5 Data for Example 18.16

	$[C_6H_5CO_2^-]$	$[C_6H_5CO_2H]$	$[OH^-]$
Initial concentration (M)	0.05	0	0
Change (M)	$(-)x$	$(+)x$	$(+)x$
Equilibrium concentration (M)	$0.05 - x$	x	x

Therefore:

$$K_aK_h = K_w \quad \text{or} \quad K_h = \frac{K_w}{K_a}$$

Example 18.16

Calculate the hydroxide ion concentration, the degree of hydrolysis, and the pH of a 0.05 M solution of sodium benzoate. Assume that sodium benzoate salt ionizes completely in water.

$$K_h(C_6H_5CO_2^-) = 1.5 \times 10^{-10} \text{ at } 25°\text{C}.$$

Solution

1. First write an equation for the reaction and an expression for K_h.

$$C_6H_5O_2^-(aq) + H_2O(l) \rightleftharpoons HC_6H_5O_2(aq) + OH^-(aq)$$

$$K_h = \frac{[HC_6H_5O_2]\left[OH^-\right]}{\left[C_6H_5O_2^-\right]} = 1.5 \times 10^{-10}$$

2. Determine the equilibrium concentrations of all species in solution including $[H^+]$ (see table 18-5).
   Substituting in the equilibrium expression yields:

$$1.5 \times 10^{-10} = \frac{(x)(x)}{(0.050 - x)} = \frac{x^2}{(0.050 - x)}$$

Since $K_h << 0.050$ (the initial concentration), we can assume x will be very small, hence we can ignore it relative to 0.05.

$$1.5 \times 10^{-10} = \frac{x^2}{(0.050)}$$

$$x = \sqrt{(1.5 \times 10^{-10})(0.05)} = 2.7 \times 10^{-6}\ \text{M} = \left[OH^-\right] = [C_6H_5CO_2H]$$

$$[C_6H_5CO_2^-] = 0.05 - 0.0000027 = 0.049997\ \text{M} \approx 0.050\ \text{M}$$

3. Calculate the degree of hydrolysis:

$$\text{Degree of hydrolysis} = \frac{[C_6H_5CO_2H]}{[C_6H_5CO_2^-]} \times 100$$

$$= \frac{2.7 \times 10^{-6}}{0.05} \times 100 = 5.4 \times 10^{-3}\%$$

4. Calculate the pH from $[OH^-]$.

$$pOH = -\log\left[OH^-\right] = -\log\left(2.7 \times 10^{-6}\right) = 5.57$$

$$pH = 14.00 - 5.57 = 8.43$$

Example 18.17

How many grams of ammonium bromide are needed per liter of solution to give a pH of 5.0? $K_a = 5.7 \times 10^{-10}$.

Solution

1. First we write the equation for the reaction and an expression for K_a.

$$NH_4^+(aq) + H_2O(l) \rightleftharpoons NH_3(aq) + H_3O^+(aq)$$

$$K_a = \frac{[NH_3]\left[H_3O^+\right]}{\left[NH_4^+\right]} = 5.7 \times 10^{-10}$$

2. Determine the equilibrium concentrations of ammonia and H^+ in solution from the given pH.

$$pH = 5.5 = -\log[H_3O^+]$$

$$[H_3O^+] = 10^{-5.5} = 3.1 \times 10^{-6}\ M$$

$$[H_3O^+] = [NH_3] = 3.1 \times 10^{-6}\ M$$

3. Calculate the concentration of ammonium ion.

$$K_a = \frac{[NH_3]\left[H_3O^+\right]}{\left[NH_4^+\right]} = 5.7 \times 10^{-10}$$

$$\frac{(3.1 \times 10^{-6})\,(3.1 \times 10^{-6})}{\left[NH_4^+\right]} = 5.7 \times 10^{-10}$$

$$\left[NH_4^+\right] = 0.0169\ M$$

$$\text{Mass of } NH_4Br \text{ needed per liter} = 0.0169\ \frac{mol}{L} \times 98.0\frac{g}{mol} = 1.65\ g.$$

Example 18.18

The pH of 0.02 M $Al(NO_3)_3$ is 3.3. Calculate the acidic ionization constant for the metal ion hydrolysis equilibrium reaction:

$$Al(H_2O)_6^{3+}(aq) \rightleftharpoons Al(OH)(H_2O)_5^{2+}(aq) + H^+(aq)$$

Solution

1. First write an equation for the reaction and an expression for K_a.

$$Al(H_2O)_6^{3+}(aq) \rightleftharpoons Al(OH)(H_2O)_5^{2+}(aq) + H^+(aq)$$

$$K_a = \frac{\left[Al(OH)(H_2O)_5^{2+}\right][H^+]}{\left[Al(H_2O)_6^{3+}\right]}$$

2. Determine the equilibrium concentrations of $Al(H_2O)_6^{3+}$, $Al(OH)(H_2O)_5^{2+}$ and $[H^+]$ in solution from the given pH.

$$pH = 3.3 = -\log[H_3O^+]$$

$$[H_3O^-] = 10^{-3.3} = 5.01 \times 10^{-4}\ M$$

$$[H_3O^-] = [Al(OH)(H_2O)_5^{2+}] = 5.01 \times 10^{-4}\ M$$

$$[Al(H_2O)_6^{3+}] = 0.02\ M - 0.000501\ M = 0.0195\ M$$

3. Calculate the equilibrium constant, K_a.

$$K_a = \frac{\left[Al(OH)(H_2O)_5^{2+}\right][H^+]}{\left[Al(H_2O)_6^{3+}\right]} = \frac{(5.01 \times 10^{-4})(5.01 \times 10^{-4})}{(0.0195)} = 1.3 \times 10^{-5}$$

18.11. The Common-Ion Effect

The *common-ion effect* is an application of Le Chatelier's principle. The dissociation of a weak acid or base is decreased when a solution of strong electrolyte that has an ion in common with the weak acid or base is added. For example, the addition of the strong electrolyte CH_3COONa to the weak acid CH_3COOH has the effect of decreasing the dissociation of the acid in solution. CH_3COOH is a weak acid and undergoes partial dissociation as follows:

$$CH_3COOH(aq) \rightleftharpoons CH_3COO^-(aq) + H^+(aq)$$

Sodium acetate, on the other hand, is a strong electrolyte and dissociates completely according to:

$$CH_3COONa(aq) \longrightarrow CH_3COO^-(aq) + Na^+(aq)$$

When sodium acetate is added to a solution of acetic acid, the acetate ion from the sodium acetate will increase the concentration of acetate ion. This will drive the equilibrium:

$$CH_3COOH(aq) \rightleftharpoons CH_3COO^-(aq) + H^+(aq)$$

to the left in accordance with Le Chatelier's principle. Therefore the degree of dissociation of CH_3COOH will be reduced. Such a shift in equilibrium due to the addition of an ion already involved in the reaction is referred to as the common-ion effect.

The pH of a solution containing a weak acid and an added common ion can be calculated in a manner similar to solutions containing only weak acids as previously described.

Example 18.19

Find the acetate ion ($C_2H_3O_2^-$) concentration and the pH of a solution containing 0.10 mol of HCl and 0.20 mol acetic acid ($HC_2H_3O_2$) in 1.0 L of solution.

Solution

1. First we write an equation for the reaction and an expression for K_a.

$$HC_2H_3O_2(aq) + H_2O(l) \rightleftharpoons C_2H_3O_2^-(aq) + H_3O^+(aq)$$

$$K_a = \frac{[C_2H_3O_2^-][H_3O^+]}{[HC_2H_3O_2]}$$

2. Determine the equilibrium concentrations of all species in solution including $[H^+]$ (see table 18-6).

Table 18-6 Data for Example 18.19

	$[HC_2H_3O_2]$	$[H_3O^+]$	$[C_2H_3O_2]$
Initial concentration (M)	0.20 M	0.10 M	0
Change (M)	$(-)x$	$(+)x$	$(+)x$
Equilibrium concentration (M)	$(0.20 - x)$	$(0.1 + x)$	x

Substituting in the equilibrium expression yields:

$$1.8 \times 10^{-5} = \frac{(0.10 + x)(x)}{(0.20 - x)}$$

Since HCl is much stronger than $HC_2H_3O_2$, almost all the H^+ in solution will be from HCl. Therefore $[H^+]$ will be approximately equal to 0.10 M, and x will be very small.
Hence:

$$1.8 \times 10^{-5} = \frac{(0.10)(x)}{(0.20)} \quad \text{or} \quad x = 3.6 \times 10^{-5}\ \text{M} = \left[C_2H_3O_2^-\right]$$

$$\left[H_3O^+\right] = (0.10 + x)\ \text{M} \approx 0.10\ \text{M}$$

Calculate the pH from $[H^+]$.

$$\text{pH} = -\log\left[H_3O^+\right] = -\log(0.10) = 1.0$$

18.12. Buffers and pH of Buffer Solutions

A buffer is a solution that resists changes in pH upon the addition of small amounts of acid or base. That is, a buffer solution keeps its pH almost constant. Buffers play a significant role in helping to maintain a constant pH in many biological and chemical processes. Human blood has a pH of about 7.4. If this value drops below 7.0, death can result by a process known as acidosis. On the other hand, if the value rises above 7.7, death can also result by a process known as alkalosis. The good news is that our blood has a buffer system that maintains the pH at the proper level.

A buffer can be prepared either by mixing a weak acid with a salt of the weak acid or by mixing a weak base with a salt of the weak base. Some examples of buffers are:

1. Acetic acid and sodium acetate (CH_3COOH/CH_3COONa).
2. Carbonic acid and sodium carbonate (H_2CO_3/Na_2CO_3).
3. Ammonium hydroxide and ammonium chloride (NH_4OH/NH_4Cl).

The reason buffers resist changes in pH is that they contain both an acidic species to neutralize added OH^- ions and a basic species to neutralize added H^+ ions.

18.12.1. How does a buffer work?

To understand how a buffer works, consider the buffer system of a weak acid, HA, and its salt, MA. The dissociation equilibrium is:

$$HA(aq) \rightleftharpoons H^+(aq) + A^-(aq)$$

For this reaction, K_a is:

$$K_a = \frac{[H^+][A^-]}{[HA]} \quad \text{and} \quad [H^+] = K_a \frac{[HA]}{[A^-]}$$

This expression indicates that the $[H^+]$, and hence the pH, depends on K_a and the ratio $[HA]/[A^-]$.

If a small amount of a base (OH^-) is added to the buffer, the OH^- ions will react with the acid, HA. As some of the HA is consumed, more of the conjugate base $[A^-]$ will be produced:

$$OH^-(aq) + HA(aq) \rightleftharpoons H_2O(l) + A^-(aq)$$

The hydrogen ion concentration, $[H^+]$, and the pH will not change as long as the value of the ratio $[HA]/[A^-]$ does not deviate greatly from the original value.

Similarly, if a small amount of acid (H^+) is added to the buffer, the H^+ will react with the conjugate base, A^-, and increase the concentration of the weak acid:

$$H^+(aq) + A^-(aq) \rightleftharpoons HA(aq)$$

However, the $[H^+]$ and hence the pH will not change as long as the value of the ratio $[HA]/[A^-]$ stays close to the original value.

18.12.2. Buffer capacity and pH

No buffer has unlimited capacity to take abuse before it is rendered useless. At some point a buffer loses its ability to function. The *buffer capacity* of a buffer is the amount of acid or a base that the solution can absorb before a significant change in pH will occur. The capacity of a buffer is directly related to the number of moles of weak acid and weak base in the system. The capacity is destroyed once these species (weak acid and weak base) are consumed.

In mathematical terms we can define buffer capacity (β) as:

$$\text{Buffer capacity } (\beta) = \frac{\Delta C_b}{\Delta \text{pH}} = -\frac{\Delta C_a}{\Delta \text{pH}}$$

where C_a and C_b are the concentrations of acid and base, respectively. The minus sign is indicative of the fact that the addition of an acid causes a decrease in pH. The unit of buffer capacity is mole per liter (M). A quick and handy formula for buffer capacity is:

$$\beta = \frac{2.303\, C_{HA} C_{A^-}}{(C_{HA} + C_{A^-})}$$

The derivation of this formula is beyond the scope of this book.

Example 18.20

What is the buffer capacity of a solution containing 0.20 M benzoic acid and 0.20 M sodium benzoate? If KOH is added such that the solution becomes 0.01 M in KOH, what will be the change in pH?

Solution

For the first part of the question substitute directly into the equation:

$$\beta = \frac{2.303 C_{HA} C_{A^-}}{(C_{HA} + C_{A^-})} = \frac{2.303\,(0.2)\,(0.2)}{(0.2 + 0.2)} = 0.23 \text{ M}.$$

For the second part of the question calculate the change in pH using the equation

$$\text{Buffer capacity } (\beta) = \frac{\Delta C_b}{\Delta \text{pH}}$$

$$0.23 = \frac{\Delta C_b}{\Delta \text{pH}} = \frac{0.01}{\Delta \text{pH}}$$

$$\Delta \text{pH} = \frac{0.01}{0.230} = 0.043$$

Buffer capacity depends on the number of moles of the weak acid and conjugate base present in the solution.

18.12.3. The Henderson–Hasselbalch equation

The pH of a buffer solution depends on the K_a of the acid as well as the relative concentrations of the weak acid and its conjugate base in the solution. We can derive an equation that shows the relationship between conjugate acid–base concentrations, pH and K_a.

Consider the buffer consisting of a weak acid HA and its salt, MA. The dissociation equilibrium is:

$$HA(aq) \rightleftharpoons H^+(aq) + A^-(aq)$$

For this reaction, K_a is:

$$K_a = \frac{[H^+][A^-]}{[HA]} \quad \text{and} \quad [H^+] = K_a \frac{[HA]}{[A^-]}$$

In general form:

$$[H^+] = K_a \frac{[\text{Acid}]}{[\text{Conjugate base}]}$$

Taking the negative log of both sides of the above equation gives the so-called Henderson–Hasselbalch equation:

$$\text{pH} = \text{p}K_a + \log\frac{[\text{Base, A}^-]}{[\text{Acid, HA}]}$$

or:

$$\text{pH} = \text{p}K_a - \log\frac{[\text{Acid, HA}]}{[\text{Base, A}^-]}$$

The Henderson–Hasselbalch equation can also be written in terms of pOH and $\text{p}K_b$:

$$\text{pOH} = \text{p}K_b + \log\frac{[\text{Conjugate cation}]}{[\text{Base}]}$$

or:

$$\text{pOH} = \text{p}K_b - \log\frac{[\text{Base}]}{[\text{Conjugate cation}]}$$

Example 18.21

What is the pH of a solution containing 0.10 M H_3BO_3 and 0.20 M NaH_2BO_3? For boric acid, $K_a = 7.3 \times 10^{-10}$.

Solution

There are two ways of solving this problem. We can use the Henderson–Hasselbalch equation, or solve for $[H^+]$ and then the pH.

Method 1

We will first determine the $\text{p}K_a$ from the given K_a.

$$\text{p}K_a = -\log K_a = -\log\left(7.3 \times 10^{-10}\right) = 9.14$$

Now calculate the pH by substituting the $\text{p}K_a$ value and known concentrations into the Henderson–Hasselbalch equation:

$$\text{pH} = \text{p}K_a + \log\frac{[\text{Base, } H_2BO_3^-]}{[\text{Acid, } H_3BO_3]}$$

$$\text{p}K_a = 9.14, \quad [H_2BO_3^-] = 0.20\text{ M}, \quad [H_3BO_3] = 0.10\text{ M}$$

$$\text{pH} = 9.14 + \log\frac{(0.20)}{(0.10)} = 9.14 + 0.30 = 9.44$$

Table 18-7 Data for Example 18.21

	$[H_3BO_3]$	$[H_3O^+]$	$[H_2BO_3^-]$
Initial concentration (M)	0.1	0	0.2
Change (M)	$(-)x$	$(+)x$	$(+)x$
Equilibrium concentration (M)	$(0.10-x)$	x	$(0.2+x)$

Method 2

1. First we the write equation for the reaction and an expression for K_a.

$$H_3BO_3(aq) + H_2O(l) \rightleftharpoons H_2BO_3^-(aq) + H_3O^+(aq)$$

$$K_a = \frac{[H_2BO_3^-][H_3O^+]}{[H_3BO_3]}$$

2. Determine the equilibrium concentrations of all species in solution including $[H^+]$ (see table 18-7).

$$K_a = \frac{(x)(x+0.2)}{(0.10-x)} = 7.3 \times 10^{-10}$$

Since we have a common ion in the solution, x will be very small compared to 0.10 or 0.20 M. Therefore we can use the approximation method to solve for x.

$$\frac{(x)(0.2)}{(0.10)} = 7.3 \times 10^{-10} \quad \text{or} \quad x = 3.6 \times 10^{-10}$$

(So the assumption is justified.)

$$[H_3O^+] = 3.6 \times 10^{-10}\text{ M}$$

3. Calculate the pH from $[H^+]$:

$$\text{pH} = -\log[H_3O^+] = -\log\left(3.6 \times 10^{-10}\right) = 9.4$$

Example 18.22

The pH of a solution containing an HCN and NaCN buffer system is 9.50. What is the ratio of $[CN^-]/[HCN]$ in the buffer? The K_a for HCN is 6.2×10^{-10}.

Solution

We will first determine the pK_a from the given K_a.

$$\text{p}K_a = -\log K_a = -\log\left(6.2 \times 10^{-10}\right) = 9.2$$

Now calculate the [base]/[acid] ratio by substituting the pH and pK_a values into the Henderson–Hasselbalch equation:

$$pH = pK_a + \log\frac{[\text{Base}, CN^-]}{[\text{Acid}, HCN]}$$

$$pH = 9.5, pK_a = 9.2$$

$$9.5 = 9.2 + \log\frac{[CN^-]}{[HCN]}$$

$$\log\frac{[CN^-]}{[HCN]} = 0.30$$

$$\frac{[CN^-]}{[HCN]} = 10^{0.30} = 2.0$$

18.13. Polyprotic Acids and Bases

A polyprotic acid such as phosphoric acid or tartaric acid has more than one ionizable hydrogen atom per molecule. In solution, polyprotic acids undergo successive dissociation, or react with a base stepwise for each proton. For example, phosphoric acid contains three protons (triprotic acid), and it dissociates as follows:

$$H_3PO_4(aq) \rightleftharpoons H^+(aq) + H_2PO_4^-(aq) \quad K_{a_1} = \frac{[H^+][H_2PO_4^-]}{[H_3PO_4]} = 7.5 \times 10^{-3}$$

$$H_2PO_4^-(aq) \rightleftharpoons H^+(aq) + HPO_4^{2-}(aq) \quad K_{a_2} = \frac{[H^+][HPO_4^{2-}]}{[H_2PO_4^-]} = 6.2 \times 10^{-8}$$

$$HPO_4^{2-}(aq) \rightleftharpoons H^+(aq) + PO_4^{3-}(aq) \quad K_{a_3} = \frac{[H^+][PO_4^{3-}]}{[HPO_4^{2-}]} = 4.8 \times 10^{-13}$$

Here, K_{a_1}, K_{a_2}, and K_{a_3} represent the first, second, and third ionization constants. For any polyprotic acid, the ionization constants always decrease in the order $K_{a_1} > K_{a_2} > K_{a_3}$.

The overall K_a value for a polyprotic acid can be found by multiplying the K_a values for the individual ionization steps. For phosphoric acid:

$$K_a = K_{a_2}K_{a_2}K_{a_3} = \frac{[H^+]\cancel{[H_2PO_4^{2-}]}}{[H_3PO_4]} \times \frac{[H^+]\cancel{[HPO_4^{2-}]}}{\cancel{[H_2PO_4^-]}} \times \frac{[H^+][PO_4^{3-}]}{\cancel{[HPO_4^{2-}]}}$$

$$K_a = \frac{[H^+]^3[PO_4^{3-}]}{[H_3PO_4^{2-}]} = 2.23 \times 10^{-22}$$

This represents the overall reaction:

$$H_3PO_4(aq) + H_2O(l) \rightleftharpoons 3\ H_3O^+(aq) + PO_4^{3-}(aq)$$

Similarly, a diprotic base such as Na_2CO_3 (or CO_3^{2-}) will have two ionization constants:

$$HCO_3^-(aq) + H_2O(l) \rightleftharpoons H_2CO_3(aq) + OH^-(aq)$$

$$K_{b_1} = \frac{[H_2CO_3][OH^-]}{[HCO_3^-]} = 2.33 \times 10^{-8}$$

$$CO_3^{2-}(aq) + H_2O(l) \rightleftharpoons HCO_3^-(aq) + OH^-(aq)$$

$$K_{b_2} = \frac{[HCO_3^-][OH^-]}{[CO_3^{2-}]} = 1.78 \times 10^{-4}$$

Example 18.23

Calculate the concentrations of H^+ and S^{2-} in a solution of 0.100 M H_2S. K_{a_1} and K_{a_2} for H_2S are 1.0×10^{-7} and 1.2×10^{-13}, respectively.

Solution

1. Write equations for the dissociation reactions and expressions for the equilibrium dissociation constants:

$$H_2S \rightleftharpoons H^+ + HS^- \quad K_{a_1} = \frac{[H^+][HS^-]}{[H_2S]} = 1.0 \times 10^{-7}$$

$$HS^- \rightleftharpoons H^+ + S^{2-} \quad K_{a_2} = \frac{[H^+][S^{2-}]}{[HS^-]} = 3.0 \times 10^{-13}$$

2. To calculate the hydrogen ion concentration we assume that all the hydrogen ions come from the first or primary reaction since K_{a1} is much larger than K_{a_2}. At equilibrium, let $[H^+] = [HS^-] = x$. Then $[H_2S] = 0.10 - x$. Using the approximation method, $[H_2S] \approx 0.10$ M since K_{a_1} is very small.
3. Substitute these values in the K_1 expression and solve for x.

$$K_{a_1} = \frac{[H^+][HS^-]}{[H_2S]} = \frac{x^2}{(0.10)} = 1.0 \times 10^{-7}$$

$$x = \sqrt{(1.0 \times 10^{-7})} = 0.00010 \quad \text{or} \quad 1.0 \times 10^{-4}\ \text{M}$$

Thus the concentration of H^+ in the solution is 1.0×10^{-4} M.

4. To calculate the concentration of S^{2-} in the solution, use the second equilibrium equation above.

$$K_2 = \frac{[H^+][S^{2-}]}{[HS^-]} = 3.0 \times 10^{-13}$$

From step 1 above:

$$[H^+] = [HS^-] = 1.0 \times 10^{-4}\ M$$

$$3.0 \times 10^{-13} = \frac{1.0 \times 10^{-4}[S^{2-}]}{1.0 \times 10^{-4}}$$

$$[S^{2-}] = 3.0 \times 10^{-13}\ M$$

Thus, concentration of S^{2-} in the solution is 3.0×10^{-13} M.

Example 18.24

Calculate the concentrations of all species present at equilibrium in 0.10 M H_3PO_4. K_{a_1}, K_{a_2}, and K_{a_3} for H_3PO_4 are 7.5×10^{-3}, 6.2×10^{-8}, and 4.8×10^{-13}, respectively.

Solution

(a) Write chemical equations for all the dissociation reactions:

$$H_3PO_4(aq) \rightleftharpoons H^+(aq) + H_2PO_4^-(aq) \quad K_{a_1} = \frac{[H^+][H_2PO_4^-]}{[H_3PO_4]} = 7.5 \times 10^{-3}$$

$$H_2PO_4^-(aq) \rightleftharpoons H^+(aq) + HPO_4^{2-}(aq) \quad K_{a_2} = \frac{[H^+][HPO_4^{2-}]}{[H_2PO_4^-]} = 6.2 \times 10^{-8}$$

$$HPO_4^{2-}(aq) \rightleftharpoons H^+(aq) + PO_4^{3-}(aq) \quad K_{a_3} = \frac{[H^+][PO_4^{3-}]}{[HPO_4^{2-}]} = 4.8 \times 10^{-13}$$

(b) Assume that H^+ comes mainly from the initial stage of the dissociation. Also assume that the concentration of any anion formed at one step of dissociation is not decreased by subsequent dissociation steps. Therefore we can say from the first equation:

$$H_3PO_4(aq) \rightleftharpoons H^+(aq) + H_2PO_4^-(aq) \quad K_{a_1} = \frac{[H^+][H_2PO_4^-]}{[H_3PO_4]} = 7.5 \times 10^{-3}$$

$$[H^+] = [H_2PO_4^-] = x$$

Substitute in the K_{a_1} expression:

$$7.5 \times 10^{-3} = \frac{x^2}{(0.10 - x)} \approx \frac{x^2}{(0.10)} \quad \text{or} \quad x = 0.0274\ \text{M}$$

This is more than 20% of the original $[H_3\ PO_4]$, which means our simplifying assumption may be invalid.

If we solve for x in the equation $x^2 + 7.5 \times 10^{-4}x - 7.5 \times 10^{-4} = 0$ using the quadratic formula

$$x = \frac{-b \pm \sqrt{b^2 - 4ac}}{2a}$$

with $a = 1$, $b = 7.5 \times 10^{-4}$, and $c = -7.5 \times 10^{-4}$ we get the same answer, i.e. $x = 0.027$ M. Thus:

$$[H^+] = [H_2PO_4^-] = 0.027\ \text{M}$$

(c) Use the values for $[H^+]$ and $[H_2PO_4^-]$ above to solve for $[HPO_4^{2-}]$:

$$H_2PO_4^-(aq) \rightleftharpoons H^+(aq) + HPO_4^{2-}(aq) \quad K_{a_2} = \frac{[H^+][HPO_4^{2-}]}{[H_2PO_4^-]} = 6.2 \times 10^{-8}$$

$$[HPO_4^{2-}] = K_{a_2}\frac{[H_2PO_4^-]}{[H^+]} = \frac{(6.2 \times 10^{-8})(0.027)}{(0.027)} = 6.2 \times 10^{-8}\text{M}$$

(d) Finally, use the values for $[H^+]$ and $[HPO_4^{2-}]$ above to solve for $[PO_4^{3-}]$:

$$HPO_4^{2-}(aq) \rightleftharpoons H^+(aq) + PO_4^{3-}(aq) \quad K_{a_3} = \frac{[H^+][PO_4^{3-}]}{[HPO_4^{2-}]} = 4.8 \times 10^{-13}$$

$$[PO_4^{3-}] = K_{a_3}\frac{[HPO_4^{2-}]}{[H^+]} = \frac{(4.8 \times 10^{-13})(6.2 \times 10^{-8})}{(0.027)} = 1.10 \times 10^{-18}\ \text{M}$$

18.14. More Acid–Base Titration

The progressive addition of a measured volume of an acid of known concentration (standard acid) to a solution of a base of unknown concentration (or vice versa) is called a *titration* (see chapter 14). When a solution of an acid such as HCl is titrated with a base such NaOH, the reaction that occurs is:

$$HCl(aq) + NaOH(aq) \longrightarrow NaCl(aq) + H_2O(l)$$

or simply:

$$H^+(aq) + OH^-(aq) \longrightarrow H_2O(l)$$

The *equivalence point* is the point at which the moles of base added are exactly equal to the moles of hydrogen ion in the original solution. It terms of chemical equivalence, it can be defined as the point in a titration at which stoichiometrically equivalent quantities of reactants (acid and base) have been mixed.

The *end point* is the point in a titration at which the indicator changes color. It is the point at which the titration is terminated. Although the end point sometimes coincides with the equivalence point, they are not necessarily the same thing. However, indicators for a given titration are selected to ensure the end point is as close as possible to the equivalence point.

18.14.1. Acid–base indicators

Indicators are organic compounds used to detect equivalence points, i.e., the point at which the reaction is complete in an acid–base titration. Indicators change colors over a characteristic short range of pH. Table 18-8 lists some common acid–base indicators, their colors in acidic and basic media, pH ranges, and pK_a values.

Table 18-8 is a guide to selecting a suitable indicator for an acid–base titration experiment. The first thing you want to do is to determine or estimate the pH of the

Table 18-8 Some Acid–Base Indicators

Indicator	*Acid color*	*Base color*	*pH range*	*pK_a*
Methyl orange	Red	Yellow	3.2–4.4	3.4
Bromophenol blue	Yellow	Blue	3.0–4.6	3.9
Methyl red	Red	Yellow	4.8–6.0	5
Bromthymol blue	Yellow	Blue	6.0–7.6	7.1
Litmus	Red	Blue	5.0–8.0	6.5
Thymol blue	Yellow	Blue	8.0–9.6	8.9
Phenolphthalein	Colorless	Pink	8.2–10.0	9.4
Alizarin yellow	Yellow	Red	10.1–12.0	11.2

solution (titration mixture) at the equivalence point. Then select an indicator that changes color at or close to this pH.

Acid–base indicators are weak organic acids, HIn, or to a lesser degree weak organic bases, InOH. The "In" in the formula represents the complex organic group. For a weak indicator HIn the following equilibrium is established with its conjugate base at equilibrium:

$$\underset{\text{(color A)}}{\mathrm{HIn\,(aq)}} + \mathrm{H_2O(l)} \rightleftharpoons \mathrm{H_3O^+(aq)} + \underset{\text{(color B)}}{\mathrm{In^-(aq)}}$$

The acid (HIn) and its conjugate base (In^-) have different colors. At low pH values, the concentration of H_3O^+ is high. Consequently the equilibrium position lies on the left-hand side of the equation and the solution exhibits color A. At high pH values, the concentration of H_3O^+ is low and as a result the equilibrium position lies to the right with the solution exhibiting color B.

The equilibrium constant for the above acid–base indicator equilibrium is:

$$K_{\mathrm{In}} = \frac{[\mathrm{H_3O^+}][\mathrm{In^-}]}{[\mathrm{HIn}]}$$

K_{In} is known as the *indicator dissociation constant*. We will assume that the change between the two colored forms HIn and In^- occurs when we have equal amounts of each. The transition or turning point is when $[\mathrm{HIn}] = [\mathrm{In^-}]$. At this point, substituting into the K_{In} expression yields:

$$K_{\mathrm{In}} = \frac{[\mathrm{H_3O^+}]\cancel{[\mathrm{In^-}]}}{\cancel{[\mathrm{HIn}]}} = [\mathrm{H_3O^+}]$$

$$\mathrm{p}K_{\mathrm{In}} = -\log K_{\mathrm{In}} = -\log[\mathrm{H_3O^+}] = \mathrm{pH}$$

Thus the pH of the solution at its turning point is known as the $\mathrm{p}K_{\mathrm{In}}$. At this pH, half of the indicator is in its acid form and the other half in the form of its conjugate base.

The equilibrium constant expression

$$K_{\mathrm{In}} = \frac{[\mathrm{H_3O^+}][\mathrm{In^-}]}{[\mathrm{HIn}]}$$

can be rearranged in ratio form as follows:

$$\frac{[\mathrm{In^-}]}{[\mathrm{HIn}]} = \frac{K_{\mathrm{In}}}{[\mathrm{H_3O^+}]}$$

Thus the ratio $[\mathrm{In^-}]/[\mathrm{HIn}]$ is inversely proportional to $[\mathrm{H_3O^+}]$ and is responsible for the color exhibited by the indicator.

The detailed behavior of acid–base indicators depends on several considerations besides the pH of the solution, including:

- The nature of the solvent
- Ionic strength
- Temperature
- Colloidal particulates, which may interfere through surface adsorption of the indicator

18.14.2. Perception of indicator color change

The human eye typically responds to only dramatic color changes. Those that are less than 10% are hardly seen. Thus, for the color change of an indicator to be seen clearly, 90% of the indicator must be in one form. At a pH equal to an indicator's pK_a, the indicator is 50% ionized, that is, the two forms are present in equal amounts. If the pH of the solution is 1 unit above the pK_a, 90% of the ionizable indicator will be in its basic form. On the other hand, if the pH of the solution is 1 unit below the pK_a, 90% of the ionizable indicator will be in its acidic form. Hence indicators show a color transition 1 pH unit above or below their pK_a. This is why indicators are generally selected based on the proximity of their pK_a to pH at the titration endpoint. For the solution to show the color due to HIn:

$$\frac{[In^-]}{[HIn]} \leq 0.10$$

and to see the In^- color,

$$\frac{[In^-]}{[HIn]} \geq 0.10$$

Example 18.25

A certain indicator, HIn, has $K_{In} = 5.25 \times 10^{-10}$. Calculate the pH of the solution when the ratio $[In^-]/[HIn]$ is $\frac{1}{2}$.

Solution

1. Write an equation for the ionization of the indicator and an expression for the hydrogen ion concentration in terms of K_{In}.

$$HIn + H_2O(l) \rightleftharpoons H_3O^+ + In^-$$

$$K_{In} = \frac{[H_3O^+][In^-]}{[HIn]} \quad \text{or} \quad [H_3O^+] = K_{In}\frac{[HIn]}{[In^-]}$$

2. Solve for the hydrogen ion concentration by making appropriate substitutions:

$$[H_3O^+] = K_{In}\frac{[HIn]}{[In^-]} = \left(5.25 \times 10^{-10}\right)\left(\frac{2}{1}\right) = 1.05 \times 10^{-9}M$$

3. Calculate the pH.

$$pH = -\log[H_3O^+] = -\log\left(1.05 \times 10^{-9}\right) = 9.0$$

18.15. pH Titration Curves

An acid–base titration curve is a plot of the pH of a solution of acid (or base) versus the volume of base (or acid) added. Depending on the strength and concentrations of the acid and base involved, pH curves can have different characteristic shapes. For example, the titration curve for the titration of a strong acid by a strong base is different from the curve for the titration of a weak acid by a strong base. Titration curves can be used in selecting a good indicator for a given titration process. While a pH meter can determine changes in pH, it is also possible to calculate the pH of each point on the titration curve.

18.15.1. Titration of a strong acid against a strong base

The net reaction between a strong acid and a strong base is the formation of water:

$$H^+(aq) + OH^-(aq) \rightarrow H_2O(l)$$

The hydrogen ion concentration at any given point in the titration can be calculated from the amount of hydrogen ion remaining at that point divided by the total volume of the solution.

- The pH of the solution before the addition of the strong base is determined by the initial concentration of the strong acid.
- As the strong base is added to the strong acid, the pH values between the initial pH and the equivalence point are determined by the concentration of excess acid (unneutralized acid).
- The pH at the equivalence point is easy to determine. No calculation is required because at the equivalence point, equal amounts or numbers of moles of the acid and the base have reacted, resulting in the formation of a neutral salt. The pH is 7.0.
- After the equivalence point has been reached, the pH of the solution is determined by the concentration of excess OH^- in the solution.

The example below illustrates the steps involved in the calculation.

Example 18.26

A 50-mL sample of 0.10 M HCl is titrated with 0.10 M NaOH. Calculate the pH:

1. At the start of the titration
2. After 10 mL of NaOH has been added
3. After 25 mL of NaOH has been added
4. After 50 mL of NaOH has been added
5. After 70 mL of NaOH has been added.

Solution

(a) At the start of the titration, before NaOH is added, we have:

$$HCl + H_2O \xrightarrow{100\%} H_3O^+ + Cl^-$$

$$\left[H_3O^+\right] = 0.10\ \text{M}, \quad pH = 1.0$$

(b) After 10 mL of NaOH has been added:

$$HCl + NaOH \longrightarrow NaCl + H_2O$$

$$\text{Moles OH}^- \text{ added} = 0.010\,\text{L} \times 0.10\frac{\text{mol}}{\text{L}} = 0.0010\ \text{mol}$$

$$\text{Original moles of H}^+ = 0.050\,\text{L} \times 0.10\frac{\text{mol}}{\text{L}} = 0.0050\ \text{mol.}$$

All the OH^- is completely reacted leaving excess H^+. Note that the volume of solution has increased by 10 mL from the original 50 mL.

$$\text{Excess H}^+ = (0.005 - 0.001) = 0.004\ \text{mol}$$

$$\left[H^+\right] = \frac{0.004\ \text{mol}}{60\ \text{mL}} \times \frac{1000\ \text{mL}}{1\,\text{L}} = 0.0667\ \text{M}$$

$$pH = -\log(0.0667) = 1.18$$

(c) Repeat the calculation as above remembering to account for the changed volume of solution.

$$HCl + NaOH \longrightarrow NaCl + H_2O$$

$$\text{Moles OH}^- \text{ added} = 0.025\ \text{L} \times 0.10\frac{\text{mol}}{\text{L}} = 0.0025\ \text{mol}$$

$$\text{Original moles of H}^+ = 0.050\ \text{L} \times 0.10\frac{\text{mol}}{\text{L}} = 0.0050\ \text{mol}$$

All the OH^- is completely reacted leaving excess H^+. Excess H^+ = $(0.005 - 0.0025) = 0.0025$ mol, so

$$[H^+] = \frac{0.0025\text{ mol}}{75\text{ mL}} \times \frac{1000\text{ mL}}{1\text{ L}} = 0.033\text{ M}$$
$$pH = -\log(0.033) = 1.48$$

(d) After 50 mL of NaOH has been added:

$$HCl + NaOH \longrightarrow NaCl + H_2O$$

$$\text{Moles } OH^- \text{ added } = 0.050\text{ L} \times 0.10\frac{\text{mol}}{\text{L}} = 0.005\text{ mol}$$

$$\text{Original moles of } H^+ = 0.050\text{ L} \times 0.10\frac{\text{mol}}{\text{L}} = 0.0050\text{ mol}$$

At this point both OH^- and H^+ are completely consumed.

$$\text{Excess } H^+ = (0.005 - 0.005) = 0$$

The equivalence point is reached when 50 mL of NaOH has been added.

NaCl will be formed and pH = pOH = 7.

(e) After 70 mL of NaOH has been added, the excess NaOH will be calculated as follows:

$$\text{Excess } OH^- = 20\text{ mL} \times \frac{0.10\text{ mol}}{1\text{ L}} \times \frac{1\text{ L}}{1000\text{ mL}} = 0.0020\text{ mol}$$

This amount is present in 120 mL of solution:

$$[OH^-] = \frac{0.0020\text{ mol}}{0.120\text{ L}} = 0.0167\text{ M}$$
$$pOH = -\log[OH^-] = -\log(0.0167) = 1.78$$
$$pH = 12.22$$

18.15.2. Titration of a weak acid against a strong base

- At the start of the titration of a weak acid HA versus a strong base such as NaOH, the initial pH is the pH of the weak acid solution. As the strong base is gradually added to the weak acid, partial neutralization of the acid occurs, resulting in a mixture of the acid HA and its conjugate base A^-.
- To calculate the pH prior to the equivalence point, we first determine the concentrations of HA and A^-. From the values of K_a, [HA], and $[A^-]$, we can determine $[H^+]$, and hence the pH. This is simple with the Henderson–Hasselbalch equation.

- The equivalence point is reached after equal numbers of moles of NaOH and HA have reacted, resulting in the formation of sodium salt of the acid. Since A^- is a weak base, the pH at the equivalence point is normally above 7.
- After the equivalence point, the $[OH^-]$ derived from the hydrolysis of A^- is negligible compared to the $[OH^-]$ from excess NaOH. Hence the pH is determined by the concentration of OH^- from the excess NaOH.

The following example will illustrate the steps involved in the calculations.

Example 18.27

50 mL of 0.100 M acetic acid, $HC_2H_3O_2$, is titrated with 0.100 M NaOH. Calculate the pH of the solution when:

(a) 40 mL of 0.10 M NaOH has been added
(b) 50 mL of 0.10 M NaOH has been added
(c) 70 mL of 0.10 M NaOH has been added
(K_a for acetic acid $= 1.8 \times 10^{-5}$.)

Solution

(a) Acetic acid is a weak acid. Use the following procedure to calculate the pH:
1. Write the equation for the reaction:

$$HC_2H_3O_2 + OH^- \rightleftharpoons C_2H_3O_2^- + H_2O$$

2. Calculate the moles of reactants before reaction.

$$\text{Moles NaOH} = (40\ \text{mL})\left(\frac{0.100\ \text{mol}}{1\ \text{L}}\right)\left(\frac{1\ \text{L}}{1000\ \text{mL}}\right) = 4.0 \times 10^{-3}\text{mol}$$

$$\text{Moles } HC_2H_3O_2 = (50\ \text{mL})\left(\frac{0.100\ \text{mol}}{1\ \text{L}}\right)\left(\frac{1\ \text{L}}{1000\ \text{mL}}\right) = 5.0 \times 10^{-3}\text{mol}$$

Once the NaOH solution is added, 0.004 mol of acetic acid is consumed by 0.004 mol of NaOH, and 0.004 mol of the acetate ion is produced. The excess acid is calculated as follows:

$$\text{Excess acid} = (0.005 - 0.004) = 0.001\ \text{mol}$$

$$\text{Total vol} = 40\ \text{mL} + 50\ \text{mL} = 90\ \text{mL} = 0.090\ \text{L}$$

$$[HC_2H_3O_2] = \left(\frac{0.001\ \text{mol}}{0.090\ \text{L}}\right) = 0.011\ \text{M}$$

$$\left[C_2H_3O_2^-\right] = \left(\frac{0.004\ \text{mol}}{0.090\ \text{L}}\right) = 0.044\ \text{M}$$

3. Using equilibrium conditions, calculate hydrogen ion concentration and then the pH.

$$K_a = 1.8 \times 10^{-5} = \frac{[H^+][C_2H_3O_2^-]}{[HC_2H_3O_2]} \quad \text{or}$$

$$[H^+] = \left(1.8 \times 10^{-5}\right) \frac{[HC_2H_3O_2]}{[C_2H_3O_2^-]}$$

$$[H^+] = \left(1.8 \times 10^{-5}\right) \left(\frac{0.011}{0.044}\right) = 4.5 \times 10^{-6} \text{ M}$$

$$pH = -\log\left(4.5 \times 10^{-6}\right) = 5.35$$

(b) Calculate the pH after 50 mL NaOH has been added.

At the equivalence point, moles of acetic acid = moles of NaOH, and the solution will contain essentially 5.0×10^{-3} mol sodium acetate. The following hydrolysis reaction will occur:

$$C_2H_3O_2^- + H_2O \rightleftharpoons HC_2H_3O_2 + OH^- \quad \text{and}$$

$$K_b = \frac{K_w}{K_a} = \frac{[OH^-][HC_2H_3O_2]}{[C_2H_3O_2^-]}$$

$$K_b = \frac{[OH^-][HC_2H_3O_2]}{[C_2H_3O_2^-]} = \frac{1 \times 10^{-14}}{1.8 \times 10^{-5}} = 5.56 \times 10^{-10}$$

$$[HC_2H_3O_2] = [OH^-] = x$$

$$[C_2H_3O_2^-] = \frac{0.005 \text{ mol}}{0.100 \text{ L}} = 0.050 \text{ M}$$

$$5.56 \times 10^{-10} = \frac{x^2}{0.050} \quad \text{or} \quad x = 5.27 \times 10^{-6}\text{M}$$

Thus $[OH^-] = 5.27 \times 10^{-6}$M and pOH = 5.28

$$pH = 8.72$$

(c) Calculate the pH after 70 mL NaOH has been added.

After the addition of 70 mL of 0.100 M NaOH:

$$\text{mol of NaOH added} = \left(\frac{70 \text{ mL} \times 0.100 \text{ mol}}{1 \text{ L}}\right)\left(\frac{1 \text{ L}}{1000 \text{ mL}}\right) = 0.0070 \text{ mol.}$$

This OH^- concentration is more than enough to neutralize the 0.0050 mol of acid initially present. The total volume of the solution is 50 mL + 70 mL = 120 mL = 0.120 L.

$$[CH_3COO^-] = \frac{0.005 \text{ mol}}{0.12 \text{ L}} = 0.042 \text{ M}$$

$$[OH^-] = \frac{(0.007 \text{ mol} - 0.005 \text{ mol})}{0.12 \text{ L}} = 0.067 \text{ M}$$

At this stage of the titration, $[OH^-]$ from the hydrolysis of CH_3COO^- is negligible compared with $[OH^-]$ from the excess NaOH. Therefore, the $[H_3O^+]$, and hence the pH, of the solution after the equivalence point can be calculated from the $[OH^-]$derived from excess NaOH.

$$[H_3O^+][OH^-] = K_w$$

$$[H_3O^+] = \frac{K_w}{[OH^-]} = \frac{1.0 \times 10^{-14}}{0.0167} = 6.0 \times 10^{-13} \text{M}$$

$$\text{pH} = -\log [H_3O^+] = 12.2$$

18.16. Problems

18.16.1. pH, pOH, and percent ionization

1. Calculate the pH of each of the following solutions:

   a. $[H^+] = 1.0 \times 10^{-5}$ M
   b. $[H^+] = 5.5 \times 10^{-2}$ M
   c. $[H^+] = 2.1 \times 10^{-11}$ M
   d. $[H^+] = 9.8 \times 10^{-7}$ M

2. Calculate the pH of each of the following solutions:

   a. $[OH^-] = 1.0 \times 10^{-5}$ M
   b. $[OH^-] = 5.5 \times 10^{-2}$ M
   c. $[OH^-] = 2.1 \times 10^{-11}$ M
   d. $[OH^-] = 9.8 \times 10^{-7}$ M

3. Calculate the hydrogen ion concentration in solutions with the following pH values:
   (a) 11.8 (b) 2 (c) 7.9 (d) 6.5 (e) 1

4. Calculate the hydroxide ion concentration in solutions with the following pH values:
   (a) 11.8 (b) 2 (c) 7.9 (d) 6.5 (e) 1

5. What is the pH of the following?

   (a) 6.3 g of HNO_3 in 1 L of solution
   (b) 0.56 g of KOH in 500 mL of solution
   (c) 4.9 g of H_2SO_4 in 250 mL of solution
   (d) 14.99 g of CsOH in 2 L of solution

6. Calculate the range of hydrogen ion concentrations in the following substances from their typical pH ranges:

   (a) Gastric juice (1.6–1.8)
   (b) Human urine (4.8–8.8)
   (c) Cow's milk (6.3–6.6)
   (d) Lime juice (1.8–2.1)
   (e) Human blood (7.35–7.45)
   (f) Milk of magnesia (9.9–10.1)
   (g) Tomatoes (4.0–4.4)

7. The hydrogen ion concentration in a sample of black coffee is 1.2×10^{-5} M. What is the hydroxide ion concentration in this coffee?

8. Commercial concentrated sulfuric acid labeled 98% (w/w) H_2SO_4 has a density of 1.84 g/mL.

   (a) Calculate the molarity of this (stock) solution
   (b) Calculate the pH of a solution prepared by diluting 1.50 mL of the stock solution to 2.00 L, assuming 100% ionization.

9. A chemistry student from Ambrose Alli University in a final year project decided to investigate the temperature dependence of pH in water. He collected water samples from Ikpoba River for his experiment. The hydrogen ion concentration in the water samples was determined at 25 and 100°C, respectively. To his surprise, at 100°C the hydrogen ion concentration in the water sample was found to be 1.05×10^{-6} M. This is about 10 times higher than the result he obtained at 25°C. What can he conclude from these data?

10. Determine the pH and pOH of water at 60°C given that at 60°C K_w has a value of 9.62×10^{-14}.

11. Calculate the pH of the following solutions (assume all have a density of 1.0 g/mL):

    (a) 5% (w/w) of HNO_3
    (b) 0.03% (w/w) of HBr
    (c) 1.0% (w/w) of NaOH
    (d) 0.035% (w/w) of RbOH

12. What are the equilibrium concentrations of benzoic acid ($C_6H_5CO_2H$), the benzoate ion, and H_3O^+ for a 0.020 M benzoic acid solution ($K_a = 6.3 \times 10^{-5}$)? What is the pH of the solution?

13. The pH of 0.00090 M formic acid (HCO_2H) is 3.5. Determine the K_a and the equilibrium concentration of HCO_2H.

14. Calculate the pH of 0.25 M aqueous acetic acid ($CH_3CO_2H, K_a = 1.8 \times 10^{-5}$).
15. Calculate the pH and pOH of a 0.10 M nitrous acid (HNO_2) solution ($K_a = 4.6 \times 10^{-4}$).
16. Calculate the pH of 0.20 M $CH_3NH_2 (K_b = 4.2 \times 10^{-4})$.
17. A solution is prepared by dissolving 0.34 g of ammonia gas in 1 L of water. Calculate the pH and pOH given that K_b for ammonia is 1.8×10^{-5}.
18. Calculate the concentration of OH^- in a 0.01 M solution of pyridine, C_5H_5N ($K_b = 1.7 \times 10^{-9}$).
19. What is the percent ionization of a 0.0010 M solution of phenol (C_6H_5OH, $K_a = 1.3 \times 10^{-10}$)?
20. Calculate the molar concentration of a benzoic acid ($C_6H_5CO_2H$) solution, if it is 5% ionized ($K_a = 6.3 \times 10^{-5}$).
21. In a 9.0×10^4 M HCO_2H solution, the concentration of HCO_2H is 3.16×10^4 M. Determine the degree of ionization.
22. At 25°C a 0.20 M solution of vitamin C (ascorbic acid) is 2.0 % dissociated. Calculate the equilibrium constant, K_a, for the reaction:

$$C_6H_8O_6 + H_2O \rightleftharpoons C_6H_7O_6^- + H_3O^+$$

18.16.2. Salt hydrolysis

23. Calculate the pH of a 0.10 M solution of sodium acetate. $K_a = 1.8 \times 10^{-5}$. (Note: you will need to determine the hydrolysis constant K_h from $K_h = K_w/K_a$.)
24. Calculate the pH and the degree of hydrolysis of a 0.20 M solution of sodium benzoate ($C_6H_5CO_2Na, K_h = 1.5 \times 10^{-10}$ at 25°C).
25. Calculate the pH of a 0.0350 M solution of $Al(NO_3)_3$ given that K_a for $Al(H_2O)_6^{3+}$ is 1.35×10^{-5}. Assume that the only reactions of importance are:

    (1) $Al(NO_3)_3(s) \rightarrow Al^{3+}(aq) + 3\ NO_3^-(aq)$
    (2) $Al(H_2O)_6^{3+}(aq) + H_2O(l) \rightleftharpoons Al(OH)(H_2O)_5^{2+}(aq) + H_3O^+(aq)$

26. The pH of a 0.30 M $Co(NO_3)_2$ solution is 3.0. Assuming the only significant hydrolysis is:

$$Co(H_2O)_6^{2+}(aq) + H_2O(l) \rightleftharpoons Co(OH)(H_2O)_5^+(aq) + H_3O^+(aq)$$

what is the hydrolysis constant?

18.16.3. Common ion effect

27. What is the hydrogen ion concentration in a solution made by adding 6.66 g of HCO_2Na to 1.00 L of 0.100 M formic acid (HCO_2H, $K_a = 1.8 \times 10^{-4}$)?

28. Calculate the pH of a solution containing 0.50 M acetic acid (CH_3CO_2H) and 0.50 M sodium acetate (CH_3CO_2Na). For acetic acid, CH_3CO_2H, $K_a = 1.8 \times 10^{-5}$.
29. Calculate the pH and the percent ionization of ammonia in a solution that is 0.10 M in NH_3 and 0.30 M NH_4Cl. $K_{b,\,NH_3} = 1.8 \times 10^{-5}$.
30. Calculate the concentration of all the species and the pH of a solution made by adding 0.065 mol $CH_3NH_3^+Cl^-$ to 1.0 L of 0.055 M CH_3NH_2. Note: $pK_b(CH_3NH_2) = 3.44$.

18.16.4. Buffers

31. What is the pH of the buffer prepared by making up a solution that is 0.25 M CH_3CO_2H and 0.30 M CH_3CO_2Na?$K_a(CH_3CO_2H) = 1.8 \times 10^{-5}$.
32. What $[CH_3CO_2H]/[CH_3CO_2^-]$ ratio is required for the buffer solution in problem 18.3 to maintain a pH of 7?
33. Provide a recipe for preparing a phosphate buffer with a pH of approximately 7.50. Which of the following equations will offer the most effective buffer system and why?

$$H_3PO_4(aq) \rightleftharpoons H^+(aq) + H_2PO_4^-(aq), \quad pK_{a1} = 2.12$$

$$H_2PO_4^-(aq) \rightleftharpoons H^+(aq) + HPO_4^{2-}(aq), \quad pK_{a2} = 7.21$$

$$HPO_4^{2-}(aq) \rightleftharpoons H^+(aq) + PO_4^{3-}(aq), \quad pK_{a3} = 12.32$$

34. A buffer solution was prepared by mixing equal volumes of a 0.2 M acid and its 0.75 M salt. The pH of the resulting buffer was measured to be 6.1. (a) Calculate K_a for the acid. (b) Is this a good buffer system for this pH?
35. The pK_a of benzoic acid is 4.18. How many moles of the salt sodium benzoate must be added to 1.0 L of 0.05 M benzoic acid to bring the pH of the solution to 3.8? (Assume no volume change by the addition of the salt.)
36. Calculate the pH of a solution prepared by adding 7.5 g of NaOH to 1.0 L of 0.50 M phenylboric acid ($C_6H_5H_2BO_3$). The pK_a of phenylboric acid is 8.86.
37. Determine the pH, $[HCO_3^-]$, and $[CO_3^{2-}]$ of a 0.05 M aqueous CO_2 solution at 25°C. The equilibrium reactions are:

$$CO_2(aq) + 2\,H_2O(l) \rightleftharpoons H_3O^+(aq) + HCO_3^-(aq), \quad K_{a1} = 2.0 \times 10^{-4}$$

$$HCO_3^{-1}(aq) + H_2O(l) \rightleftharpoons H_3O^+(aq) + CO_3^{2-}(aq), \quad K_{a2} = 5.6 \times 10^{-11}$$

38. The ionization of hydrogen sulfide, H_2S, is shown below:

$$H_2S \rightleftharpoons H^+ + HS^-, \quad K_1 = 1.0 \times 10^{-7}$$

$$HS^- \rightleftharpoons H^+ + S^{2-}, \quad K_2 = 1.0 \times 10^{-19}$$

What is the pH, $[HS^-]$, and $[S^{2-}]$ of 0.20 M aqueous H_2S at 25°C?

39. A solution saturated with tellurous acid undergoes the following equilibrium reactions:

$$H_2TeO_3 \rightleftharpoons H^+ + HTeO_3^-, \quad K_1 = 2.0 \times 10^{-3}$$

$$HTeO_3^- \rightleftharpoons H^+ + TeO_3^{2-}, \quad K_2 = 1.0 \times 10^{-8}$$

(a) What are the approximate values of $[H^+]$ and $[HTeO_3^-]$ in a 0.200 M solution of H_2TeO_3?
(b) Calculate the concentration of TeO_3^{2-} present at equilibrium.

18.16.5. Titration

40. Using table 18-1 as a guide, recommend an indicator for the following titrations:

(a) 0.10 M NaOH titrated with 0.10 M HCl
(b) 0.10 M NaOH titrated with formic acid 0.10 M (HCO_2H, $K_a = 1.8 \times 10^{-4}$)
(c) 0.10 M NH_3 titrated with 0.10 M HCl

41. At the equivalence point in the titration of 50.00 mL of 0.100 M (CH_3CO_2H) with 0.100 M NaOH, 0.0050 mol of CH_3CO_2Na) ($K_b = 5.6 \times 10^{-10}$) are formed in 100.00 mL of solution. (a) Calculate the pH for this point on the titration curve. (b) Recommend a suitable indicator for the titration.

42. For Alizarin yellow (HAY), $K_a = 6.3 \times 10^{-12}$. When the ratio $[AY^-]/[HAY] \leq \frac{1}{10}$, the human eye can just detect the red color. Calculate the highest pH at which the solution of the acid form (HAY) appears red.

43. In a titration experiment, 25.00 mL of 0.10 M HCl is titrated with 0.10 M NaOH. Calculate the pH of the solution formed when:

(a) 0.00 mL of NaOH is added
(b) 5.00 mL of NaOH is added
(c) 15.00 mL of NaOH is added
(d) 25.00 mL of NaOH is added
(e) 35.00 mL of NaOH is added.

44. In the titration of 25.00 mL of 0.10 M HCl by a 0.10 M KOH solution, calculate the pH of the titration mixture (a) before the addition of any acid, (b) after the addition of 20 mL HCl.

45. In the titration of 25 mL of 0.10 M acetic acid with 0.10 M NaOH, what is the pH of the titration mixture after

(a) 0.00 mL of 0.10 M NaOH is added
(b) 10.00 mL of 0.10 M NaOH is added
(c) 25.00 mL of 0.10 M NaOH is added
(d) 35.00 mL of 0.10 M NaOH is added.

46. In the titration of 25 mL of 0.10 M NH_3 with 0.10 M HCl, what is the pH of the titration mixture after:

    (a) 0.00 mL of 0.10 M HCl is added
    (b) 10.00 mL of 0.10 M HCl is added
    (c) 20.00 mL of 0.10 M HCl is added
    (d) 35.00 mL of 0.10 M HCl is added.

19

Solubility and Complex-Ion Equilibria

19.1. Solubility Equilibria

Solubility equilibria are heterogeneous equilibria that describe the dissolution and precipitation of slightly soluble ionic compounds. Examples are commonly encountered in many chemical and biological processes. This chapter deals with the application of chemical equilibrium to heterogeneous systems involving saturated solutions of slightly soluble ionic compounds.

19.2. The Solubility Product Principle

When a slightly soluble salt is dissolved in water, it eventually reaches a point of saturation where equilibrium is established between the undissolved salt and its solution. For example, a saturated solution of $BaSO_4$ involves this equilibrium:

$$BaSO_4(s) \rightleftharpoons Ba^{2+}(aq) + SO_4^{2-}(aq)$$

We can apply the law of chemical equilibrium to this system and obtain the expression

$$K_{eq} = \frac{\left[Ba^{2+}\right]\left[SO_4^{2-}\right]}{[BaSO_4]}$$

Since the concentration of solid $BaSO_4$ remains essentially constant, the product $K_{eq} \times [BaSO_4]$ will also be constant. Therefore, we can write:

$$K_{eq}[BaSO_4] = \left[Ba^{2+}\right]\left[SO_4^{2-}\right] = K_{sp}$$

The equilibrium constant K_{sp} is called the *solubility product constant*, or simply *solubility product.*

For the general solubility equilibrium

$$A_xB_y(s) \rightleftharpoons xA^{n+}(aq) + yB^{m-}(aq)$$

The equilibrium constant expression is:

$$K_{sp} = [A^{n+}(aq)]^x [B^m - (aq)]^y$$

19.2.1. Definition

The solubility product constant is equal to the product of the concentrations of the constituent ions involved in the equilibrium, each raised to the power corresponding to its coefficient in the balanced equilibrium equation.

Example 19.1

Write the expression for the solubility product constant, K_{sp}, for the following salts:

(a) $CaF_2(s)$
(b) $Ag_2CrO_4(s)$
(c) $Bi_2S_3(s)$

Solution

(a) $CaF_2(s) \rightleftharpoons Ca^{2+}(aq) + 2F^-(aq) \qquad K_{sp} = [Ca^{2+}][F^-]^2$

(b) $Ag_2CrO_4(s) \rightleftharpoons 2Ag^+(aq) + CrO_4^{2-}(aq) \qquad K_{sp} = [Ag^+]^2[CrO_4^{2-}]$

(c) $Bi_2S_3(s) \rightleftharpoons 2Bi^{3+}(aq) + 3S^{2-}(aq) \qquad K_{sp} = [Bi^{3+}]^2[S^{2-}]^3$

19.3. Determining K_{sp} from Molar Solubility

The value of K_{sp} for a salt whose ions do not react appreciably with water can be determined if the solubility of the salt is known. Sometimes the concentration of the ions in solution or solubility of a salt is expressed in mass per unit volume of water. All mass concentrations must be converted to moles per liter before substituting in the K_{sp} expression.

Example 19.2

The solubility of magnesium carbonate at room temperature is 0.2666 g/L. Calculate the solubility product for this salt.

Solution

1. Write the chemical equation and K_{sp} expression for the dissolution of $MgCO_3$:

$$MgCO_3(s) \rightleftharpoons Mg^{2+}(aq) + CO_3^{2-}(aq)$$

$$K_{sp} = \left[Mg^{2+}\right]\left[CO_3^{2-}\right]$$

2. A solution of $MgCO_3$ contains equal amounts of Mg^{2+} and CO_3^{2-}, which implies that when the solution is saturated, $\left[Mg^{2+}\right] = \left[CO_3^{2-}\right]$ = molar solubility, x.
3. Calculate the solubility of $MgCO_3$ in moles per liter.

$$\text{Solubility in moles per liter} = \left(\frac{0.2666\text{ g } MgCO_3}{1\text{ L}}\right)\left(\frac{1\text{ mol } MgCO_3}{84.3\text{ g } MgCO_3}\right)$$

$$= 3.2 \times 10^{-3}\text{ M}$$

Hence, $\left[Mg^{2+}\right] = \left[CO_3^{2-}\right] = 3.2 \times 10^{-3}$ M.

4. Now substitute for x in the K_{sp} expression:

$$K_{sp} = \left[Mg^{2+}\right]\left[CO_3^{2-}\right] = (x)(x) = \left(x^2\right)$$

$$= \left(3.2 \times 10^{-3}\right)^2 = 1.0 \times 10^{-5}$$

Example 19.3

A saturated solution of BaF_2 prepared by dissolving solid BaF_2 in water has $\left[F^-\right] = 1.0 \times 10^{-3}$M. What is the value of K_{sp} for BaF_2?

Solution

1. Write the chemical equation and K_{sp} expression for the dissolution of BaF_2.

$$BaF_2(s) \rightleftharpoons Ba^{2+}(aq) + 2F^-(aq)$$

$$K_{sp} = \left[Ba^{2+}\right]\left[F^-\right]^2$$

2. Two moles of fluoride ions are formed for each mole of BaF_2 that dissolves. Thus:

$$\left[F^-\right] = 2\left[Ba^{2+}\right]$$

Since $\left[F^-\right] = 1.0 \times 10^{-3}$ M, $\left[Ba^{2+}\right] = 5.0 \times 10^{-4}$ M.

3. Substitute these values in the K_{sp} expression:

$$K_{sp} = \left[Ba^{2+}\right][F^-]^2 = \left(5.0 \times 10^{-4}\right)\left(1.0 \times 10^{-3}\right)^2$$
$$= 5.0 \times 10^{-10}$$

Example 19.4

The solubility of silver sulfate is 0.020 M at 25°C. Calculate the solubility product constant.

Solution

1. Write the chemical equation and K_{sp} expression for the dissolution of Ag_2SO_4:

$$Ag_2SO_4(s) \rightleftharpoons 2Ag^+(aq) + SO_4^{2-}(aq)$$
$$K_{sp} = [Ag^+]^2\left[SO_4^{2-}\right]$$

2. Two moles of silver ions are formed for each mole of Ag_2SO_4 that dissolves. Thus:

$$[Ag^+] = 2[Ag_2SO_4] = 2\left[SO_4^{2-}\right]$$
$$\text{Since } \left[SO_4^{2-}\right] = 2.0 \times 10^{-2}\ \text{M}, [Ag^+] = 4.0 \times 10^{-2}\ \text{M}$$

3. Substitute these values in the K_{sp} expression.

$$K_{sp} = [Ag^-]^2\left[SO_4^{2-}\right] = \left(4.0 \times 10^{-2}\right)^2\left(2.0 \times 10^{-2}\right)$$
$$= 3.2 \times 10^{-5}$$

19.4. Calculating Molar Solubility from K_{sp}

The solubility of a slightly soluble salt in moles per liter can be readily calculated if the solubility product constant is known.

Example 19.5

The solubility product constant of lead iodide in water is 7.1×10^{-9}. Calculate the solubility in (a) moles per liter, (b) grams per liter.

Solution

1. Write the chemical equation and K_{sp} expression for the dissolution of PbI_2:

$$PbI_2(s) \rightleftharpoons Pb^{2+}(aq) + 2\,I^-(aq)$$

$$K_{sp} = \left[Pb^{2+}\right]\left[I^-\right]^2 = 7.1 \times 10^{-9}$$

2. Let x be the solubility of PbI_2 in moles per liter. Note that 2 mol iodide ions and 1 mol of lead ions are formed for each mole of the salt that dissolves. Therefore:

$\left[Pb^{2+}\right] = x$ and $\left[I^-\right] = 2x$

$PbI_2(s) \rightleftharpoons Pb^{2+}_{x}(aq) + 2\,I^-_{2x}(aq)$

$$K_{sp} = \left[Pb^{2+}\right]\left[I^-\right]^2 = (x)(2x)^2 = 4x^3 = 7.1 \times 10^{-9}$$

$$x^3 = 1.18 \times 10^{-9}$$

$$x = 1.21 \times 10^{-3}$$

(a) The solubility of PbI_2 is 1.21×10^{-3} mol/L.

(b) Solubility in g/L $= \left(\frac{1.21 \times 10^{-3}\ \text{mol } PbI_2}{1\ \text{L}}\right)\left(\frac{334.2\ \text{g } PbI_2}{1\ \text{mol } PbI_2}\right)$

$= 0.40$ g/L

Example 19.6

Magnesium hydroxide is used as an antacid and laxative. (a) How many moles of $Mg(OH)_2$ would dissolve in 1.0 L of water? (b) If $Mg(OH)_2$ is the only active ingredient in the antacid, what is the pH of the resulting medicinal emulsion? The K_{sp} for $Mg(OH)_2$ is 8.8×10^{-12}.

Solution

(a) Write the equation for the dissolution reaction, and an expression for K_{sp}.

$$Mg(OH)_2(s) \rightleftharpoons Mg^{2+}(aq) + 2\,OH^-(aq)$$

$$K_{sp} = \left[Mg^{2+}\right]\left[OH^-\right]^2 = 8.8 \times 10^{-12}$$

Let x = moles of $Mg(OH)_2$ which dissolve.

$$\left[Mg^{2+}\right] = x \quad \text{and} \quad \left[OH^-\right] = 2x$$

$$K_{sp} = (x)(2x)^2 = 4x^3 = 8.8 \times 10^{-12} \quad \text{or} \quad x^3 = 2.2 \times 10^{-12}$$

$$x = \sqrt[3]{(2.2 \times 10^{-12})} = 1.0 \times 10^{-4}\text{M}.$$

Therefore the solubility of $Mg(OH)_2$ in moles per liter is 1.3×10^{-4}M. Also, $\left[OH^-\right] = 2.6 \times 10^{-4}$M.

(b) Calculate the hydrogen ion concentration and pH using the known hydroxide ion concentration:

$$\left[H^+\right]\left[OH^-\right] = 10^{-14}$$

$$\left[H^+\right] = \frac{10^{-14}}{\left[OH^-\right]} = \frac{1.0 \times 10^{-14}}{2.6 \times 10^{-4}} = 3.85 \times 10^{-11}\text{M}$$

$$\text{pH} = -\log\left[H^+\right] = -\log\left(5 \times 10^{-11}\right) = 10.4$$

19.5. K_{sp} and Precipitation

Generally, a precipitate will form whenever the solubility limit of a substance is exceeded. To predict whether a precipitate will form when solutions are mixed, we calculate the quantity known as ion product, Q, and compare its value with the K_{sp} of a given solid. The ion product Q is defined just like K_{sp} except that initial concentrations are used for Q instead of equilibrium concentrations. For example, the Q expression for Ag_2CrO_4 is:

$$Q = \left[Ag^+\right]_0^2\left[CrO_4^{2-}\right]_0$$

If $Q > K_{sp}$, the solution is supersaturated and precipitation will occur until $Q = K_{sp}$.
If $Q = K_{sp}$, the solution is saturated and is already in equilibrium.
If $Q < K_{sp}$, the solution is unsaturated, so more solid will dissolve until $Q = K_{sp}$.

Example 19.7

A solution is prepared by mixing 100 mL of 0.010 M $CaCl_2$ solution and 50 mL of 0.005 M NaF solution at 25°C. Is it possible for CaF_2 to precipitate? K_{sp} for CaF_2 is 1.5×10^{-10}.

Solution

1. First calculate the $[Ca^{2+}]$ and $[F^-]$ in the mixture, assuming no precipitation.

   Total after mixing = 150 mL or 0.150 L

$$\left[Ca^{2+}\right] = \frac{(0.100\,L \times 0.010\,M)}{0.150\,L} = 0.0067\ M$$

$$\left[F^-\right] = \frac{(0.050\,L \times 0.005\,M)}{0.150\,L} = 0.0017\ M$$

2. Now calculate Q after mixing:

   The ion product $Q = [Ca^{2+}][F^-]^2$

$$Q = (0.0067\ M)(0.0017\ M)^2$$
$$= 1.94 \times 10^{-8}$$

3. Compare Q to K_{sp} and decide if precipitation will occur.

   Since K_{sp} is 1.5×10^{-10}, Q is greater than K_{sp}, so CaF_2 will precipitate.

Example 19.8

A student had a solution of lead dichromate in water; he wanted to bring the Pb^{2+} concentration to 0.20 M exactly. What concentration of CrO_4^{2-} (added as chromic acid) would accomplish this? (K_{sp} for $PbCrO_4 = 2.0 \times 10^{-16}$.)

Solution

1. Write an equation for the dissolution reaction:

$$PbCrO_4 \rightleftharpoons Pb^{2+} + CrO_4^{2-}$$

2. When the solution is saturated with $PbCrO_4$, the expression $\left[Pb^{2+}\right]\left[CrO_4^{2-}\right] = K_{sp} = 2.0 \times 10^{-16}$ holds.
3. Calculate the chromate ion concentration from the expression for K_{sp}:

$$\left[Pb^{2+}\right]\left[CrO_4^{2-}\right] = K_{sp} = 2.0 \times 10^{-16}$$

$$\left[CrO_4^{2-}\right] = \frac{2.0 \times 10^{-16}}{\left[Pb^{2+}\right]} = \frac{2.0 \times 10^{-16}}{0.20} = 1.0 \times 10^{-15}\ M$$

   The concentration of chromate ion which will bring the lead ion concentration in the solution to 0.20 M is 1.0×10^{-15} M.

Example 19.9

A solution contains 0.010 M silver ion (Ag^+) and 0.0010 M lead ion (Pb^{2+}). If an iodide ion solution is slowly added, which solid will precipitate first, AgI or PbI_2? (K_{sp} for AgI = 1.5×10^{-16}, and $PbI_2 = 7.1 \times 10^{-8}$).

Solution

1. Write equations for both reactions:

$$AgI \rightleftharpoons Ag^+ + I^-, \quad K_{sp} = [Ag^+][I^-] = 1.5 \times 10^{-16}$$

$$PbI_2 \rightleftharpoons Pb^{2+} + 2I^-, \quad K_{sp} = [Pb^{2+}][I^-]^2 = 7.1 \times 10^{-8}$$

2. PbI_2 will precipitate when $[Pb^{2+}][I^-]^2 = K_{sp} = 7.1 \times 10^{-8}$. From this expression, calculate the iodide ion concentration required for precipitation to occur.

$$[Pb^{2+}][I^-]^2 = K_{sp} = 7.1 \times 10^{-8}$$

$$[Pb^{2+}] = 0.0010 \text{ M}$$

$$[I^-] = \sqrt{\frac{7.1 \times 10^{-8}}{[Pb^{2+}]}} = \sqrt{\frac{7.1 \times 10^{-8}}{0.0010}} = 0.0084 \text{ M}$$

Thus PbI_2 will precipitate when $[I^-] = 0.0084$ M.

3. Similarly, AgI will precipitate when $[Ag^+][I^-] = K_{sp} = 1.50 \times 10^{-16}$. From this expression, calculate the iodide ion concentration required for precipitation to occur:

$$[Ag^+][I^-] = 1.50 \times 10^{-16}$$

$$[Ag^+] = 0.010 \text{ M}$$

$$[I^-] = \frac{1.50 \times 10^{-16}}{[Ag^+]} = \frac{1.50 \times 10^{-16}}{0.010} = 1.50 \times 10^{-14} \text{ M}$$

So AgI will precipitate when the iodide ion concentration equals 1.50×10^{-14} M.

4. The solid that requires the lowest $[I^-]$ will be the first to precipitate. Therefore AgI will precipitate first since 1.50×10^{-14} M is reached before 3.70×10^{-3} M.

19.6. Complex-Ion Equilibria and Formation of Complex Ions

A complex ion is one which contains a positive metal ion (usually a transition metal) bonded to one or more uncharged small molecules (e.g., H_2O or NH_3) or negative ions (e.g., OH^-, CN^-, or Cl^-) called *ligands*. The metal ion is an electron acceptor (or Lewis acid), while the ligands are electron-donating groups (or Lewis bases). The ligands surrounding the central ion need not all be the same. Also, solvent molecules can occupy some positions. Ligands are linked to the central metal ion by coordinate (dative covalent) bonds. The number of coordinate bonds formed by the central metal ion (or the number of ligands bonded to the central metal ion) is known as the *coordination number*.

The cations in many slightly soluble electrolytes can be dissolved through complex ion formation. For example, consider the dissolution of AgCl precipitate by the addition of aqueous ammonia. This results in the formation of the complex $[Ag(NH_3)_2]^+$. The net reaction is shown by equation 3 below.

$$AgCl(s) \rightleftharpoons Ag^+(aq) + Cl^-(aq) \tag{1}$$

$$Ag^+ + 2NH_3(aq) \rightleftharpoons Ag(NH_3)_2^+(aq) \tag{2}$$

$$AgCl(s) + 2NH_3(aq) \rightleftharpoons Ag(NH_3)_2^+ + Cl^-(aq) \tag{3}$$

According to Le Chatelier's principle, the added ammonia removes Ag^+ ions from solution and shifts the position of equilibrium. More AgCl(s) will dissolve to furnish new Ag^+ ions in solution to restore the equilibrium. These are further converted to $[Ag(NH_3)_2]^+$ by the NH_3 present in the solution. The entire precipitate can be dissolved if excess ammonia is present in the solution.

An equilibrium constant expression can be written for the formation of the complex ion (equation 2) as:

$$K = \frac{[Ag(NH_3)^+{}_2]}{[Ag^+][NH_3]^2} = K_f$$

This equilibrium constant is known as the *formation* or *stability constant*, K_f, which measures how tightly or loosely the ligands hold the metal ions.

The dissociation of the complex ion $[Ag(NH_3)_2]^+$ may be represented as:

$$Ag(NH_3)_2^+ \rightleftharpoons Ag^+ + 2NH_3$$

The equilibrium constant for this reaction is known as the *dissociation constant* K_d, and is given by the expression:

$$K_d = \frac{[Ag^+][NH_3]^2}{[Ag(NH_3)_2^+]}$$

Like polyprotic acids, complex ions usually exhibit stepwise dissociation equilibria. Equilibrium constants can be written for each reaction step. The net dissociation constant K_d will be the product of the K values for the successive steps.

K_d is related to K_f by:

$$K_d = \frac{1}{K_f}$$

This constant can be used to calculate the solubility and concentration of an aqueous metal ion in equilibrium with a complex ion.

Example 19.10

A solution was prepared by adding 0.10 mol of $AgNO_3$ to 1.0 L of 2.5 M NH_3. Calculate the concentrations of Ag^+ and $Ag(NH_3)_2^+$ ions in the solution. The K_f value for $Ag(NH_3)_2^+$ is 1.70×10^7.

Solution

1. First write a balanced equation for the complex ion formation:

$$Ag^+(aq) + 2NH_3(aq) \rightleftharpoons Ag(NH_3)_2^+(aq)$$

2. Write the equilibrium or K_f expression for the reaction:

$$K_f = \frac{[Ag(NH_3)_2^+]}{[Ag^+][NH_3]^2}$$

3. Because K_f is large, we will assume all Ag^+ is in the form of $Ag(NH_3)_2^+$, and then we will see what happens if a small amount of this dissociates again (see table 19-1).
4. Substitute the equilibrium concentrations into the K_f expression and solve for x:

$$K_f = 1.70 \times 10^7 = \frac{[Ag(NH_3)_2^+]}{[Ag^+][NH_3]^2} = \frac{(0.10 - x)}{(x)(2.3 + 2x)}$$

Table 19-1 Data for Example 19.10

	$Ag^+(aq)$	$NH(aq)$	$Ag(NH_3)_2^-$
Initial concentration (M)	0.1	2.5	0
After complete reaction of Ag^+ (M)	0	2.3	0.1
Equilibrium concentration (M)	x	$2.3 + 2x$	$0.1 - x$

The expression can be simplified by making the approximation that x is small compared to 0.10 and 2.30.

$$1.70 \times 10^7 = \frac{(0.10)}{(x)(2.3)} \quad \text{or} \quad x = 2.56 \times 10^{-9}$$

Hence, $[Ag^+] = 2.56 \times 10^{-9}$ M, and

$$[Ag(NH_3)_2^+] = (0.10 - x) = \left(0.10 - 2.56 \times 10^{-9}\right) = 0.10\ \text{M}$$

5. The concentrations of Ag^+ and $Ag(NH_3)_2^+$ ions are 2.56×10^{-9} M and 0.10 M, respectively.

Example 19.11

The dissociation constant of $Ag(NH_3)_2^+$ is 6.2×10^{-8}. For a 0.15 M, solution prepared by dissolving solid $[Ag(NH_3)_2]NO_3$ in water, calculate the equilibrium concentrations of Ag^+ and NH_3.

Solution

1. First write a balanced equation for the dissociation of the complex ion.

$$Ag(NH_3)_2^+(aq) \rightleftharpoons Ag^+(aq) + 2NH_3(aq)$$

2. Write the equilibrium or K_d expression for the reaction:

$$K_d = \frac{[Ag^+][NH_3]^2}{[Ag(NH_3)_2^-]}$$

3. For the dissociation of $Ag(NH_3)_2^+$, make a table showing initial and equilibrium concentrations of all species (see table 19-2).
4. Substitute the equilibrium concentrations into the K_d expression and solve for x:

$$K_d = 6.20 \times 10^{-8} = \frac{[Ag^+][NH_3]^2}{[Ag(NH_3)_2^+]} = \frac{x(2x)^2}{(0.15 - x)} = \frac{4x^3}{(0.15 - x)}$$

Table 19-2 Data for Example 19.11

	$[Ag(NH_3)_2^+]$	$[Ag^+]$	$[NH_3]$
Initial concentration (M)	0.15 *M*	0	0
Change (M)	$(-)x$	$(+)x$	$(+)2x$
Equilibrium concentration (M)	$(0.15 - x)$	x	$2x$

The expression can be simplified by making the approximation that x is small compared to 0.15.

$$6.20 \times 10^{-8} = \frac{4x^3}{(0.15)} \quad \text{or} \quad x = 1.30 \times 10^{-3}$$

Hence, $[Ag^+] = 1.3 \times 10^{-3}$ M, $[NH_3] = 2.6 \times 10^{-3}$M and

$$[Ag(NH_3)_2^-] = (0.15 - x) = (0.15 - 1.30 \times 10^{-3}) \approx 0.15\,\text{M}$$

5. The concentrations of Ag^+ ions and NH_3 at equilibrium are 1.30×10^{-3} M, and 2.60×10^{-3} M respectively.

19.7. Problems

1. Write the solubility product expressions, K_{sp}, for each of the following: (a) Ag_2CrO_4 (b) CaF_2 (c) $CaCO_3$ (d) $Ca_3(PO_4)_2$ (e) $Pb_3(AsO_4)_2$
2. Write the solubility product expressions for the following salts: (a) AgI (b) CaC_2O_4 (c) Hg_2Cl_2 (d) PbI_2 (e) $Fe(OH)_3$
3. At 25°C, a 1.0 L solution saturated with magnesium oxalate, MgC_2O_4, is evaporated to dryness, resulting in 1.0345 g of MgC_2O_4 residue. Calculate the K_{sp} at this temperature.
4. A saturated solution prepared by dissolving excess MgF_2 in water has $[Mg^{2+}] = 2.55 \times 10^{-4}$ M at 25°C. Calculate the K_{sp}.
5. Calculate the molar solubility of aluminum hydroxide, $Al(OH)_3$, in water at 25°C. $K_{sp} = 1.9 \times 10^{-33}$.
6. The solubility product constant, K_{sp}, of calcium phosphate, $Ca_3(PO_4)_2$, is 1.0×10^{-26} at 25°C. Calculate the solubility of $Ca_3(PO_4)_2$ (a) in mol/L and (b) in g/L.
7. The solubility product constant, K_{sp}, of CaF_2 is 3.9×10^{-11}. What is the solubility of CaF_2 in a 0.030 M LiF solution?
8. A student mixed 50 mL of 0.200 M $AgNO_3$ with 70 mL of 0.10 M K_2CrO_4. Calculate the concentration of each ion (Ag^+, NO_3^-, K^+, and CrO_4^{2-}) in solution at equilibrium if K_{sp} of $Ag_2CrO_4 = 1.2 \times 10^{-13}$.
9. The concentration of Ba^{2+} ions in a given aqueous solution is 5.0×10^{-4} M. What SO_4^{2-} ion concentration must be exceeded before $BaSO_4$ can precipitate? The K_{sp} of $BaSO_4$ at 25°C is 1.5×10^{-9}.
10. 100 mL of a solution containing 0.0025 mol Ag^+ ions is mixed with 100 mL of 0.250 M HBr to precipitate AgBr. Calculate the concentration of Ag^+ ions remaining in solution. The K_{sp} of AgBr at 25°C is 5×10^{-13}.
11. A 0.01 M mercuric (I) ion solution is mixed with another solution containing 0.001 M lead (II) ions. A dilute NaI solution is slowly added. Which solid will precipitate first—Hg_2I_2 ($K_{sp} = 4.5 \times 10^{-29}$) or PbI_2 ($K_{sp} = 1.4 \times 10^{-8}$)?

12. A solution of $AgNO_3$ is gradually added to one containing 0.025 mol of NaBr and 0.0025 mol of NaI. Determine which salt, AgBr ($K_{sp} = 5.0 \times 10^{-13}$) or AgI ($1.5 \times 10^{-16}$), will precipitate first.
13. Determine the maximum concentration of hydrogen ion at which FeS ($K_{sp} = 3.7 \times 10^{-19}$) will not precipitate from a solution which is 0.01 M in $FeCl_2$ and saturated with H_2S ($K_{sp} = 1.3 \times 10^{-21}$).
14. If 450 mL of 0.05 M $Pb(NO_3)_2$ is added to 250 mL of 0.125 M Na_2CO_3, would you expect $PbCO_3$ ($K_{sp} = 1.5 \times 10^{-15}$) to precipitate from this solution?
15. A solution contains 0.0010 M Ba^{2+} and 0.100 M Ca^{2+}. A dilute NaF solution is slowly added. Will BaF_2 ($K_{sp} = 2.4 \times 10^{-5}$) or CaF_2 (4.0×10^{-11}) precipitate first? Indicate the concentration of fluoride ion necessary to begin precipitation of each salt.
16. Calculate the number of moles of ammonia that must be added to 1.0 L of 0.50 M $AgNO_3$ so as to reduce the silver ion concentration, $[Ag^+]$, to 2.50×10^{-6} M. The formation constant, K_f, for the amine complex ion at 25°C is 1.7×10^7.
17. Determine the number of moles of ammonia that must be added to 1.0 L of 0.025 M $Cu(NO_3)_2$ so as to reduce the copper ion concentration, $[Cu^{2+}]$, to 2.50×10^{-12} M. The dissociation constant K_d for $Cu(NH_3)_4^{2+}$ at 25°C is 4.35×10^{-13}. (Remember $K_f = 1/K_d$.)
18. What is the molar solubility of AgCl at 25°C in 1.0 M NH_3?
19. The solubility product constant, K_{sp}, of $Zn(OH)_2$ is 1.8×10^{-14}. Calculate the concentration of aqueous ammonia required to initiate the precipitation of $Zn(OH)_2$ from a 0.050 M solution of $Zn(NO_3)_2$.
20. Nickel forms a complex with ammonia according to the following equation:

$$Ni(NH_3)_4^{2+} \rightleftharpoons Ni^{2+} + 4NH_3 \qquad K_d = 1.0 \times 10^{-8}$$

If excess powdered NiS is added to 1.0 M NH_3 solution, only a small amount will dissolve. Calculate the exact amount of NiS that will dissolve in 1.0 M NH_3 solution. The K_{sp} of NiS is 1×10^{-22}.

20

Thermochemistry

20.1. Introduction

All chemical reactions involve energy changes. Some reactions liberate heat to the surroundings; others absorb heat from the surroundings. The breaking of chemical bonds in reactants and the formation of new ones in the products is the source of these energy changes.

20.2. Calorimetry and Heat Capacity

Calorimetry is the experimental determination of the amount of heat transferred during a chemical reaction. This measurement is carried out in a device called a *calorimeter*, which allows all the heat entering or leaving the reaction to be accounted for. This is done by observing the temperature change within the calorimeter as the reaction takes place; if we know how much energy is needed to change the calorimeter's temperature by a given amount, we can calculate the amount of energy involved in the reaction.

The relation between energy and temperature change for the calorimeter or for any other physical object is known as its *heat capacity* (C), which is the amount of heat energy required to raise the temperature of that object by 1°C (or 1 K). This can be expressed in mathematical terms as:

$$C = \frac{q}{\Delta T}$$

where q is the quantity of heat transferred and ΔT is the change in temperature, calculated as $\Delta T = T_f - T_i$. The larger the heat capacity of a body, the larger the amount of heat required to produce a given rise in temperature.

The heat capacity of 1 mol of a substance is known as the *molar heat capacity*. Also, the heat capacity of 1 g of a substance is known as the *specific heat* . To determine the specific heat of a substance, measure the temperature change, ΔT, that a known mass, m, of a substance undergoes as it gains or loses a known quantity of heat, q. That is:

$$\text{Specific heat } (c) = \frac{\text{Quantity of heat gained or lost}}{\text{Mass of substance (in grams)} \times \text{Temperature change } (\Delta T)} \quad \text{or}$$

$$c = \frac{q}{m \times \Delta T}$$

The unit of specific heat is J/g-K or J/g°C.

Example 20.1

A 20-g sample of water absorbs 250.5 J of heat in a calorimeter. If the temperature rises from 25°C to 28°C, what is the specific heat of water?

Solution

$$\text{Specific heat, } c = \frac{q}{m \times \Delta T}$$

$$m = 20\text{ g}; \quad q = 250.5\text{ J}; \quad \Delta T = 28°\text{C} - 25°C = 3°\text{C} \times \frac{1\text{K}}{1°\text{C}} = 3\text{ K}$$

$$c = \frac{q}{m \times \Delta T} = \frac{250.5\text{ J}}{20\text{ g} \times 3\text{ K}} = 4.18\text{ J/g-K}$$

Example 20.2

The specific heat of NaCl is 0.864 J/g-K. Calculate the quantity of heat necessary to raise the temperature of 100 g of NaCl by 7.5 K.

Solution

Substitute known values into the equation and solve for q directly:

$$\text{Specific heat, } c = \frac{q}{m \times \Delta T}$$

or

$$q = mc\Delta T$$

$$m = 100\text{ g}; \quad c = 0.864\text{ J/g-K}; \quad \Delta T = 70.5\text{ K}$$

$$q = mc\Delta T = 100\text{ g} \times 0.864\ \frac{\text{J}}{\text{g-K}} \times 70.5\text{ K} = 6{,}090\text{ J}$$

Example 20.3

The specific heat of gold is 0.129 J/g-K. What will be the final temperature of 1500 g of gold at 60°C after 4.35 kJ of heat is transferred away from it?

Solution

Calculate ΔT and add the value to the initial temperature. This will give you the final temperature. Note that q is negative since heat is transferred away from the system.

$$\Delta T = \frac{q}{m \times c}$$

$$m = 1500\text{ g}; \quad q = -4.35\text{ kJ} \quad \text{or} \quad -4{,}350\text{ J}; \quad c = 0.129\text{ J/g-K}$$

$$\Delta T = \frac{-4{,}350\text{ J}}{1500\text{ g} \times 0.129\ \frac{\text{J}}{\text{g-K}}} = -22.5\text{ K} \quad \text{or} -22.5°\text{C}\left(\text{i.e.,} -22.5\text{ K} \times \frac{1°\text{C}}{1\text{ K}}\right)$$

$$T_f = T_i + \Delta T = 60°\text{C} - 22.5°\text{C} = 38.5°\text{C}$$

20.3. Enthalpy

Enthalpy (H) is defined as *the total energy bound up in a substance.* Note that we cannot measure this quantity directly. We can only measure the changes that occur during reactions. We use the symbol ΔH to refer to these changes.

The change in the enthalpy or heat content in a chemical reaction is given by the equation:

$$\Delta H = H_2 - H_1$$

H_1 is the heat content before the reaction, and H_2 is the heat content after the reaction.

An *exothermic reaction* is one in which *energy is given off by the reacting substances*. This generally appears as an increase in temperature as internal energy is converted to heat (though it can take other forms). In these reactions, ΔH is negative because the products have less potential energy than the reactants. Once an exothermic reaction is initiated, it will continue until the reactants are completely consumed. Figure 20-1 illustrates the energy changes in an exothermic reaction.

An *endothermic reaction* is one in which *heat energy is absorbed from the surroundings*. Here, ΔH is positive because the potential energy of the products is higher than that of the reactants, as illustrated by figure 20-2. Consequently, the temperature of the reaction mixture decreases as heat is converted into chemical bonds. Endothermic reactions are not self-sustaining. Once the source of energy is removed, the reaction will terminate.

The *heat of reaction* is the heat change produced when the number of moles of the reactants represented by the chemical equation has reacted completely. Heat of reaction has the unit of kJ/mol and is indicated by ΔH.

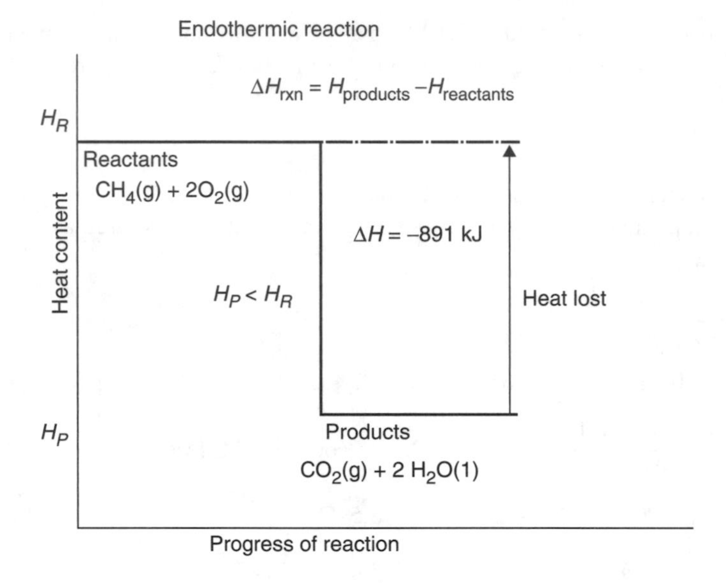

Figure 20-1 Change in heat content during the combustion of methane gas.

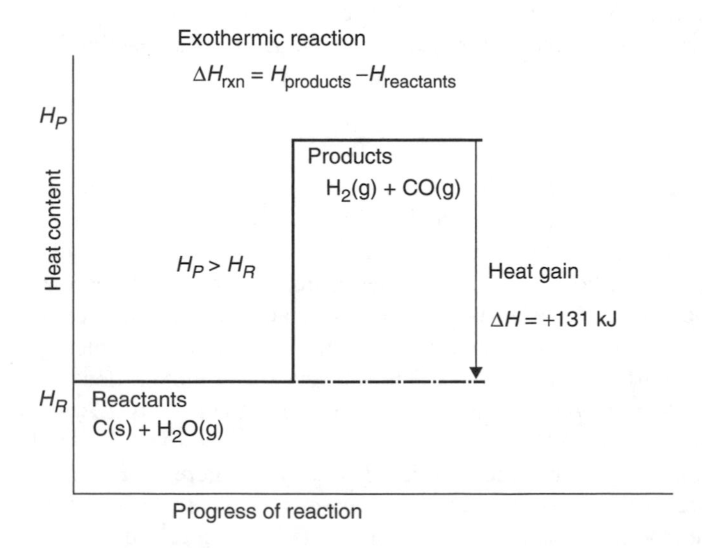

Figure 20-2 Change in heat content during the water gas reaction (formation of carbon monoxide).

20.3.1. Calculating ΔH of reaction

Consider the following general reaction:

$$aA + bB \longrightarrow cC + dD$$
$$\text{reactants} \qquad \text{products}$$

$$\Delta H \text{ of reaction} = \text{Total heat of formation of products} - \text{Total heat of formation of reactants}$$

$$\Delta H \text{ of reaction} = \sum \Delta H_{\text{products}} - \sum \Delta H_{\text{reactants}}$$

$$\Delta H \text{ of reaction} = (c.\Delta H_C + d.\Delta H_D) - (a.\Delta H_A + b.\Delta H_B)$$

The *heat of formation* of a compound is defined as the amount of heat liberated or absorbed when 1 mol of it is formed from its constituent elements. For example, the heat of formation of ammonia gas is given as:

$$N_2(g) + 3\ H_2(g) \longrightarrow 2\ NH_3(g), \qquad \Delta H = -100\ \text{kJ}$$

The *standard molar heat of formation* is the amount of heat absorbed or liberated when 1 mol of a substance in a specified state is formed from its elements in their standard states. It is given the symbol ΔH_f^0. The superscript zero in the ΔH_f^0 indicates standard temperature (298 K) and pressure (1 atm). Under international convention, it is generally accepted that the standard heat of formation of all elements in their most stable form is zero. Table 20-1 provides some common heats of formation.

Example 20.4

Syngas (synthesis gas) consists primarily of carbon monoxide and hydrogen and has less than half the energy density of natural gas. Methanol can be produced by the direct conversion of Syngas according to the following equation:

$$CO(g) + 2\ H_2(g) \longrightarrow CH_3OH(l)$$

Calculate the heat of reaction, given the heats of formation of reactants and product as follows:

$$\Delta H_{CO(g)} = -110.5\ \text{kJ/mol}$$
$$\Delta H_{H_2(g)} = 0$$
$$\Delta H_{CH_3OH(l)} = -239\ \text{kJ/mol}^{-1}$$

Solution

$$\Delta H \text{ of reaction} = \sum \Delta H_{\text{products}} - \sum \Delta H_{\text{reactants}}$$
$$\Delta H \text{ of reaction} = (-239\ \text{kJ/mol}) - (-110.5\ \text{kJ/mol} + 0)$$
$$\Delta H \text{ of reaction} = -128.5\ \text{kJ/mol}$$

Table 20-1 Selected Standard Heat of Formation of Some Compounds

Substance and state	ΔH_f^0 *(kJ/mol)*	*Substance and state*	ΔH_f^0 *(kJ/mol)*	*Substance and state*	ΔH_f^0 *(kJ/mol)*
Al(s)	0	Cl_2(g)	0	HNO_3(aq)	−207
Al_2O_3(s)	−1676	Cl_2(aq)	−23	HNO_3(l)	−174
$Al(OH)_3$(s)	−1277	Cl^-(aq)	−167	O_2(g)	0
Ba(s)	0	HCl(g)	−92	O(g)	249
$Ba(OH)_2$(s)	−946	F_2(g)	0	O_3(g)	143
$BaSO_4$(s)	−1465	F^-(aq)	−333	H_3PO_4(s)	−1279
Br_2(l)	0	HF(g)	−271	H_3PO_4(l)	−1267
Br_2(g)	31	H_2(g)	0	H_3PO_4(aq)	−1288
Br_2(aq)	−3	H(g)	217	K(s)	0
Br^-(aq)	−121	H^+(aq)	0	KCl(s)	−436
HBr(aq)	−36	OH^-(aq)	230	KOH(s)	−425
Ca(s)	0	H_2O(l)	−286	KOH(aq)	−481
$CaCO_3$(s)	−63	H_2O(g)	−242	Na(s)	0
CaO(s)	−635	I_2(s)	0	Na^+(aq)	−240
$Ca(OH)_2$(s)	−987	I_2(g)	62	NaCl(s)	−411
C(s) (graphite)	0	I_2(aq)	23	NaOH(s)	−427
C(s) (diamond)	2	I^-(aq)	−55	NaOH(aq)	−470
CO(g)	−110.5	N_2(g)	0	S(s) (rhombic)	0
CO_2(g)	−393.5	NH_3(g)	−46	S(s) (monoclinic)	0.3
CH_4(s)	−75	NH_3(aq)	−80	S_8(g)	102
CH_3OH(g)	−201	NH_4^+(aq)	−132	SF_6(g)	−1209
CH_3OH(l)	−239	NO(g)	90	H_2S(g)	−21
C_2H_2(g)	227	NO_2(g)	94	SO_2(g)	−297
C_2H_4(g)	52	N_2O(g)	82	SO_3(g)	−396
C_2H_5OH(l)	−278	N_2O_4(g)	10	SO_4^{2-}(aq)	−909
C_2H_6(g)	−84.7	N_2O_4(l)	−20	H_2SO_4(l)	−814
$C_6H_{12}O_6$(s)	−1275	N_2O_5(s)	−42	H_2SO_4(aq)	−909

Heat of combustion is defined as the heat evolved when 1 mol of a substance is burned completely in oxygen at constant temperature. For example, the heat of combustion of methane is:

$$CH_4(g) + 2\,O_2(g) \longrightarrow CO_2(g) + 2\,H_2O(l) \qquad \Delta H = -891\text{ kJ}$$

Combustion is always an exothermic process.

The *heat of neutralization* is the amount of heat liberated when 1 mol of hydrogen ion, H^+, from an acid reacts with 1 mol of hydroxide ion, OH^-, from a base to form 1 mol of water.

$$H^+(aq) + OH^-(aq) \longrightarrow H_2O(l) \qquad \Delta H = -57.4\text{ kJ}$$

The heat of neutralization is always negative.

The *heat of solution* of a substance is the quantity of heat liberated or absorbed when 1 mol of it is dissolved in a large quantity of solvent. For example, the heat of solution of sodium chloride is:

$$NaCl(s) + H_2O(l) \longrightarrow NaCl(aq) \qquad \Delta H + 4\,kJ$$

20.4. Hess's Law of Heat Summation

Hess's law states that if a reaction is carried out in a series of steps, the overall enthalpy change is equal to the sum of the enthalpy changes for the individual steps in the reaction. This implies that the overall enthalpy change for the process is independent of the number or type of steps through which the reaction is carried out.

When solving problems using Hess's law, it is important to know that reactants and products in the individual steps can be added or subtracted just like algebraic terms to obtain the overall equation. In general, the law may be represented as:

$$\Delta H^0_{rxn} = \Delta H^0_1 + \Delta H^0_2 + \Delta H^0_3 + \cdots$$

where 1, 2, 3, ... represent the equations that can be added to give the target equation for the reaction.

20.4.1. Hints for using Hess's law

1. The enthalpy change for any reaction is equal in magnitude but opposite in sign to that for the backward reaction. When an equation is reversed, the associated ΔH must be multiplied by -1.
2. An equation can be multiplied by any necessary coefficient; when this is done, the ΔH value must be multiplied by the same coefficient.

The following worked examples will illustrate Hess's law.

Example 20.5

A mixture of CO and H_2, known as "Syngas" or "water gas," is prepared by passing steam over red-hot charcoal at 1000°C according to the following equation:

$$C(s) + H_2O(g) \longrightarrow CO(g) + H_2(g) \qquad \Delta H_{rxn} = ?$$

Use the following data to calculate ΔH^0 in kJ for the reaction.

$$C(s) + O_2(g) \longrightarrow CO_2(g) \qquad \Delta H_1 = -393.5\,kJ \qquad (1)$$

$$2\,H_2(g) + O_2(g) \longrightarrow 2\,H_2O(g) \qquad \Delta H_2 = -483.6\,kJ \qquad (2)$$

$$2\,CO(g) + O_2(g) \longrightarrow 2\,CO_2(g) \qquad \Delta H_3 = -566.0\,kJ \qquad (3)$$

Solution

The target equation is

$$C(s) + H_2O(g) \longrightarrow CO(g) + H_2(g) \quad \Delta H_{rxn} = ?$$

We might notice that C appears only in equation 1, on the left side, which is where we need it for the target equation. Since it also has a coefficient of 1, as in the target, we can use equation 1 exactly as written. We next note that H_2O appears only in equation 2. In this case, it is on the wrong side, and has a coefficient of 2 instead of 1. Thus we will reverse equation 2, and multiply all the coefficients (as well as the ΔH) by $\frac{1}{2}$. Neither of these equations accounts for the CO in the target equation. This appears in equation 3—once again on the wrong side and with a coefficient of 2 instead of 1. We reverse equation 3 and multiply by $\frac{1}{2}$. Now we can add the resulting equations to obtain the target equation:

$$2C(s) + \cancel{O_2}(g) \longrightarrow \cancel{CO_2}(g) \qquad \Delta H_1 = -393.5\,\text{kJ} \tag{1}$$

$$H_2O(g) \longrightarrow H_2(g) + \frac{1}{2}\cancel{O_2}(g) \quad \Delta H_2 = -\frac{1}{2} \times -483.6\,\text{kJ} = 241.8\,\text{kJ} \tag{2}$$

$$\cancel{CO_2}(g) \longrightarrow CO(g) + \frac{1}{2}\cancel{O_2}(g) \quad \Delta H_3 = -\frac{1}{2} \times -566.0\,\text{kJ} = 283\,\text{kJ} \tag{3}$$

$$C(s) + H_2O(g) \longrightarrow CO(g) + H_2(g)\Delta H_{rxn} = \left[\Delta H + \tfrac{1}{2}(-\Delta H_2) + \tfrac{1}{2}(-\Delta H_3)\right]$$

$$= 131.3\,\text{kJ}$$

Thus the heat of reaction for water gas synthesis is +131.3 kJ.

Example 20.6

Ethanol can be produced by reacting ethylene with water according to the following equation:

$$C_2H_4(g) + H_2O(l) \longrightarrow C_2H_5OH(l)$$

From the following data, calculate the heat of reaction at 298 K.

$$C_2H_5OH(l) + 3\ O_2(g) \longrightarrow 2\ CO_2(g) + 3\ H_2O(l) \quad \Delta H = -1367\,\text{kJ} \tag{1}$$

$$C_2H_4(g) + 3\ O_2(g) \longrightarrow 2\ CO_2(g) + 2\ H_2O(l) \quad \Delta H = -1411\,\text{kJ} \tag{2}$$

Solution

To solve this problem, reverse equation 1 and multiply its ΔH by -1. Add the result to equation 2:

$$2\,\cancel{CO_2}(g) + \cancel{3}\,H_2O(l) \longrightarrow C_2H_5OH(l) + \cancel{3O_2}(g) \qquad \Delta H = 1367\,\text{kJ} \qquad (1)$$

$$C_2H_4(g) + \cancel{3O_2}(g) \longrightarrow 2\,\cancel{CO_2}(g) + \cancel{2\,H_2O}(l) \qquad \Delta H = -1411\,\text{kJ} \qquad (2)$$

$$C_2H_4(g) + H_2O(l) \longrightarrow C_2H_5OH(l) \qquad \Delta H^0_{rxn} = -44\,\text{kJ}$$

Thus ΔH^0_{rxn} per mole of $C_2H_5OH(l)$ formed is -44 kJ.

20.5. Bond Energies and Enthalpy

Chemical reactions involve breaking and forming chemical bonds. To break a bond, energy must be added to the system; hence this is an endothermic process. The formation of a bond liberates energy, that is, it is exothermic. *The average amount of energy necessary to dissociate 1 mol of bonds in a covalent substance in the gaseous state into atoms in the gaseous state is known as the bond energy (BE).* Bond energy (BE) always has a positive sign. The larger the bond energy, more energy is needed to break the bond, or, conversely, the more energy is liberated when the bond forms. Thus bond-energy data can serve as a measure of bond stability. For example, the H—F bond, with energy of 569 kJ/mol, has higher bond stability than C—C, with a bond energy of 339 kJ/mol. Some average bond energies are given in table 20-2.

The enthalpy change for a reaction can be written as:

$$\Delta H^0_{rxn} = \sum \text{BE (bonds broken)} - \sum \text{BE (bonds formed)}$$

(Energy needed) (Energy liberated)

where $\sum$ represents the sum of terms while BE represents the bond energy per mole of bonds.

Since bonds are always broken in reactants and formed in products, the above expression can also be written as:

$$\Delta H^0_{rxn} = \sum \text{BE (reactants)} - \sum \text{BE (products)}$$

Consider the reaction between hydrogen and iodine to form hydrogen iodide:

H—H + I—I ⟶ 2 H—I

Bonds broken Bond formed

Table 20-2 Average Bond Energies (kJ/mol) of Some Bonds

Bond	*Bond energy (kJ/mol)*	*Bond*	*Bond energy (kJ/mol)*	*Bond*	*Bond energy (kJ/mol)*
H—H	432	F—F	154	S—Br	218
H—F	565	Cl—Cl	239	S—S	266
H—Cl	427	Cl—Br	218	O—H	467
H—Br	365	Br—Br	193	O—O	146
H—I	295	I—I	149	O=O	495
C—H	413	N—H	391	O—F	190
C—C	347	N—N	160	O—Cl	203
C=C	614	N=N	418	O—I	234
C≡C	839	N≡N	941	O—P	351
C—N	305	N—F	272	O—N	201
C≡N	891	N—Cl	200	Si—H	323
C—O	358	N—Br	243	Si—C	301
C=O	799	N—P	209	Si—O	368
C≡O	1072	S—H	347	Si—Si	226
C—Cl	339	S—F	327		
C—Br	276	S—Cl	253		

One H—H bond and one I—I bond are broken. On the other hand, two H—I bonds are formed. The approximate overall heat of reaction can be calculated as follows:

$$\Delta H^0_{\text{rxn}} = \sum \text{BE (bonds broken)} - \sum \text{BE (bonds formed)}$$
$$= (\text{BE}_{\text{H}_2} + \text{BE}_{\text{I}_2}) - (2 \times \text{BE}_{\text{HI}})$$

Using BE data from table 20-2:

$$= \left(1 \text{ mol} \times 432 \frac{\text{kJ}}{\text{mol}}\right) + \left(1 \text{ mol} \times 149 \frac{\text{kJ}}{\text{mol}}\right) - \left(2 \text{ mol} \times 295 \frac{\text{kJ}}{\text{mol}}\right)$$
$$= 581 \text{ kJ} - 590 \text{ kJ}$$
$$= -9 \text{ kJ}$$

Thus the reaction is slightly exothermic.

Example 20.7

Using the bond energies listed in table 20-2, estimate ΔH^0 for the reaction:

$$\text{H}_2(\text{g}) + \text{Br}_2(\text{g}) \longrightarrow 2\ \text{HBr}(\text{g})$$

Is the reaction endothermic or exothermic?

Solution

To solve this problem, we can use the following expression:

$$\Delta H^0_{rxn} = \sum BE\ (\text{reactants}) - \sum BE\ (\text{products})$$

1. Obtain the bond dissociation values for all the bonds involved:

   $D(H{-}H) = 432\,kJ/mol$; $D(Br{-}Br) = 193\,kJ/mol$; $D(H{-}Br) = 365\,kJ/mol$

2. Write the expression for calculating the heat of reaction for this reaction and substitute the bond energy values. Remember to multiply each BE by the corresponding number of moles.

$$\begin{aligned}\Delta H_{rxn} &= [D(H{-}H) + D(Br{-}Br)] - [2 \times D(H{-}Br)]\\ &= \{(1\ mol)(432\ kJ/mol) + (1\ mol)(193\ kJ/mol)\} - \{(2\ mol)(365\ kJ/mol)\}\\ &= -105\ kJ\end{aligned}$$

   The reaction is exothermic since ΔH_{rxn} is negative. This suggests that the bonds formed are more stable than those that are broken.

Example 20.8

Calculate the approximate heat of reaction for the hydrogenation of ethylene using the bond energies listed in table 20-2.

$$H_2C{=}CH_2(g) + H_2(g) \longrightarrow CH_3CH_3(g)$$

Is the reaction endothermic or exothermic?

Solution

To solve this problem, we can use the following expression:

$$\Delta H^0_{rxn} = \sum BE\ (\text{reactants}) - \sum BE\ (\text{products})$$

1. Obtain the bond dissociation values for all the bonds involved:

   Bonds broken

   $1\ D(H{-}H) = 432\ kJ/mol$; and $1\ D(C{=}C) = 614\ kJ/mol$

   Bonds formed

   $1\ D(C{-}C) = 347\ kJ/mol$, and $2\ D(C{-}H) = 413\ kJ/mol$

2. Write the expression for calculating the heat of reaction for this reaction and substitute the bond energy values. Remember to multiply each BE by the corresponding number of moles.

$$\begin{aligned}\Delta H_{rxn} &= [D(H{-}H) + D(C{=}C)] - [2 \times D(C{-}H) + D(C{-}C)] \\ &= \{(1\text{ mol})(432\text{ kJ/mol}) + (1\text{ mol})(614\text{ kJ/mol})\} \\ &\quad - \{(2\text{ mol})(413\text{ kJ/mol}) + (1\text{ mol})(347\text{ kJ/mol})\} \\ &= -127\text{ kJ}\end{aligned}$$

The reaction is exothermic since ΔH_{rxn} is negative.

20.6. Problems

1. Determine the amount of heat absorbed when 500 g of ethyl alcohol is heated from 30°C to 65°C. The specific heat of ethyl alcohol is 2.2202 J/g-°C.
2. An industrial chemistry student was asked to measure the heat of combustion of benzene, C_6H_6, a commercially important hydrocarbon used to enhance the octane rating of gasoline. To determine the heat of combustion, he ignited a 1.2575-g sample of benzene in a bomb calorimeter in the presence of excess oxygen. Calculate the heat of combustion per mole for benzene if the temperature increase of the calorimeter is 17°C. The heat capacity of the bomb calorimeter is 8.23 kJ/°C.
3. Determine the heat capacity of a 25 g metal sample that has absorbed 4200 J of heat over a temperature change of 50°C.
4. A 75-g sample of an iron alloy was heated to 100°C and dropped into a beaker containing 150 g of water at 25°C. Calculate the final temperature of the water if the specific heat of the metal is 0.55 J/°C (specific heat capacity for water is 4.2 J/g-°C).
5. Find the amount of heat required to raise the temperature of 550 g of nickel (Ni) from 15°C to 110°C. The specific heat of Ni is 0.4452 J/g-°C.
6. The thermochemical equation for the hydrogenation of ethene is

$$C_2H_4(g) + H_2(g) \longrightarrow C_2H_6(g) \qquad \Delta H^0_{rxn} = -138\text{ kJ}$$

   (a) Is the reaction endothermic or exothermic?
   (b) Calculate the quantity of heat liberated when 5.0 mol of ethane react with excess hydrogen.
   (c) Estimate the mass of ethane required to produce 420 kJ of heat.

7. The thermochemical equation for the combustion of ethane is:

$$C_2H_6(g) + \tfrac{1}{2}O_2(g) \longrightarrow 2\ CO_2(g) + 2\ H_2O(g) \qquad \Delta H^0_{rxn} = -1560\text{ kJ}$$

   Find the amount of heat, in kJ, given off by burning 0.50 kg of ethane.

8. Consider the reaction

$$2\ CH_4(g) + 2\ NH_3(g) + 3\ O_2(g) \longrightarrow 2\ HCN(g) + 6\ H_2O(g) \quad \Delta H^0_{rxn} = -930\,kJ$$

   (a) Is the reaction endothermic or exothermic?
   (b) Re-write the above reaction showing the heat energy appropriately as a reactant or a product.
   (c) What is the heat released per mole of CH_4?
   (d) Find the amount of heat liberated when 48.0 g of CH_4 are burned in excess oxygen.

9. Classify each of the following reactions as endothermic or exothermic.

   (a) $2\ NH_3(g) + CO_2(g) \longrightarrow (H_2N)_2CO(g) + H_2O(g) + 133.6\ kJ$
   (b) $C_{(graphite)} + 1.895\ kJ \longrightarrow C_{(diamond)}$
   (c) $Sn(s) + 2\ Cl_2(g) \longrightarrow SnCl_4(l) + 545.2\ kJ$
   (d) $HCl(aq) + NaOH(aq) \longrightarrow NaCl(aq) + H_2O(l) + 57.3\ kJ$
   (e) $Ca^{2+}(aq) + CO_3^{2-}(aq) + 26.6\ kJ \longrightarrow CaCO_3(s)$

10. The formation of limestone, $CaCO_3$, is represented by the reaction

$$CaO(s) + CO_2(g) \longrightarrow CaCO_3(s)$$

   (a) Determine the heat of reaction, given that

$$\Delta H_{f(CaO)} = -635\ kJ/mol$$

$$\Delta H_{f(CO_2)} = -393.5\ kJ/mol$$

$$\Delta H_{f(CaCO_3)} = -1205.9\ kJ/mol$$

   (b) Is the reaction endothermic or exothermic?
   (c) What is the effect of adding heat to the reaction vessel?

11. Dissolving sulfur trioxide gas in water can produce sulfuric acid.

$$H_2O(l) + SO_3(g) \longrightarrow H_2SO_4(aq)$$

   (a) Determine the heat of reaction from the following data:

$$\Delta H_{f(H_2O)} = -286\ kJ/mol$$

$$\Delta H_{f(SO_3)} = -396\ kJ/mol$$

$$\Delta H_{f(H_2SO_4)} = -909\ kJ/mol$$

   (b) Is the reaction endothermic or exothermic?
   (c) What is the effect of adding heat to the reaction vessel?

12. The enthalpy change at 25°C for the reaction:

$$2\,N_2O(g) + 3\,O_2(g) \longrightarrow 2\,N_2O_4(g)$$

ΔH_f^0(kJ/mol): 81.6 0 ?

is -143.9 kJ. Determine the standard heat of formation of $N_2O_4(g)$.

13. The combustion of glucose is represented by the equation:

$$C_6H_{12}O_6(s) + 6\,O_2(g) \longrightarrow 6\,CO_2(g) + 6\,H_2O(g)$$

Calculate ΔH^0_{rxn} at 25°C for the combustion of 1 mol of glucose from the following standard heat of formation data:

$$\Delta H_{f(H_2O)} = -241.6 \text{ kJ/mol}$$

$$\Delta H_{f(CO_2)} = -393.5 \text{ kJ/mol}$$

$$\Delta H_{f(O_2)} = 0$$

$$\Delta H_{f(C_6H_{12}O_6)} = -1275.1 \text{ kJ/mol}$$

14. The enthalpy change for the reaction:

$$2\,KClO_3(s) \longrightarrow 2\,KCl(s) + 3\,O_2(g)$$

is –89.3 kJ. What is the enthalpy of formation of $KClO_3$?

$$\Delta H^0_{f(KClO_3)} = ?$$

$$\Delta H^0_{f(KCl)} = -435.5 \text{ kJ/mol}$$

$$\Delta H^0_{f(O_2)} = 0$$

15. The first step in the manufacture of nitric acid, HNO_3, is:

$$4\,NH_3(g) + 7\,O_2(g) \longrightarrow 4\,NO_3(g) + 6\,H_2O(l)$$

The ΔH^0_{rxn} is -1396 kJ. What is the enthalpy of formation of NO_2?

$$\Delta H_{f(H_2O(l))} = -286 \text{ kJ/mol}$$

$$\Delta H_{f(NO_2)} = ? \text{ kJ/mol}$$

$$\Delta H_{f(O_2)} = 0$$

$$\Delta H_{f(NH_3)} = -46 \text{ kJ/mol}$$

16. Calculate the heat of reaction for:

$$2\ CO(g) + O_2(g) \longrightarrow 2\ CO_2(g)$$

from the enthalpy changes for the following reactions:

$$C(s) + O_2(g) \longrightarrow CO_2(g) \qquad \Delta H = -393.3\ kJ$$
$$CO(g) \longrightarrow C(s) + \tfrac{1}{2}O_2(g) \qquad \Delta H = +110.5\ kJ$$

17. Calculate the enthalpy of formation of solid PCl_5 from the enthalpy changes for the following reactions:

$$P_4(s) + 6\ Cl_2(g) \longrightarrow 4\ PCl_3(l) \qquad \Delta H = -1270.7\ kJ$$
$$PCl_3(l) + Cl_2(g) \longrightarrow PCl_5(s) \qquad \Delta H = -137.1\ kJ$$

18. The formation of diborane from its constituent elements is represented by:

$$2\ B(s) + 3\ H_2(g) \longrightarrow B_2H_6(g)$$

Use the following thermochemical data to calculate the enthalpy change in kJ for this reaction:

$$B_2H_6(g) + 3\ O_2(g) \longrightarrow B_2O_3(g) + 3\ H_2O(g) \qquad \Delta H = \ \ 1941\ kJ$$
$$2\ B(s) + \tfrac{3}{2}O_2(g) \longrightarrow B_2O_3(g) \qquad \Delta H = -2368\ kJ$$
$$H_2(g) + \tfrac{1}{2}O_2(g) \longrightarrow H_2O(g) \qquad \Delta H = -242\ kJ$$

19. Methylene chloride is an important industrial solvent, commonly prepared by the chlorination of methane gas:

$$CH_4(g) + 2\ Cl_2(g) \longrightarrow CH_2Cl_2(g) + 2\ HCl(g)$$

From the following thermochemical data, calculate the enthalpy change for this chlorination reaction:

$$CH_4(g) + Cl_2(g) \longrightarrow CH_3Cl(g) + HCl(g) \qquad \Delta H = -98.3\ kJ$$
$$CH_3Cl(g) + Cl_2(g) \longrightarrow CH_2Cl_2(g) + HCl(g) \qquad \Delta H = -104\ kJ$$

20. In the laboratory, acetylene gas, C_2H_2, is prepared by the action of cold water on calcium carbide:

$$CaC_2(s) + 2H_2O(l) \longrightarrow Ca(OH)_2(aq) + C_2H_2(g)$$

(a) Calculate the enthalpy change for the reaction using the following thermochemical data:

$$Ca(s) + 2\ C_{(graphite)} \longrightarrow CaC_2(s) \qquad \Delta H = -62.8\ kJ$$

$$Ca(s) + \tfrac{1}{2}O_2(g) \longrightarrow CaO(s) \qquad \Delta H = -635.5\ kJ$$

$$CaO(s) + H_2O(l) \longrightarrow Ca(OH)_2(aq) \qquad \Delta H = -653.1\ kJ$$

$$C_2H_2(g) + \tfrac{5}{2}O_2(g) \longrightarrow 2\ CO_2(g) + H_2O(l) \qquad \Delta H = -1300.1\ kJ$$

$$C_{(graphite)} + O_2(g) \longrightarrow CO_2(g) \qquad \Delta H = -393.5\ kJ$$

(b) Is the reaction endothermic or exothermic?

21. For the reaction:

$$H_2(g) + F_2(g) \longrightarrow 2\ HF(g)$$

(a) Estimate ΔH^0_{rxn} using the bond energy values given below:

H—H	436.4 kJ/mol
F—F	156.9 kJ/mol
H—F	568.2 kJ/mol

(b) Calculate ΔH^0_{rxn} from the standard enthalpy of formation data:

$$\Delta H^0_{f(H_2)} = 0$$

$$\Delta H^0_{f(F_2)} = 0$$

$$\Delta H^0_{f(HF)} = -268.6\ kJ/mol$$

22. Using the bond dissociation energy data below, calculate the ΔH^0 for the following reactions:

(a) $$H_3C{-}CH_3 \longrightarrow H_2C{=}CH_2 + H_2$$

(b) $$H_3C{-}CH_2{-}CH_2{-}CH_3 + 13/2\ O_2 \longrightarrow 4\ CO_2 + 5H_2O$$

(c) $2\,H_2(g) + O_2(g) \longrightarrow 2\,H_2O(g)$

Bond	kJ/mol
H—H	436.4
C—C	331
C=C	590
C—H	414
O=O	498
C=O	803
O—H	464

23. Esters have pleasant aromas and are commonly used in perfumes, and also as flavoring agents in the food industry. They are typically formed by the reaction of a carboxylic acid with an alcohol:

```
   H  O            H  H                  H  O     H  H
   |  ||           |  |                  |  ||    |  |
H—C—C—OH  +  HO—C—C—H  ———→  H—C—C—O—C—C—H  + H2O
   |               |  |                  |        |  |
   H               H  H                  H        H  H
```

Using bond dissociation energy data, calculate the standard enthalpy change for this reaction:

Bond	kJ/mol
C—C	331
C—H	414
C—O	351
C=O	803
O—H	464

24. The hydrogenation of ethylene is represented by the reaction:

```
           H  H                H  H
           |  |                |  |
H2  +  H—C—C—H  ———→  H—C—C—H
                               |  |
                               H  H
```

ΔH^0 for the reaction is -184 kJ. Calculate the bond energy of the carbon-to-carbon double bond in ethylene, given that the C—C = 331 kJ/mol, C—H = 414 kJ/mol, and H—H = 436.4 kJ/mol.

21

Chemical Thermodynamics

21.1. Definition of Terms

Chemical thermodynamics is the study of the energy changes and transfers associated with chemical and physical transformations.

Energy is the ability to do work or to transfer heat.

A *spontaneous process* is one that can occur on its own without any external influence. A spontaneous process always moves a system in the direction of equilibrium.

When a process or reaction cannot occur under the prescribed conditions, it is *nonspontaneous*. The reverse of a spontaneous process or reaction is always nonspontaneous.

Heat (q) is the energy transferred between a system and its surroundings due to a temperature difference.

Work (w) is the energy change when a force (F) moves an object through a distance (d). Thus $W = F \times d$.

A *system* is a specified part of the universe (e.g., a sample or a reaction mixture we are studying).

Everything outside the system is referred to as the *surroundings*.

The *universe* is the system plus the surroundings.

A *state function* is a thermodynamic quantity that defines the present state or condition of the system. Changes in state function quantities are independent of the path (or process) used to arrive at the final state from the initial state. Examples of state functions include enthalpy change (ΔH), entropy change, (ΔS) and free energy change, (ΔG).

The *internal energy* of a system is the sum of the kinetic and potential energies of the particles making up the system. While it is not possible to determine the absolute internal energy of a system, we can easily measure changes in internal energy (which correspond to energy given off or absorbed by the system). The change in internal energy, ΔE, is:

$$\Delta E = E_{\text{final}} - E_{\text{initial}}$$

21.2. The First Law of Thermodynamics

The first law of thermodynamics, also called the *law of conservation of energy*, states that *the total amount of energy in the universe is constant,* that is, *energy can neither be created nor destroyed*. It can only be converted from one form into another. In mathematical terms, the law states that the change in internal energy of a system, ΔE, equals $q + w$. That is,

$$\Delta E = q + w$$

In other words, the change in E is equal to the heat absorbed (or emitted) by the system, plus work done on (or by) the system.

Example 21.1

What is the change in internal energy in a process in which the system absorbs 550 J of heat energy, and at the same time 300 J of work are done on the system?

Solution

We know that 300 J of work are done on the system and 550 J of heat are absorbed. Therefore w has a positive value ($w = +300$) and $q = +550$ J. We can calculate the change in internal energy as follows:

$$\Delta E = q + w = 550 + 300 = 850 \text{ J}$$

Example 21.2

In a certain chemical process, the system lost 1300 J of heat while doing work by expanding against the surrounding atmosphere. What is the work done if the change in internal energy is – 2500 J?

Solution

Here heat is transferred from the system to the surroundings, and work is done by the system. Therefore q is negative and the value of w should be negative.

$$\Delta E = q + w$$
$$w = \Delta E - q = (-2500 \text{ J}) - (-1300 \text{ J}) = -1200 \text{ J}$$

21.3. Expansion Work

When a force causes a mass to move, or when energy is transferred from one point to another, work is done. Consider a cylinder which has a movable piston, and is filled with a gas. Work is done whenever the gas is compressed or expands due to the influence of external pressure. The work done by a constant pressure during expansion or contraction is given by:

$$\text{Work } (w) = \text{Pressure } (P) \times \text{change in volume } (\Delta V)$$

or:

$$w = P\Delta V = P(V_2 - V_1)$$

If the piston pushes back the surrounding atmosphere, the system (the gas within the piston) is giving up energy to do so—that is, if V increases ($\Delta V > 0$), energy is lost by the system, so that $w < 0$. On the other hand, if V decreases ($\Delta V < 0$), positive work is done on the system. Hence we define:

$$w_{\text{expansion}} = -P\Delta V$$

Now, if we substitute $-P\Delta V$ for w in the expression $\Delta E - q + w$, we obtain:

$$\Delta E = q - P\Delta V$$

In a constant-volume process, $\Delta V = 0$, and no $P\Delta V$ work is done. Hence:

$$\Delta E = q_v$$

Thus, in a constant-volume process, no work is done and the heat evolved is equal to the change in internal energy. Note that q_v refers to the heat generated at constant volume.

When a gas is compressed, ΔV is a negative quantity due to the decrease in volume. This makes w a positive quantity.

Liquids and solids are relatively incompressible, and so for these components of a system $P\Delta V$ is usually zero (i.e., $\Delta E = q$).

For reactions involving a change in the number of moles of gases, the work done can be calculated from the following expression:

$$w = P\Delta V = (\Delta n)RT$$

Here Δn equals the total moles of gaseous products minus the total moles of gaseous reactants, R is the ideal gas constant, and T is the absolute temperature.

Example 21.3

During a chemical decomposition reaction, the volume of a reaction vessel expanded from 20.0 L to 35.0 L against an external pressure of 3.5 atm. What is the work done in kJ? (1 L-atm $= 101$ J.)

Solution

1. Work done in expansion or contraction is $w = -P\Delta V$. Here, $P = 3.5$ atm, and: $\Delta V = V_f - V_i = (35.0 - 20.0) = 15.0\,\text{L}$

$$w = -P\Delta V = -(3.5\ \text{atm})(15.0\ \text{L}) = -52.5\ \text{L-atm}$$

2. Convert L-atm to kJ:

$$(-52.5\ \text{L-atm})\left(101\frac{\text{J}}{\text{L-atm}}\right)\left(\frac{1\ \text{kJ}}{1000\ \text{J}}\right) = -5.3\ \text{kJ}$$

Example 21.4

The reaction of nitric oxide with oxygen to form nitrogen dioxide is an important step in the industrial synthesis of nitric acid via the Ostwald process.

$$2\ NO(g) + O_2(g) \rightleftharpoons 2\ NO_2(g)$$

Calculate the work done if the volume contracts from 25.5 L to 18.0 L at a fixed pressure of 20 atm.

Solution

1. Since the volume is changing at constant pressure, use the formula $w = -P\Delta V$ to calculate work done. $P = 20.0$ atm. and $\Delta V = V_f - V_i = (18.0 - 25.5) = -7.5$ L.

$$w = -P\Delta V = -(20.0\ \text{atm})(-7.5\ \text{L}) = 150.0\ \text{L-atm}$$

2. Convert L-atm to kJ:

$$(150.0\ \text{L-atm})\left(101\frac{\text{J}}{\text{L-atm}}\right)\left(\frac{1\ \text{kJ}}{1000\ \text{J}}\right) = 15.15\ \text{kJ}$$

The sign is positive because work is done on the system, that is, energy flows into the system.

Example 21.5

A reaction between gaseous carbon monoxide and oxygen to form carbon dioxide generates 566.0 kJ of heat.

$$2CO(g) + O_2(g) \longrightarrow 2\ CO_2(g) \quad \Delta H = -566.0\ \text{kJ}$$

How much P-V work is done if the reaction is carried out at a constant pressure of 30 atm and the volume change is −2.5 L? Calculate the change in internal energy, ΔE, of the system.

Solution

1. First calculate the work done in kJ by using the formula:

$$w = -P\Delta V = -(30.0\text{ atm})(-2.5\text{ L}) = 75.0\text{ L-atm}$$

$$w = (75.0\text{ L-atm})\left(101\frac{\text{J}}{\text{L-atm}}\right)\left(\frac{1\text{ kJ}}{1000\text{ J}}\right) = 7.58\text{ kJ}$$

2. Calculate the value of ΔE by using the equation

$$\Delta E = q + w \quad \text{or} \quad \Delta E = q - P\Delta V$$

From the question, $q = -566.0$ kJ, and $w = 7.58$ kJ:

$$\Delta E = q + w = -566.0 + 7.58 = -558.4\text{ kJ}$$

Thus we can see that the system has given off energy.

21.4. Entropy

Entropy is a measure of the degree of disorder or randomness of a system, and is denoted by the symbol S. Entropy can also be regarded as a measure of the statistical probability of a system. The larger the value of the entropy, the greater is the disorder or randomness of the atoms, ions, or molecules of the system. For example, the particles in a solid are more closely packed than those in a liquid or gas, hence they are less disordered; in other words, they have lower entropy. As a system changes phase from solid to liquid to gas, the entropy of the system increases.

A change in entropy is represented by ΔS and can be evaluated from the expression

$$\Delta S = S_{\text{final}} - S_{\text{initial}}$$

If $S_{\text{final}} > S_{\text{initial}}$, ΔS is positive, indicating the system has become more disordered.

If $S_{\text{final}} < S_{\text{initial}}$, ΔS is negative, indicating the system has become less disordered.

A process is likely to be favored, that is, spontaneous, if S for the system increases; it is probably nonspontaneous if S decreases. The unit of entropy is joules per kelvin (J/mol-K).

21.5. The Second Law of Thermodynamics

The second law states that *in any spontaneous process the universe tends toward a state of higher entropy.* This implies that every spontaneous chemical or physical change will increase the entropy of the universe. In any process, spontaneous or not, the following condition holds:

$$\Delta S_{\text{universe}} = \Delta S_{\text{system}} + \Delta S_{\text{surrounding}}$$

If $\Delta S_{\text{univ.}} > 0$ the process is spontaneous. If $\Delta S_{\text{univ.}} = 0$ the process can go equally well in either direction (it is reversible). If $\Delta S_{\text{univ.}} < 0$, the process is nonspontaneous; it will not proceed as written, but it will be spontaneous in the reverse direction.

21.6. Calculation of Entropy Changes in Chemical Reactions

The standard state entropy change, ΔS^0, for a chemical reaction can be calculated from the absolute standard state entropies of the reactants and products. Values of S_f^0 at 298 K are tabulated in units of J/mol-K instead of kJ/mol as used for enthalpy changes. Table 21-1 gives the standard molar enthalpies, entropies, and free energies of formation of some common substances. A comprehensive list can be found in most general chemistry textbooks, or a chemistry handbook.

Table 21-1 Standard Free Energies, Enthalpies and Entropies of Formation of Selected Substances at 25°C (298 K)

Substance	*Formula*	ΔH_f^0 *(kJ/mol)*	ΔG_f^0 *(kJ/mol)*	S_f^0 *(J/mol-K)*
Aluminum	$Al(s)$	0	0	28.32
	$Al_2O_3(s)$	−1668.8	−1576.5	51
Bromine	$Br_2(l)$	0	0	152.3
	$Br_2(g)$	30.71	3.14	245.3
	$Br^-(aq)$	−121	−102.8	80.71
	$HBr(aq)$	−36.23	−53.2	198.5
Carbon	C(s) (graphite)	0	0	5.69
	C(s) (diamond)	2	2.84	2.43
	$CO(g)$	−110.5	−137.2	198
	$CO_2(g)$	−393.5	−394.4	213.6
	$CH_4(s)$	−75	−50.8	186.3
	$CH_3OH(g)$	−201	−162	237.6
	$C_2H_4(g)$	52.3	68.1	219.4
	$C_2H_6(g)$	−84.7	−32.9	229.5
Chlorine	$Cl_2(g)$	0	0	223
	$Cl^-(aq)$	−167	−131.2	56.5
	$HCl(g)$	−92.3	−95.3	186.7
Fluorine	$F_2(g)$	0	0	202.7
	$HF(g)$	−271.1	−273.2	173.7
Hydrogen	$H_2(g)$	0	0	130.6
	$H^+(aq)$	0	0	0

Table 21-1 Continued

Substance	*Formula*	ΔH_f^0 *(kJ/mol)*	ΔG_f^0 *(kJ/mol)*	S_f^0 *(J/mol-K)*
Iron	$Fe(s)$	0	0	27.15
	$FeCl_2(S)$	−342	−302.2	118
	$FeCl_3(S)$	−400	−334	142.3
	$Fe_2O_3(s)$	−822.2	−741	90
	$Fe_3O_4(s)$	−1117.1	−1014.2	146
Iodine	$I_2(s)$	0	0	116.73
	$I_2(g)$	62.25	19.37	260.6
	$I^-(aq)$	−55.2	−51.6	111.3
	$HI(g)$	25.94	1.3	206.3
Nitrogen	$N_2(g)$	0	0	191.5
	$NH_3(g)$	−46	−16.7	192.5
	$N_2H_4(g)$	95.4	159.4	238.5
	$NO(g)$	90.37	86.7	210.6
	$NO_2(g)$	33.8	51.84	240.45
	$N_2O_4(g)$	10	98.28	304.3
Oxygen	$O_2(g)$	0	0	205
	$O_3(g)$	143	163.4	237.6
	$OH^-(aq)$	−230	−157.3	−10.7
	$H_2O(l)$	−286	−237.2	69.9
	$H_2O(g)$	−242	−228.6	188.8
Sulfur	S(s, rhombic)	0	0	31.8
	$H_2S(g)$	−21	−33.6	205.7
	$SO_2(g)$	−297	−300.4	248.5
	$SO_3(g)$	−396	−370.4	256.2

The entropy change of a reaction is given by the sum of the entropies of the products minus the sum of the entropies of the reactants. This calculation works just like the one for ΔH^0 of reaction. That is:

$$\Delta S^0 = \sum m S^0 \text{ (products)} - \sum n S^0 \text{ (reactants)}$$

The coefficients m and n represent the stoichiometric coefficients in the balanced chemical equation.

Unlike the enthalpies of formation, the standard molar entropies of formation of the elements are not zero and must be included in all calculations.

Example 21.6

Considering the reaction

$$C(s, \text{diamond}) + O_2(g) \longrightarrow CO_2(g)$$

calculate the standard state entropy change for the formation of gaseous carbon dioxide, given the following standard state entropies:

$$S^0_{C} = 2 \text{ J/mol-K}; \quad S^0_{O_2} = 205 \text{ J/mol-K}; \quad S^0_{CO_2} = 214 \text{ J/mol-K}$$

Solution

For the formation of 1 mol of CO_2 from 1 mol of C and 1 mol of O_2:

$$\begin{aligned}\Delta S^0 &= \sum n\Delta S^0_{\text{product}} - \sum n\Delta S^0_{\text{rectant}} \\ &= (1 \text{ mol})(214 \text{ J/mol-K}) - \{(1 \text{ mol})(2 \text{ J/mol-K}) \\ &\quad + (1 \text{ mol})(205 \text{ J/mol-K})\} \\ &= 7 \text{ J/K}\end{aligned}$$

This reaction is spontaneous if considered from the standpoint of entropy alone.

Example 21.7

Using data from the standard entropy table, calculate ΔS^0 for the following reactions:

1. $N_2(g) + 3H_2(g) \rightleftharpoons 2\,NH_3(g)$
2. $C_2H_6(g) \longrightarrow C_2H_4(g) + H_2(g)$
3. $H_2(g) + F_2(g) \longrightarrow 2\,HF(g)$
4. $2H_2(g) + 2C(s, \text{diamond}) + O_2(g) \longrightarrow CH_3COOH(l)$

In each case, decide if the reaction is spontaneous on the basis of entropy alone. (Note: $S^0_{CH_3COOH} = 159.8$ J/K.)

Solution

For each problem, estimate ΔS^0 by using the expression:

$$\Delta S^0 = \sum nS^0(\text{products}) - \sum nS^0\ (\text{reactants})$$

Then substitute the appropriate ΔS^0 values from table 21-1 to obtain ΔS^0.

1. $N_2(g) + 3H_2(g) \rightleftharpoons 2\,NH_3(g)$

$$\Delta S^0 = 2S^0_{NH_3} - \left[S^0_{N_2} + 3S^0_{H_2}\right]$$

$$\Delta S^0 = (2\text{ mol})(192.5\text{ J/mol-K}) - \{(1\text{ mol})(191.5\text{ J/mol-K}) + (3\text{ mol})(130.6\text{ J/mol-K})\}$$

$$= -198.3\text{ J/K}$$

The reaction is nonspontaneous from entropy considerations alone.

2. $C_2H_6(g) \longrightarrow C_2H_4(g) + H_2(g)$

$$\Delta S^0 = \left[S^0_{C_2H_4} + S^0_{H_2}\right] - \left[S^0_{C_2H_6}\right]$$

$$= \{(1\text{ mol})(219.4\text{ J/mol-K}) + (1\text{ mol})(130.6\text{ J/mol-K})\} - (1\text{ mol})(229.5\text{ J/mol-K})$$

$$= +120.5\text{ J/K}$$

The reaction is spontaneous from entropy considerations alone.

3. $H_2(g) + F_2(g) \longrightarrow 2\,HF(g)$

$$\Delta S^0 = 2S^0_{HF} - \left[S^0_{H_2} + 3S^0_{F_2}\right]$$

$$\Delta S^0 = (2\text{ mol})(173.7\text{ J/mol-K}) - \{(1\text{ mol})(130.6\text{ J/mol-K}) + (1\text{ mol})(202.7\text{ J/mol-K})\}$$

$$= +14.1\text{ J/K}$$

The reaction is spontaneous, considering only entropy changes.

4. $2H_2(g) + 2C(s) + O_2(g) \longrightarrow CH_3COOH(l)$

$$\Delta S^0 = S^0_{CH_3COOH} - \left[2S^0_{H_2} + 2S^0_{C} + S^0_{O_2}\right]$$

$$\Delta S^0 = (1\text{ mol})\left(159.8\text{Jmol}^{-1}\text{K}^{-1}\right) - \{(2\text{ mol})(130.6\text{ J/mol-K}) + (2\text{ mol})(5.69\text{ J/mol-K}) + (1\text{ mol})(205\text{ J/mol-K})\}$$

$$= -317.8\text{ J/K}$$

The reaction is nonspontaneous, considering only entropy changes.

21.7. Free Energy

The Gibbs free energy (G) of a system is a thermodynamic measure which incorporates enthalpy and entropy changes. Neither of these alone is enough to determine absolutely whether a process is spontaneous; G combines them in a way that provides a single simple answer to the question "Is this process spontaneous?" In mathematical terms, G is defined by:

$$G = H - TS$$

Like enthalpy and entropy, the Gibbs free energy of a system is a state function. For a given reaction at constant temperature, changes in H and S result in a change in G, which can be expressed by the equation:

$$\Delta G = \Delta H - T\Delta S$$

The sign of ΔG is a general criterion for the spontaneity of a chemical reaction. If:

$\Delta G < 0$ (negative), the reaction is spontaneous

$\Delta G = 0$, the reaction is at equilibrium—no change

$\Delta G > 0$ (positive), the reaction is not spontaneous

If a reaction is not spontaneous as written, the reverse reaction is always spontaneous. A larger magnitude for ΔG implies that the process is more strongly spontaneous (if negative) or nonspontaneous (if positive).

21.8. The Standard Free Energy Change

The standard free energy change, ΔG^0, for a given reaction is the free energy change that occurs as reactants in their standard states are converted to products in their standard states. The standard conditions include 1 atm pressure, 1 M concentration for solutions, and 25°C (298 K) temperature.

21.8.1. Calculating the standard free energy change

There are several ways to calculate ΔG^0 for a given chemical reaction. Three of the common methods are described here.

1. By direct substitution into the equation $\Delta G^0 = \Delta H^0 - T\Delta S^0$. If we know the values of ΔH^0 and ΔS^0, then we can calculate ΔG^0 directly.

Example 21.8

Calculate ΔG^0 in kJ per mole of SO_3 formed for the following reaction at 298 K, given that the values of ΔH^0 and ΔS^0 are – 198 kJ and – 200 J/K.

$$2\ SO_2(g) + O_2(g) \longrightarrow 2\ SO_3(g)$$

Is the reaction spontaneous under these conditions?

Solution

Substitute the given values directly into the expression for ΔG^0 (watch the units: ΔH^0 is usually in kJ but ΔS^0 is usually in J).

$$\begin{aligned}\Delta G^0 &= \Delta H^0 - T\Delta S^0\\ &= (-198{,}000\ \text{J}) - (298\ \text{K})\left(-200\frac{\text{J}}{\text{K}}\right) = -138{,}400\ \text{J}\\ &= -138.4\ \text{kJ}\end{aligned}$$

$$\Delta G^0 \text{ in kJ/mol } SO_3 = \frac{-138.4\ \text{kJ}}{2\ \text{mol } SO_3} = -69.2\ \text{kJ/mol}$$

The reaction is spontaneous since ΔG^0 is negative.

2. By using Hess's law: ΔG^0 is a state function, so if we know the values of ΔG^0 for a series of reactions, we can use Hess's law to combine the ΔG^0 values and the equations to get the net ΔG^0 for the overall reaction.

Example 21.9

Using the following equations at 298 K:

$$2\ FeO(s) + \tfrac{1}{2}\ O_2(g) \longrightarrow Fe_2O_3(s) \qquad \Delta G^0 = -252.5\ \text{kJ} \qquad (1)$$

$$3\ FeO(s) + \tfrac{1}{2}\ O_2(g) \longrightarrow Fe_3O_4(s) \qquad \Delta G^0 = -281.2\ \text{kJ} \qquad (2)$$

Calculate ΔG^0 for the reaction:

$$3\ Fe_2O_3(s) \longrightarrow 2\ Fe_3O_4(s) + \tfrac{1}{2}\ O_2(g)$$

Solution

To solve this problem, we multiply equation 1 by 3 and equation 2 by 2. Then we reverse the first equation to make iron (III) oxide a reactant and then add both equations. (Remember that the ΔG^0 values for both equations must be multiplied by the coefficients as well; and when an equation is reversed, the sign for its ΔG^0 must be changed.)

$$3 \times \left[2\ \mathrm{FeO(s)} + \tfrac{1}{2}\ \mathrm{O_2(g)} \longrightarrow \mathrm{Fe_2O_3(s)} \qquad \Delta G^0 = -252.5\ \mathrm{kJ}\right] \qquad (3)$$

$$2 \times \left[3\ \mathrm{FeO(s)} + \tfrac{1}{2}\ \mathrm{O_2(g)} \longrightarrow \mathrm{Fe_3O_4(s)} \qquad \Delta G^0 = -281.2\ \mathrm{kJ}\right] \qquad (4)$$

Reversed equation 3:

$$3\ \mathrm{Fe_2O_3(s)} \longrightarrow \cancel{6\ \mathrm{FeO(s)}} + \cancel{\tfrac{3}{2}}\tfrac{1}{2}\mathrm{O_2(g)} \qquad \Delta G^0 = +757.5\ \mathrm{kJ} \qquad (5)$$

$$\cancel{6\ \mathrm{FeO(s)}} + \cancel{\mathrm{O_2(g)}} \longrightarrow 2\ \mathrm{Fe_3O_4(s)} \qquad \Delta G^0 = -562.4\ \mathrm{kJ} \qquad (6)$$

$$3\ \mathrm{Fe_2O_3(s)} \longrightarrow 2\ \mathrm{Fe_3O_4(s)} + \tfrac{1}{2}\ \mathrm{O_2(g)} \qquad \Delta G^0 = +195.1\ \mathrm{kJ} \qquad (7)$$

Since ΔG^0 is positive, the transformation from hematite (Fe_2O_3) to magnetite (Fe_3O_4) is not spontaneous at 298 K.

3. By calculation from tabulated values, using the following expression:

$$\Delta G^0 = \sum \Delta G_f^0\ \text{(products)} - \sum \Delta G_f^0\ \text{(reactants)}$$

   If we know the standard free energies of formation of reactants and products, then we can calculate ΔG^0.

Example 21.10

Is it feasible to synthesize $KClO_3$ from solid KCl and gaseous O_2 under standard conditions?

$$2\ \mathrm{KCl(s)} + 3\mathrm{O_2(g)} \longrightarrow 2\mathrm{KClO_3(s)}$$

The values of ΔG_f^0(in kJ/mol) are:

$$\mathrm{KCl(s)} = -408.8, \quad \mathrm{O_2(g)} = 0, \quad \text{and} \quad \mathrm{KClO_3(s)} = -289.9$$

Solution

To determine whether the given reaction is feasible (or spontaneous) under standard conditions, we will need to determine ΔG^0 for the reaction, then use the

ΔG^0 criteria for spontaneity of a chemical reaction. Calculate ΔG^o by substituting the values of $\Delta G_f^0$for the reactants and product into the expression:

$$\begin{aligned}\Delta G^0 &= \sum \Delta G_f^0(\text{products}) - \sum \Delta G_f^0(\text{reactants})\\ &= 2\Delta G_f^0(\text{KClO}_3(\text{s})) - \left[2\Delta G_f^0(\text{KCl(s)} + 3\Delta G_f^0(\text{O}_2(\text{g})))\right]\\ &= 2\text{ mol} \times \left(-289.9\frac{\text{kJ}}{\text{mol}}\right) - \left[2\text{ mol} \times \left(-408.8\frac{\text{kJ}}{\text{mol}}\right) + 3\text{ mol} \times 0\frac{\text{kJ}}{\text{mol}}\right]\\ &= -579.8\text{ kJ} + 817.6\text{ kJ}\\ &= +237.8\text{ kJ}\end{aligned}$$

The reaction is not feasible as written since ΔG^0 is positive and fairly large. However, the reverse reaction will be spontaneous.

21.9. Enthalpy and Entropy Changes During a Phase Change

A *phase change* is a process in which a substance changes physical form but not chemical composition or identity. Common phase changes include:

- Melting (fusion) — a change from solid to liquid
- Freezing — a change from liquid to solid
- Sublimation — a change from solid to gas
- Deposition — a change from gas to solid
- Vaporization — a change from liquid to gas
- Condensation — a change from gas to liquid

All phase changes are accompanied by a change in enthalpy and entropy, as well as a change in the free energy of the system. If we know the values of ΔH and ΔS for a phase transition, we can calculate the temperature at which a given phase change will occur. By definition, the phases involved in the change are at equilibrium, so $\Delta G = 0$. Knowing this allows us to solve for T. That is:

$$\begin{aligned}\Delta G &= \Delta H - T\Delta S\\ 0 &= \Delta H - T\Delta S\\ \Delta H &= T\Delta S\\ T &= \frac{\Delta H}{\Delta S}\end{aligned}$$

Hence the molar entropy change for any phase transition is given by the expression:

$$\Delta S = \frac{\Delta H}{T}$$

In this equation, ΔH represents the enthalpy change for the phase change and T is the absolute temperature of the change of state.

Example 21.11

The heat of vaporization of benzene is $\Delta H_{vap} = 30.8$ kJ/mol and the entropy change for the vaporization is $\Delta S_{vap} = 87.2$ J/K. What is the boiling point of benzene?

Solution

To determine the boiling point, use the formula:

$$T_b = \frac{\Delta H_{vap}}{\Delta S_{vap}} = \frac{30{,}800 \text{ J/mol}}{87.2 \text{ J/K-mol}} = 353.2 \text{ K} \quad \text{or} \quad 80.2°\text{C}$$

The trick here is to remember to convert your units. You cannot operate on J and kJ at the same time. You must express both units in J or in kJ before carrying out any mathematical operation involving both of them.

Example 21.12

The molar heat of fusion of ice is 333.15 J/g. What is the entropy change for the melting of 1 mol of ice to liquid water at 0°C?

Solution

1. First convert molar heat of fusion from J/g to J/mol. The molar mass of water is 18.0 g/mol:

$$\Delta H_{fus} = \left(333.15 \frac{\text{J}}{\text{g}}\right)\left(18.0 \frac{\text{g}}{\text{mol}}\right) = 5996.6 \text{ J/mol}$$

2. Calculate ΔS_{fus} using the expression $\Delta S = \Delta H/T$:

$$\Delta S = \frac{\Delta H}{T} = \frac{5996.6 \text{ J/mol}}{273 \text{ K}} = 21.97 \text{ J/mol-K}$$

Example 21.13

At 2.6°C, gallium undergoes a solid-solid phase transition. If ΔS^0 for this change is 7.62 J/mol-K, what is ΔH^0 for the transition at this temperature?

Solution

The entropy change for this phase transition is 7.62 J/mol-K.

Since $\Delta H - T\Delta S = 0$ for a phase transition:

$$\Delta S^0_{275.6\text{ K}} = \frac{\Delta H_{\text{s-s phase change}}}{T}$$

$$\Delta H_{\text{s-s phase change}} = T\Delta S^0_{275.6\text{ K}} = (275.6\text{ K})(7.62\text{ J/mol-K})$$
$$= 2100\text{ J/mol}$$

21.10. Free Energy and the Equilibrium Constant

For a reaction under standard conditions, the standard-state free energy, ΔG^0, and the equilibrium constant, K, are related by:

$$\Delta G^0 = -RT\ln K \quad \text{or} \quad \Delta G^0 = -2.303RT\log K$$

where R is the gas constant (8.314 J/K-mol) and T is the thermodynamic or absolute temperature in kelvin.

In actual reactions where the reactants and products are present at nonstandard pressures and concentrations, the free energy change ΔG is related to the standard state free energy ΔG^0 and the composition of the reaction mixture as follows:

$$\Delta G = \Delta G^0 + RT\ln Q \quad \text{or} \quad \Delta G = \Delta G^0 + 2.303RT\log Q$$

where ΔG is the free energy change under nonstandard conditions and Q is the reaction quotient. Details about Q can be found in chapter 17, dealing with chemical equilibrium. Q can be expressed in terms of partial pressure for reactions involving gases, or in terms of molar concentration for reaction involving solutes in solution. Table 21-2 describes the relationship between $\Delta G^0$and K.

Example 21.14

Calculate the value of the equilibrium constant at 298 K for the reaction:

$$N_2(g) + 3\ H_2(g) \longrightarrow 2\ NH_3(g) \quad \Delta G^0 = -33.1\text{ kJ}$$

Table 21-2 Relationship Between Standard Free Energy Change and Equilibrium Constant

ΔG^0	$\ln K$	K	*Comments on products and reactants*
$\Delta G^0 < 0$	$\ln K > 0$	$K > 1$	Products are favored over reactants at equilibrium
$\Delta G^0 = 0$	$\ln K = 0$	$K = 1$	Amount of products is approximately equal to the amount of reactants at equilibrium
$\Delta G^0 > 0$	$\ln K < 0$	$K < 1$	Reactants are favored over products at equilibrium

Solution

We can calculate K from the expression $\Delta G^0 = -2.303\,RT \log K$. Rearrange this equation to obtain $\log K$:

$$\log K_{\mathrm{p}} = \frac{-\Delta G^0}{2.303RT}$$

Now solve for K by substituting the values of R, T, and ΔG^0 into the equation.

$$\log K_{\mathrm{p}} = \frac{-\Delta G^0}{2.303RT} = \frac{-(-33.1 \times 10^3 \mathrm{J})}{(2.303) \times (8.314\ \mathrm{J/K}) \times (298\ \mathrm{K})} = 5.8$$

$$K = \text{antilog}\ (5.8) = 10^{5.8} = 6.32 \times 10^5$$

Example 21.15

At 25°C, the solubility product constant K_{sp} for $CuCO_3$ is 2.5×10^{-10}. Calculate ΔG^0 at 25°C for the reaction:

$$\mathrm{CuCO_3(s)} \rightleftharpoons \mathrm{Cu^{2+}(aq)} + \mathrm{CO_3^{2-}(aq)}$$

Solution

Calculate ΔG^0 from the expression $\Delta G^0 = -2.303RT \log K$. (Note that $K_{\mathrm{sp}} = K$.)

$$\Delta G^0 = -2.303RT \log K = (-2.303)(8.314\ \mathrm{J/K})(298\ \mathrm{K})\left(\log 2.5 \times 10^{-10}\right)$$

$$= 54{,}788\ \mathrm{J} = 54.8\ \mathrm{kJ}$$

Example 21.16

For the gaseous reaction:

$$4\ \mathrm{NH_3(g)} + 5\mathrm{O_2(g)} \rightleftharpoons 4\ \mathrm{NO(g)} + 6\ \mathrm{H_2O(g)} \qquad \Delta G^0 = -957.4\ \mathrm{kJ}$$

the equilibrium partial pressures are:

$$P_{\mathrm{NH_3}} = 0.2\ \mathrm{atm}, \quad P_{\mathrm{O_2}} = 15\ \mathrm{atm}, \quad P_{\mathrm{NO}} = 2.5, \quad \text{and} \quad P_{\mathrm{H_2O}} = 1.5\ \mathrm{atm}.$$

Calculate ΔG at 298 K for this reaction.

Solution

To calculate ΔG for this reaction we use the expression

$$\Delta G = \Delta G^0 + 2.303RT \log Q$$

First we need to calculate Q. From the given data:

$$Q = \frac{(P_{NO})^4 (P_{H_2O})^6}{(P_{NH_3})^4 (P_{O_2})^5} = \frac{(2.5)^4 (1.5)^6}{(0.2)^4 (15)^5} = 0.3662$$

Now solve for ΔG:

$$\begin{aligned}\Delta G &= \Delta G^0 + 2.303RT \log Q \\ &= (-957,400 \text{ J}) + (2.303)(8.314 \text{ J/K})(298 \text{ K})(\log 0.3662) \\ &= -9.60 \times 10^5 \text{J} = -960 \text{ kJ}\end{aligned}$$

This reaction is spontaneous since ΔG is negative.

21.11. Variation of ΔG and Equilibrium Constant with Temperature

The dependence of ΔG on the reaction temperature can be seen from the quantitative relationship between ΔG, ΔH, and ΔS, as well as between ΔG and the equilibrium constant K. If we know ΔG at one temperature, we can calculate ΔG at another temperature if we assume that ΔH and ΔS are constant. For the two temperatures in question,

$$\Delta G_1 = \Delta H - T_1 \Delta S \quad \text{and} \quad \Delta G_2 = \Delta H - T_2 \Delta S$$

Divide through the expressions for ΔG by the temperature term. This gives:

$$\frac{\Delta G_1}{T_1} = \frac{\Delta H}{T_1} - \Delta S \quad \text{and} \quad \frac{\Delta G_2}{T_2} = \frac{\Delta H}{T_2} - \Delta S$$

Rearranging:

$$\frac{\Delta H}{T_1} - \frac{\Delta G_1}{T_1} = \Delta S \quad \text{and} \quad \Delta S = \frac{\Delta H}{T_2} - \frac{\Delta G_2}{T_2}$$

Since ΔS is the same in both equations (by our assumption):

$$\frac{\Delta H^0}{T_1} - \frac{\Delta G^0}{T_1} = \frac{\Delta H^0}{T_2} - \frac{\Delta G^0}{T_2}$$

$$\frac{\Delta G_2^0}{T_2} - \frac{\Delta G_1^0}{T_1} = \Delta H \left(\frac{1}{T_2} - \frac{1}{T_1} \right)$$

or:

$$\frac{\Delta G_2^0}{T_2} = \frac{\Delta G_1^0}{T_1} + \Delta H \left(\frac{1}{T_2} - \frac{1}{T_1} \right)$$

Example 21.17

The standard free energy change at 25°C is – 57.7 kJ for the reaction:

$$FeCl_2(aq) + \tfrac{1}{2}Cl_2(g) \longrightarrow FeCl_3(aq)$$

Given that ΔH^0 is – 127.2 kJ and ΔS^0 is – 233 J/K, calculate ΔG^0 at 100°C. Comment on your answer.

Solution

Note that while the ΔS^0 value is given, it is not relevant in solving this problem. To calculate ΔG^0 at 100°C (or 373 K), use the equation:

$$\frac{\Delta G_2^0}{T_2} = \frac{\Delta G_1^0}{T_1} + \Delta H\left(\frac{1}{T_2} - \frac{1}{T_1}\right)$$

$$\frac{\Delta G_2^0}{373\text{ K}} = \frac{(-57.7\text{ kJ})}{298\text{ K}} + (-127.2\text{ kJ})\left(\frac{1}{373\text{ K}} - \frac{1}{298\text{ K}}\right)$$

$$= \left(-\,0.1936\,\frac{\text{kJ}}{\text{K}}\right) + \left(0.0858\,\frac{\text{kJ}}{\text{K}}\right)$$

$$\frac{\Delta G_2^0}{373\text{ K}} = -\,0.1078\,\frac{\text{kJ}}{\text{K}}$$

$$\Delta G_2^0 = \left(-\,0.1078\,\frac{\text{kJ}}{\text{K}}\right)(373\text{ K}) = -\,40.2\text{ kJ}$$

The value of ΔG^0 at 25°C is more negative than that at 100°C. Therefore the reaction is less favorable at elevated temperature.

21.11.1. Relationship between ΔG^0 and K at different temperatures

The expressions for ΔG^0 we have so far encountered are:

$$\Delta G^0 = \Delta H^0 - T\Delta S^0$$

$$\Delta G^0 = -2.303\,RT\log K$$

If we combine these two equations (i.e., $\Delta H^0 - T\Delta S^0 = -2.303RT\log K$) and solve for $\log K$, we will have:

$$\log K = -\frac{\Delta H^0}{2.303RT} + \frac{T\Delta S^0}{2.303RT} \quad \text{or} \quad \log K = -\frac{\Delta H^0}{2.303RT} + \frac{\Delta S^0}{2.303R}$$

Now, assuming that a reaction is carried out at two temperatures, T_1 and T_2, and that ΔH^0 and ΔS^0 are constant at these temperatures, the above equation will give:

$$\log K_1 = -\frac{\Delta H^0}{2.303RT_1} + \frac{\Delta S^0}{2.303R}$$

and:

$$\log K_2 = -\frac{\Delta H^0}{2.303RT_2} + \frac{\Delta S^0}{2.303R}$$

Subtracting the first equation from the second and rearranging gives:

$$\log\left(\frac{K_2}{K_1}\right) = \frac{-\Delta H^0}{2.303R}\left(\frac{1}{T_2} - \frac{1}{T_1}\right)$$

which shows how the equilibrium constant changes with temperature.

Example 21.18

The following reaction plays a significant role in the refining of copper; it is usually carried out at high temperatures.

$$\mathrm{CuS(s) + H_2(g) \longrightarrow Cu(s) + H_2S(g)}$$

$K = 3.14 \times 10^{-4}$ at 25^{0}C. What is K at 727°C, given that $\Delta H^0 = 33$ kJ? Assume that ΔH^0 is constant over this temperature range.

Solution

To calculate K at 727°C, use the equation:

$$\log\left(\frac{K_2}{K_1}\right) = \frac{-\Delta H^0}{2.303\,R}\left(\frac{1}{T_2} - \frac{1}{T_1}\right)$$

$$\log\left(\frac{K_2}{3.14 \times 10^{-4}}\right) = \frac{(-33,000\ \mathrm{J})}{(2.303)(8.314\ \mathrm{J/K})}\left(\frac{1}{1000} - \frac{1}{298}\right)$$

$$\log\left(\frac{K_2}{3.14 \times 10^{-4}}\right) = 4.060$$

$$K_2 = \left(3.14 \times 10^{-4}\right)(\text{antilog } 4.060) = 3.6$$

The equilibrium constant at 727°C is 3.6. Thus the process is more favorable at higher temperature.

21.12. Problems

1. 150 J of heat was added to a biological system. Determine the change in internal energy for the system if it does 250 J of work on its surroundings.
2. The reaction of nitrogen with hydrogen to make ammonia yields – 92.2 kJ.

$$N_2(g) + 3\ H_2(g) \longrightarrow 2\ NH_3(g)$$

If the change in internal energy is 80.5 kJ, calculate the maximum amount of work that can be done by this chemical process.

3. The combustion of methane gas in a reactor causes a volume expansion of 7.5 L against an external pressure of 10 atm. Calculate the work done, in kJ, during the reaction. (Note: 1 L-atm = 101 J.)
4. The highly exothermic fluorination of methane is represented by the equation:

$$CH_4(g) + 4\ F_2(g) \longrightarrow CF_4(g) + 4\ HF(g) \qquad \Delta H = -1935\,\text{kJ}$$

If the reaction is carried out at a constant pressure of 25 atm, what is the change in internal energy, ΔE, if the reaction results in a volume change of – 0.95 L?

5. The industrial production of ammonia is represented by the following reaction:

$$N_2(g) + 3\ H_2(g) \longrightarrow 2\ NH_3(g) \qquad \Delta H = -92.2\ \text{kJ}$$

Calculate the change in internal energy, ΔE, for the reaction at 298 K ($R = 8.314$ J/K-mol).

6. Oxygen is often prepared in the laboratory by the thermal decomposition of potassium chlorate in the presence of manganese dioxide:

$$2\ KClO_3(s) \xrightarrow{MnO_2} 2\ KCl(s) + 3\ O_2(g)$$

Calculate the work done when the above reaction is conducted at STP.

7. Using the appropriate S^0 values in table 21-1, calculate the standard entropy change for the following reaction:

$$N_2O_4(g) \longrightarrow 2\ NO_2(g)$$

8. Using the appropriate S^0 values in table 21-1, calculate the entropy change that accompanies the formation of 2 mol of nitrogen dioxide from nitrogen monoxide and oxygen.

$$2\ NO(g) + O_2(g) \longrightarrow 2\ NO_2(g)$$

9. Determine which of the following processes is accompanied by an increase or a decrease in the entropy of the system. No calculations are needed.

   (a) $I_2\ (s) \longrightarrow I_2(g)$
   (b) $CaCO_3(s) \longrightarrow CaO(s) + CO_2(g)$

(c) $H_2(g) + I_2(g) \longrightarrow 2\,HI(g)$
(d) $2\,SO_2(g) + O_2(g) \longrightarrow 2\,SO_3(g)$
(e) $H_2O(s) \longrightarrow H_2O(l)$
(f) $2\,H_2(g) + CO(g) \longrightarrow CH_3OH(l)$

10. Without any calculation, predict whether the entropy change in each of the following systems is positive or negative:

    (a) $3\,O_2\,(g) \longrightarrow 2\,O_3(g)$
    (b) $NH_4Cl(s) \longrightarrow NH_3(g) + HCl(g)$
    (c) $H_2(g) \longrightarrow 2\,H(g)$
    (d) $2\,HgO(s) \longrightarrow 2\,Hg(l) + O_2(g)$
    (e) $N_2O_4(g) \longrightarrow 2\,NO_2(g)$
    (f) $2\,H_2O(l, 90°C) \longrightarrow H_2O(s, 0°C)$

11. Calculate the standard entropy changes for transformations of 1 mol of the following at 25°C:
    (a) S_8 (rhombic) ($S_r^0 = 31.8$ J/K-mol) $\longrightarrow$ S_8 (monoclinic) ($S_m^0 = 32.5$ J/K-mol)

    (b) SiO_2(quartz) ($S_q^0 = 10$ J/K-mol) $\longrightarrow$ SiO_2(tridymite) ($S_t^0 = 11.2$ J/K-mol)

12. Calculate the standard entropy change for each of the following reactions:

    (a) $2\,H_2S\,(g) + SO_2(g) \longrightarrow 3\,S(s) + 2\,H_2O(g)$
    (b) $SO_3(g) + H_2O(l) \longrightarrow H_2SO_4(aq)$
    (c) $2\,H_2O_2(l) \longrightarrow 2\,H_2O(l) + O_2(g)$
    S^0 in J/K-mol: $H_2O_2(l) = 110$, $SO_2(g) = 248.5$, S (rhombic) $= 31.8$
    $SO_3(g) = 256.2$, $H_2O(l) = 69.9$, $H_2O(g) = 188.7$, $H_2SO_4(aq) = 20$,
    $O_2(g) = 205$, $H_2S(g) = 205.7$.

13. The heat of vaporization of ethanol is 39.3 kJ/mol and its normal boiling point is 78°C. Calculate the entropy change associated with the vaporization of ethanol.

14. The molar heat of sublimation of solid carbon dioxide (dry ice) is 25.2 kJ/mol and it sublimes at 194.8 K. What is the entropy change for the sublimation of dry ice?

15. The molar heat of fusion and vaporization of NaCl are 28.5 and 170.7 kJ/mol respectively. Determine the entropy changes for the solid-to-liquid and liquid-to-gas transformations. The melting and boiling point are 1081 K and 1738 K at 1 atm pressure.

16. A certain polymerization process has $\Delta H = 55.8$ kJ and $\Delta S = 335$ J/K at 298 K. (a) Calculate ΔG. (b) Is the reaction spontaneous under these conditions? (c) Is the reaction spontaneous at 200°C?

17. Calculate ΔG^0 in kJ per mole of H_2 formed for the following reaction at 298 K, using the data below:

$$H_2(g) + (Cl_2(g) \longrightarrow 2HCl(g)$$

	ΔH_f^0/kJ/mol	S^0/J/K-mol
$H_2(g)$	0.0	130.6
$Cl_2(g)$	0.0	223.0
$HCl(g)$	−92.3	186.7

Is the reaction spontaneous under these conditions?

18. An engineer without adequate chemistry background boasted he could synthesize ammonia by dissolving nitric oxide in water at 80°C according to the following equation:

$$4\ NO(g) + 6\ H_2O(g) \longrightarrow 4\ NH_3(g) + 5\ O_2(g)$$

Using the following thermodynamic data, predict if the engineer will have any luck synthesizing ammonia under these conditions.

	ΔH_f^0/kJ/mol	S^0/J/K-mol
$NO(g)$	90.4	210.6
$H_2O(g)$	−242	188.8
$O_2(g)$	0	205.0
$NH_3(g)$	−46	192.5

19. Using the given enthalpy and entropy changes, estimate the temperature range for which each of the following reactions is spontaneous:

(a) $NH_4Cl(s) \longrightarrow NH_3(g) + HCl(g)$
($\Delta H^0_{rxn} = +176$ kJ; $\Delta S^0_{rxn} = +285$ J/K)

(b) $N_2H_4(g) + 2\ H_2O_2(l) \longrightarrow N_2(g) + 4H_2O(g)$
($\Delta H^0_{rxn} = -642.2$ kJ; $\Delta S^0_{rxn} = +606$ J/K)

20. For the reaction:

$$Fe_2O_3(s) + 3\ CO(g) \longrightarrow 2\ Fe(s) + 3\ CO_2(g)$$

ΔG^0 and ΔH^0 were experimentally determined to be −29.5 kJ and − 24.9 kJ, respectively.

(a) Calculate $\Delta G^0_{CO_2}$ and $\Delta H^0_{CO_2}$ from the thermodynamic data given below
(b) Determine the change in entropy for the reaction at 25°C.

	ΔG_f^0/kJ/mol	ΔH_f^0/kJ/mol
$CO(g)$	−137.7	−110.5
$CO_2(g)$	?	?
$Fe_2O_3(s)$	−742.2	−824.2
$Fe(s)$	0	0
$NH_3(g)$	−46	192.5

21. The decomposition of hydrogen peroxide is represented by the equation:

	$H_2O_2(l)$	$\longrightarrow$	$H_2O(g)$	+	$\frac{1}{2}O_2(g)$
ΔG_f^0 (kJ/mol)	−187.9		−241.8		0

Calculate ΔG^0 for the reaction and determine if the reaction will occur as it is written (forward direction).

22. Calculate the equilibrium constant (K_p) for the reaction given in problem 21 at 25°C ($R = 8.31$ J/K).
23. Using the standard free energies of formation given below, calculate the equilibrium constant (K_p) at 25°C for the process:

	$2\ NH_3(g)$	$\rightleftharpoons$	$N_2H_4(g)$	+	$H_2(g)$
ΔG_f^0 (kJ/mol)	-16.5		159.3		0

24. Calculate the free energy change for the oxidation of ammonia at 25°C from the following equilibrium partial pressures: 2.0 atm NH_3, 1.0 atm O_2, 0.5 atm NO, and 0.75 atm H_2O.

$$4\ NH_3(g) + 5\ O_2(g) \rightleftharpoons 4\ NO(g) + 6\ H_2O(g) \qquad \Delta G^0_{rxn} = -958 \text{ kJ/mol}$$

25. The standard free energies of formation for liquid and gaseous ethanol are -174.9 and -168.6 kJ/mol respectively. Calculate the vapor pressure of ethanol (C_2H_5OH) at 25°C. (Assume that, since the liquid and the vapor are in equilibrium, $\Delta G = 0$.)
26. The vapor pressure of mercury at 25°C is 2.6×10^{-6} atm. Calculate the free energy change for the vaporization process.
27. The equilibrium constant for the formation of hydrogen iodide gas is 10.2 at 25°C, and 78 at 425°C. Calculate the enthalpy change for the reaction.
28. At 25°C, the equilibrium constant, K_c, for the process

$$N_2O_4(g) \rightleftharpoons 2\ NO_2(g) \qquad \Delta H^0_{rxn} = 58 \text{ kJ}$$

is 4.6×10^{-3}. What is the equilibrium constant at 100°C?

22

Oxidation and Reduction Reactions

22.1. Introduction

Oxidation-reduction reactions, or *redox reactions*, occur in many chemical and biochemical systems. The process involves the complete or partial transfer of electrons from one atom to another. Oxidation and reduction processes are complementary. For every oxidation, there is always a corresponding reduction process. This is because for a substance to gain electrons in a chemical reaction, another substance must be losing these electrons.

22.2. Oxidation and Reduction in Terms of Electron Transfer

Oxidation is defined as *a process by which an atom or ion loses electrons*. This can occur in several ways:

- Addition of oxygen or other electronegative elements to a substance:

$$2\ Mg(s) + O_2(g) \longrightarrow 2\ MgO(s)$$
$$2\ Mg(s) + O_2(g) \longrightarrow MgCl_2\ (s)$$

- Removal of hydrogen or other electropositive elements from a substance:

$$H_2S(g) + Cl_2(g) \longrightarrow 2\ HCl(g) + S(s)$$

  Here, H_2S is oxidized.
- The direct removal of electrons from a substance:

$$2\ FeCl_2\ (s) + Cl_2(g) \longrightarrow 2\ FeCl_3\ (s)$$
$$Fe^{2+} \longrightarrow Fe^{3+} + e^-$$

Reduction is defined as *the process by which an atom or ion gains electrons.* This can occur in the following ways:

- Removal of oxygen or other electronegative elements from a substance:

$$MgO(s) + H_2(g) \longrightarrow Mg(s) + H_2O(g)$$

- Addition of hydrogen or other electropositive elements to a substance:

$$H_2(g) + Br_2(g) \longrightarrow 2\ HBr(g)$$

$$2\ Na(s) + Cl_2(g) \longrightarrow 2\ NaCl(s)$$

Here, chlorine (Cl_2) is reduced.

- The addition of electrons to a substance:

$$Fe^{3+} + e^{-} \longrightarrow Fe^{2+}$$

22.3. Oxidation Numbers

Oxidation number or *oxidation state* is a number assigned to the atoms in a substance to describe their relative state of oxidation or reduction. These numbers are used to keep track of electron transfer in chemical reactions. Some general rules are used to determine the oxidation number of an atom in free or combined state.

22.3.1. Rules for assigning oxidation numbers

1. Any atom in an uncombined (or free) element (e.g., N_2, Cl_2, S_8, O_2, O_3, and P_4) has an oxidation number of zero.
2. Hydrogen has an oxidation number of $+1$ except in metal hydrides (e.g., NaH, MgH_2) where it is -1.
3. Oxygen has an oxidation number of -2 in all compounds except in peroxides (e.g., H_2O_2, Na_2O_2) where it is –1.
4. In simple monoatomic ions such as Na^+, Zn^{2+}, Al^{3+}, Cl^-, and C^{4-}, the oxidation number is equal to the charge on the ion.
5. The algebraic sum of the oxidation numbers in a neutral molecule (e.g., $KMnO_4$, NaClO, H_2SO_4) is zero.
6. The algebraic sum of the oxidation numbers in a polyatomic ion (e.g., SO_4^{2-}, $Cr_2O_7^{2-}$) is equal to the charge on the ion.
7. Generally, in any compound or ion, the more electronegative atom is assigned the negative oxidation number while the less electronegative number is assigned the positive oxidation number.
8. Group IA elements (H, Li, Na, K, etc.) have oxidation number of $+1$; Group IIA elements (Mg, Ca, etc.) have oxidation number of +2; group IIIA (B, Al, Ga, etc.) has oxidation number $+3$. On the other hand, elements in group VIA (e.g., O, S) and group VIIA (e.g., F, Cl, Br) have oxidation numbers of -2 and -1, respectively.

22.3.2. Oxidation numbers in formulas

The oxidation number (ON) of any atom in a compound or ion can be determined using the applicable rules and some arithmetic. For example, the oxidation number of Cr in $K_2Cr_2O_7$ can be obtained as follows:

$$2 \times \text{ON [K]} + 2 \times \text{ON [Cr]} + 7 \times \text{ON [O]} = 0$$
$$\text{ON of K} = +1$$
$$\text{ON of Cr} = ?$$
$$\text{ON of O} = -2$$
$$2 \times 1 + 2\text{Cr} - 7 \times 2 = 0$$
$$2 + 2\text{Cr} - 14 = 0$$
$$2\text{Cr} - 12 = 0$$
$$\text{Cr} = 6$$

Hence the ON of Cr in $K_2Cr_2O_7 = +6$.

We can calculate the ON of Mn in MnO_4^- in a similar fashion:

$$1 \times \text{ON [Mn]} + 4 \times \text{ON[O]} = -1$$
$$\text{ON of Mn} = ?$$
$$\text{ON of O} = -2$$
$$\text{Mn} - 4 \times -2 = -1$$
$$\text{Mn} - 8 = -1$$
$$\text{Mn} = +7$$

Therefore the ON of Mn in MnO_4^- is +7.

Example 22.1

Calculate the oxidation number for the indicated element in each of the following compounds or ions:

(a) As in Na_3AsO_4
(b) S in S_2F_{10}
(c) Br in $HBrO_3$
(d) P in $P_2O_7^{4-}$
(e) Pt in $[PtCl_6]^{2-}$
(f) Fe in $[Fe(CN)_6]^{3-}$

Solution

(a) As in Na_3AsO_4

The sum of the oxidation numbers should be zero:

Na_3AsO_4

$3(+1) + \text{As} + 4(-2) = 0$

$3 + As - 8 = 0$
$As = +5$

(b) S in S_2F_{10}

The sum of the oxidation numbers should be zero:
S_2F_{10}
$2(S) + 10(-1) = 0$
$2S - 10 = 0$
$S = +5$

(c) Br in $HBrO_3$

The sum of the oxidation numbers should be zero:
$HBrO_3$
$1(+1) + Br + 3(-2) = 0$
$1 + Br - 6 = 0$
$Br = +5$

(d) P in $P_2O_7^{4-}$

The sum of the oxidation numbers should be −4:
$P_2O_7^{4-}$
$2(P) + 7(-2) = -4$
$2P - 14 = -4$
$P = +5$

(e) Pt in $[PtCl_6]^{2-}$

The sum of the oxidation numbers should be −2:
$[PtCl_6]^{2-}$
$(Pt) + 6(-1) = -2$
$Pt - 6 = -2$
$Pt = +4$

(f) Fe in $\left[Fe(CN)_6\right]^{3-}$

The sum of the oxidation numbers should be −3:
$\left[Fe(CN)_6\right]^{3-}$
$1(Fe) + 6(-1) = -3$
$Fe - 6 = -3$
$Fe = +3$

22.3.3. Oxidation number and nomenclature

The concept of oxidation number is very important in naming complex molecular or ionic compounds. Binary chemical compounds such as CO, S_2Cl_2, NO_2, and CCl_4, are commonly named by simply indicating numbers of constituent atoms. The oxidation number is often not stated. Thus CO is named carbon *mono*oxide, S_2Cl_2 is *di*sulfur *di*chloride, NO_2 is nitrogen *di*oxide and CCl_4 is carbon*tetra*chloride.

When naming complex compounds or ions under the IUPAC convention, the oxidation number of the component element that exhibits variable oxidation number is indicated in Roman numerals. Some examples are shown in table 22.1.

Table 22-1 Oxidation Number and IUPAC Nomenclature of Some Common Compounds

Formula	*IUPAC name*	*Common name*
HNO_2	Nitric(III) acid	Nitrous acid
HNO_3	Nitric(V) acid	Nitric acid
H_2SO_3	Sulfuric(IV) acid	Sulfurous acid
H_2SO_4	Sulfuric(VI) acid	Sulfuric acid
$FeCl_2$	Iron(II) chloride	Ferrous chloride
$FeCl_3$	Iron(III) chloride	Ferric chloride

22.4. Oxidation and Reduction in Terms of Oxidation Number

In oxidation–reduction reactions, the reacting substances undergo changes in oxidation number. Oxidation results in an increase in oxidation number due to a loss of electrons. Reduction results in a decrease in oxidation number due to a gain of electrons. In the reaction below, the oxidation number of Mg has increased from 0 to +2, and magnesium is therefore said to be oxidized. Hydrogen, on the other hand, is reduced because its oxidation number has decreased from +1 to 0.

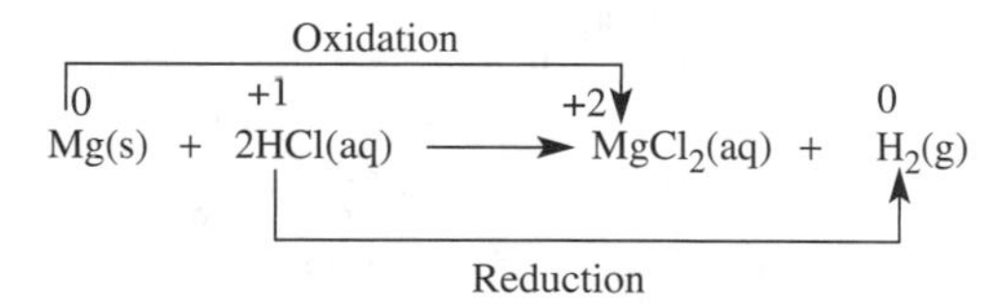

22.5. Disproportionation Reactions

Disproportionation is a special type of redox reaction in which a single reactant is simultaneously oxidized and reduced. A good example is the preparation of nitric acid from the action of nitrogen dioxide on water.

$$\overset{+4}{3NO_2(g)} + H_2O(l) \longrightarrow 2\,\overset{+5}{HNO_3(aq)} + \overset{+2}{NO(g)}$$

The nitrogen in nitrogen dioxide disproportionates from the +4 oxidation state to both +5 and +2. Similarly, copper (I) ions disproportionate in solution:

$$2\,\overset{+1}{Cu^+(aq)} \longrightarrow \overset{0}{Cu(s)} + \overset{+2}{Cu^{2+}(aq)}$$

22.6. Oxidizing and Reducing Agents

Substances which bring about oxidation are called oxidizing agents. In a redox reaction, the oxidizing agent is the:

- Substance that is reduced.
- Substance that gains electrons.
- Substance in which the oxidation number has decreased (reduced).

Substances that bring about reduction are called reducing agents. In a redox reaction, the reducing agent is the:

- Substance that is oxidized.
- Substance that loses electrons.
- Substance in which the oxidation number has increased (oxidized).

Consider the reaction below:

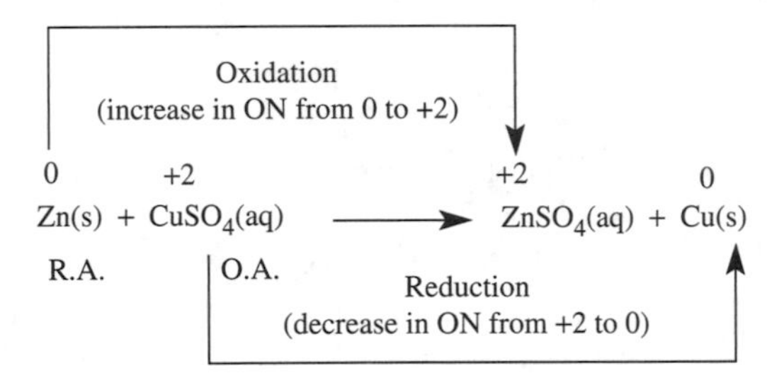

Zinc donates 2 electrons and goes into solution as zinc (Zn^{2+}) ions. Copper ions (Cu^{2+}) gain these electrons and are deposited as elemental copper. Zinc is oxidized, and hence it is the reducing agent (RA). On the other hand, copper, which is reduced, is the oxidizing agent (OA).

Reducing agents are used industrially in the extraction of metals from their ores. For example, carbon in the form of coke can be used to reduce copper (II) oxide and iron (III) oxide to the metals—copper and iron.

22.6.1. Identifying oxidizing and reducing agents

To identify oxidizing and reducing agents, calculate the oxidation numbers for every atom in the reaction, and determine which ones change. The reactant molecule with a net increase in oxidation number (loss of electrons or oxidation) is the *reducing agent*. Similarly, the reactant with a net decrease in oxidation number (gain of electrons, reduction) is the *oxidizing agent*.

Example 22.2

Assign oxidation numbers to all atoms in the equation below. Indicate which substance is oxidized and which is reduced. Identify the oxidizing and reducing agents.

$$2\,Fe_2O_3(s) + 3\,C(s) \longrightarrow 4\,Fe(s) + 3\,CO_2(g)$$

Solution

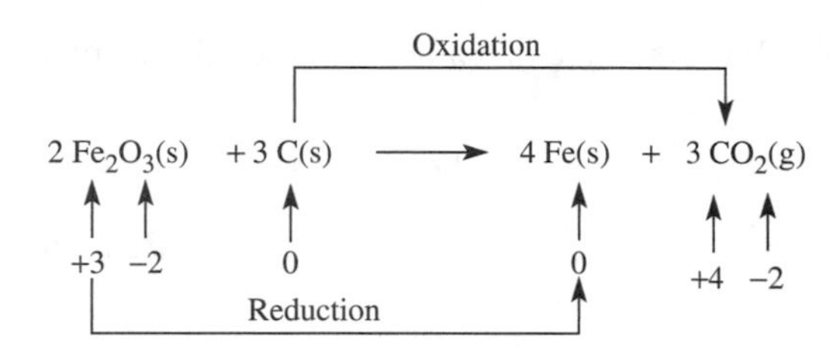

Fe_2O_3 is reduced (+3 to 0) C is oxidized (0 to +4)

Fe_2O_3 is the oxidizing agent C is the reducing agent

The uncombined C and Fe have oxidation numbers of zero; Fe in Fe_2O_3 is +3. O in Fe_2O_3 and CO_2 is −2.

C is oxidized (from 0 to +4) while Fe_2O_3 is reduced (from +3 to 0).

Fe_2O_3 is the oxidizing agent, and C is the reducing agent.

22.7. Half-Cell Reactions

The oxidation and reduction steps of a redox reaction can be described by two separate equations called *half-equations* or *half-cell reactions*. A half-cell reaction is largely a convenient calculating aid; one half-reaction never takes place in isolation. Each such equation represents half of a redox reaction, either the reduction or the oxidation.

The overall reaction is always the sum of two half-cell reactions—one for oxidation, and the other for reduction. For example, consider the redox reaction:

$$Zn(s) + Cu^{2+}(aq) \longrightarrow Zn^{2+}(aq) + Cu(s)$$

This can be separated into a half-reaction for each metal:

$$\text{Oxidation half-reaction} \quad Zn(s) \longrightarrow Zn^{2+}(aq) + 2e^-$$

$$\text{Reduction half-reaction} \quad Cu^{2+}(aq) + 2e^- \longrightarrow Cu(s)$$

$$\text{Net reaction} \quad Zn(s) + Cu^{2+}(aq) \longrightarrow -Zn^{2+}(aq) + Cu(s)$$

Example 22.3

Iron metal reacts with aqueous copper (II) ion to give iron (II) ion and copper metal according to the following equation:

$$Fe(s) + Cu^{2+}(aq) \longrightarrow Fe^{2+}(aq) + Cu(s)$$

Write the two half-cell reactions, labeling the oxidation and reduction reactions.

Solution

The two half-cell reaction are given below. When these are added, we obtain the original net reaction.

$$Fes \longrightarrow Fe^{2+}(aq) + 2e^- \quad \text{Oxidation}$$

$$Cu^{2+}(aq) + 2e^- \longrightarrow Cu(s) \quad \text{Reduction}$$

22.8. Balancing Redox Equations

Many chemical reactions are represented by simple equations that can be balanced by mere inspection. But oxidation–reduction reactions are often very complex, and it can be quite difficult and time-consuming to use inspection or trial and error to balance their equations. In general, redox equations are balanced by following the principles of conservation of charge (i.e., the total number of electrons lost by the reducing agent must be equal to the number of electrons gained by the oxidizing agent in a chemical reaction). Two common methods for balancing redox equations will be discussed here:

1. The oxidation number method.
2. The half-reaction method.

22.8.1. The oxidation-number method

1. Write the unbalanced equation.
2. Assign oxidation numbers to all the atoms in the equation. Identify the atoms that are oxidized and reduced.
3. Write two new equations, one for the oxidation step and the other for the reduction step, using only species that change in oxidation number. Add electrons to reflect the change in oxidation number.
4. Multiply both equations by small integers such that the total net increase in oxidation number for the oxidized atoms equals the total net decrease in oxidation number for the reduced atoms.
5. Assign these integers as coefficients of the oxidized and reduced species on both sides of the unbalanced equation.
6. Add water to the side with less O until O is balanced. Then, if H isn't balanced, add H^+ to the side with less H. If the reaction takes place in acidic solution,

you are done, but if the reaction takes place in basic solution, there is one more step: add as many OH^- to each side as you added free H^+. (Note that $H^+ +$ OH^- can be written as water, H_2O.) Balance the remaining species that are not oxidized or reduced.

7. Check for atom and charge balance of the final equation.

Example 22.4

Use the oxidation number method to balance the equation for the reaction of nitric acid with hydrogen sulfide.

Solution

Step 1: Write the unbalanced equation.

$$HNO_3(aq) + H_2S(g) \longrightarrow S(s) + NO(g) + H_2O(l)$$

Step 2: Assign oxidation numbers to all the atoms in the equation. Identify the atoms that are oxidized and reduced.

$$\begin{array}{ccccc} HNO_3(aq) + & H_2S(g) \longrightarrow & S(s) + & NO(g) + & H_2O(l) \\ +1+5-2 & +1-2 & 0 & +2-2 & +1-2 \end{array}$$

Nitric acid is reduced (gain of $3e^-$), sulfur is oxidized (loss of $2e^-$).

Step 3: Write two new equations, one for the oxidation step and the other for the reduction step, using only species that change in oxidation number. Add electrons to reflect the change in oxidation number.

$$\text{Oxidation} \quad S^{2-} \longrightarrow S^0 + 2e^-$$

$$\text{Reduction} \quad N^{5+} + 3e^- \longrightarrow N^{2+}$$

Step 4: Multiply both equations by the smallest possible integers such that the total net increase in oxidation number for the oxidized atoms equals the total net decrease in oxidation number for the reduced atoms.

$$\text{Oxidation} \quad 3\,S^{2-} \longrightarrow 3\,S^0 + 6e^-$$

$$\text{Reduction} \quad 2\,N^{5+} + 6e^- \longrightarrow 2\,N^{2+}$$

Step 5: Assign these integers as coefficients of the oxidized and reduced species on both sides of the unbalanced equation.

$$2\,HNO_3(aq) + 3\,H_2S(g) \longrightarrow 3\,S(s) + 2\,NO(g) + H_2O(l) \quad \text{(unbalanced)}$$

Step 6: This is not necessary for the present problem.

Step 7: Balance the remaining species that are not oxidized or reduced.

$$2\ HNO_3(aq) + 3\ H_2S(g) \longrightarrow 3\ S(s) + 2\ NO(g) + H_2O(l) \quad \text{(balanced)}$$

Step 8: The final balanced equation contains equal numbers of atoms on each side.

Example 22.5

Balance the equation below (assuming it takes place in a basic solution) by the oxidation number method.

$$MnO_4^-(aq) + SO_3^{2-}(aq) \longrightarrow MnO_4^{2-}(aq) + SO_4^{2-}(aq)$$

Solution

Step 1: Write the unbalanced equation.

$$MnO_4^-(aq) + SO_3^{2-}(aq) \longrightarrow MnO_4^{2-}(aq) + SO_4^{2-}(aq)$$

Step 2: Assign oxidation numbers to all the atoms in the equation. Identify the atoms that are oxidized and reduced.

$$MnO_4^-(aq) + SO_3^{2-}(aq) \longrightarrow MnO_4^{2-}(aq) + SO_4^{2-}(aq)$$

$$+7-2 \qquad +4-2 \qquad\qquad +6-2 \qquad +6-2$$

$MnO_4^-(aq)$ is reduced ($Mn^{7+} \longrightarrow Mn^{6+}$, gain of $1e^-$)

$SO_3^{2-}(aq)$ is oxidized ($S^{4+} \longrightarrow S^{6+}$, loss of $2e^-$)

Step 3: Write two new equations, one for the oxidation step and the other for the reduction step, using only species that change in oxidation number. Add electrons to reflect the change in oxidation number.

$$\text{Oxidation} \quad S^{4+} \longrightarrow S^{6+} + 2e^-$$

$$\text{Reduction} \quad Mn^{7+} + 1e^- \longrightarrow Mn^{6+}$$

Step 4: Multiply both equations by the smallest possible integers such that the total net increase in oxidation number for the oxidized atoms equals the total net decrease in oxidation number for the reduced atoms.

$$\text{Oxidation} \quad S^{4+} \longrightarrow S^{6+} + 2e^-$$

$$\text{Reduction} \quad 2\ Mn^{7+} + 2e^- \longrightarrow 2\ Mn^{6+}$$

Step 5: Assign these integers as coefficients of the oxidized and reduced species on both sides of the unbalanced equation.

$$2\,MnO_4^-(aq) + SO_3^{2-}(aq) \longrightarrow 2\,MnO_4^{2-}(aq) + SO_4^{2-}(aq)$$

Step 6: Balance the equation for oxygen and hydrogen as follows: add H_2O to the side with less O and add H^+ to the side with less H.

$$2\,MnO_4^-(aq) + SO_3^{2-}(aq) + H_2O(l) \longrightarrow 2\,MnO_4^{2-}(aq) + SO_4^{2-}(aq) + 2\,H^+(aq)$$

Since the reaction occurs in a basic solution, we now add $2OH^-$ to both sides of the above equation to give

$$2\,MnO_4^-(aq) + SO_3^{2-}(aq) + H_2O(l) + 2\,OH^-(aq) \longrightarrow 2\,MnO_4^{2-}(aq) + SO_4^{2-}(aq) + 2\,H_2O(l)$$

Step 7: This is not required since all the atoms are already balanced.
Step 8: The final balanced equation contains equal numbers of atoms and equal charges on both sides of the equation.

22.8.2. The half-reaction method

1. Write the unbalanced ionic equation and decide which atoms are oxidized and which atoms are reduced.
2. Write unbalanced equations for the oxidation and reduction half-reactions.
3. Balance all the elements other than hydrogen and oxygen in each half-reaction using the inspection method.
4. Balance oxygen and hydrogen. For reactions in acidic solution, add H^+ to balance hydrogen and H_2O to balance oxygen. For reactions occurring in basic solution, first treat as if the reaction is occurring in an acidic solution; then add as many OH^- ions to each side of the reaction as there are H^+ ions in the equation. Combine the H^+ and OH^- ions to form H_2O. Re-write the equation and eliminate equal numbers of water molecules appearing on opposite side of the equation.
5. Balance the charge in each half-reaction by adding electrons as reactants or products.
6. Multiply each half-reaction by an appropriate integer so that the number of electrons gained in the reduction process equals the number of electrons lost in the oxidation process.
7. Add the two half-reactions and eliminate any species that appear on both sides of the equation.

Example 22.6

Balance the following half-reactions in acidic solution:

(a) $Cr_2O_7^{2-}(aq) \longrightarrow Cr^{3+}(aq)$
(b) $Ti^{3+}(aq) \longrightarrow TiO_2(s)$

Solution

(a) Balance Cr.

$Cr_2O_7^{2-}(aq) \longrightarrow 2\,Cr^{3+}(aq)$

Balance O by adding H_2O

$Cr_2O_7^{2-}(aq) \longrightarrow 2\,Cr^{3+}(aq) + 7\,H_2O(l)$

Balance H by adding H^+

$Cr_2O_7^{2-}(aq) + 14\,H^+(aq) \longrightarrow 2\,Cr^{3+}(aq) + 7\,H_2O(l)$

Balance charge by adding electrons. This gives the balanced equation:

$Cr_2O_7^{2-}(aq) + 14\,H^+(aq) + 6e^- \longrightarrow 2\,Cr^{3+}(aq) + 7\,H_2O(l)$

(b) Ti is already balanced.

$Ti^{3+}(aq) \longrightarrow TiO_2(s)$

Balance O by adding H_2O

$Ti^{3+}(aq) + 2\,H_2O(l) \longrightarrow TiO_2(s)$

Balance H by adding H^+

$Ti^{3+}(aq) + 2\,H_2O(l) \longrightarrow TiO_2(s) + 4\,H^+(aq)$

Balance charge by adding electrons. This gives the balanced equation:

$Ti^{3+}(aq) + 2\,H_2O(l) \longrightarrow TiO_2(s) + 4\,H^+(aq) + e^-$

Example 22.7

Balance the following ionic equation in acidic solution:

$$Fe^{2+}(aq) + MnO_4{}^-(aq) \longrightarrow Fe^{3+}(aq) + Mn^{2+}(aq)$$

Solution

Step 1: Write the unbalanced ionic equation and decide which substances are oxidized and which are reduced.

$$Fe^{2+}(aq) + MnO_4{}^-(aq) \longrightarrow Fe^{3+}(aq) + Mn^{2+}(aq)$$

Oxidized Reduced

Step 2: Write unbalanced equations for the oxidation and reduction half-reactions.

Oxidation half-reaction $Fe^{2+} \longrightarrow Fe^{3+}$

Reduction half-reaction $MnO_4^- \longrightarrow Mn^{2+}$

Step 3: Balance all the elements other than hydrogen and oxygen in each half-reaction using the inspection method. Here Fe and Mn are already balanced.

Step 4: Balance oxygen and hydrogen. Since the reaction is in acidic solution, balance oxygen in the reduction half-reaction by adding 4 H_2O to the right side. Then balance hydrogen by adding $8H^+$ to the left side of the equation.

$$MnO_4{}^-(aq) + 8\,H^+(aq) \longrightarrow Mn^{2+}(aq) + 4\,H_2O(l)$$

Step 5: Balance the charge in each half-reaction by adding electrons as reactants or products. The total charge is +7 on the left side and +2 on the right side of the reduction half-reaction. Therefore add electrons to the left side. For the oxidation reaction, the total charge is +2 on the left side, and +3 on the right side. Therefore add one electron to the right side.

$$MnO_4^-(aq) + 8\,H^+(aq) + 5e^- \longrightarrow Mn^{2+}(aq) + 4\,H_2O(l)$$

$$Fe^{2+}(aq) \longrightarrow Fe^{3+}(aq) + e^-$$

Step 6: Multiply each half-reaction by an appropriate integer so that the number of electrons gained in the reduction equals the number of electrons lost in the oxidation. The reduction reaction requires the transfer of five electrons while the oxidation reaction requires the transfer of only one electron. Therefore, multiply the oxidation half-equation by 5 to equalize the electron transfer.

$$MnO_4^-(aq) + 8\,H^+(aq) + 5e^- \longrightarrow Mn^{2+}(aq) + 4\,H_2O(l)$$

$$5\,Fe^{2+}(aq) \longrightarrow 5\,Fe^{3+}(aq) + 5e^-$$

Step 7: Add the two half-reactions and eliminate any species that appear on both sides. This yields the balanced equation.

$$MnO_4^-(aq) + 8\,H^+(aq) + 5e^- \longrightarrow Mn^{2+}(aq) + 4\,H_2O(l)$$

$$5\,Fe^{2+}(aq) \longrightarrow 5\,Fe^{3+}(aq) + 5e^-$$

$$MnO_4^-(aq) + 8\,H^+(aq) + 5\,Fe^{2+}(aq) \longrightarrow Mn^{2+}(aq) + 4\,H_2O(l) + 5\,Fe^{3+}(aq)$$

Example 22.8

Balance the following net ionic reaction in acidic solution.

$$Zn(s) + NO_3^-(aq) \longrightarrow NH_4^+(aq) + Zn^{2+}(aq)$$

Solution

The unbalanced equation is $Zn(s) + NO_3^-(aq) \longrightarrow NH_4^+(aq) + Zn^{2+}(aq)$.

Step 1: Write and balance the reduction half-equation:

$NO_3^-(aq) \longrightarrow NH_4^+(aq)$

Balance O with H_2O and H with H^+:

$NO_3^-(aq) + 10\,H^+(aq) \longrightarrow NH_4^+(aq) + 3H_2O(l)$

Balance charge with electrons:

$NO_3^-(aq) + 10\,H^+(aq) + 8e^- \longrightarrow NH_4^+(aq) + 3\,H_2O(l)$

Step 2: Write and balance the oxidation half-equation:

$$Zn(s) \longrightarrow Zn^{2+}(aq) + 2e^-$$

Step 3: Multiply the redox half-equations by appropriate factors such that the number of electrons gained is equal to the number of electrons lost.

$$4 \times \left[Zn(s) \longrightarrow Zn^{2+}(aq) + 2e^-\right] = 4\,Zn(s) \longrightarrow 4\,Zn^{2+}(aq) + 8e^-$$

$$1 \times \left[NO_3^-(aq) + 10\,H^+(aq) + 8e^- \longrightarrow NH_4^+(aq) + 3\,H_2O(l)\right]$$

$$= NO_3^-(aq) + 10\,H^+(aq) + 8e^- \longrightarrow NH_4^+(aq) + 3\,H_2O(l)$$

Step 4: Add the two half-equations to get the overall balanced equation. Be sure to cancel the electrons and other species that appear on both sides of the equation, and check that the coefficients are reduced to the simplest whole numbers.

$$4\,Zn(s) \longrightarrow 4\,Zn^{2+}(aq) + 8e^-$$

$$NO_3^-(aq) + 10\,H^+(aq) + 8\,e^- \longrightarrow NH_4^+(aq) + 3\,H_2O(l)$$

$$4\,Zn(s) + NO_3^-(aq) + 10\,H^+(aq) \longrightarrow 4\,Zn^{2+}(aq) + NH_4^+(aq) + 3\,H_2O(l)$$

Example 22.9

Balance the following redox reaction, which occurs in acidic solution, using the half-reaction method.

$$I^-(aq) + Cr_2O_7^{2-}(aq) \longrightarrow Cr^{3+}(aq) + IO_3^-(aq)$$

Solution

The unbalanced equation is $I^-(aq) + Cr_2O_7^{2-}(aq) \longrightarrow Cr^{3+}(aq) + IO_3^-(aq)$.

Step 1: Write and balance the reduction half-equation:

$Cr_2O_7^{2-}(aq) \longrightarrow 2Cr^{3-}(aq)$

Balance O with H_2O and H with H^+:

$$Cr_2O_7^{2-}(aq) + 14\,H^-(aq) \longrightarrow 2\,Cr^{3+}(aq) + 7\,H_2O(l)$$

Balance charge with electrons:

$$Cr_2O_7^{2-}(aq) + 14\,H^-(aq) + 6e^- \longrightarrow 2\,Cr^{3+}(aq) + 7\,H_2O(l)$$

Step 2: Write and balance the oxidation half-equation:

$$I^-(aq) \longrightarrow IO_3^-(aq)$$

Balance O with H_2O and H with H^+:

$$I^-(aq) + 3\,H_2O(l) \longrightarrow IO_3^-(aq) + 6\,H^+(aq)$$

Balance charge with electrons:

$$I^-(aq) + 3\,H_2O(l) \longrightarrow IO_3^-(aq) + 6\,H^+(aq) + 6e^-$$

Step 3: Multiply the redox half-equations by appropriate factors such that the number of electrons gained is equal to the number of electrons lost. In this case, both half-equations contain the same number of electrons.

Step 4: Add the two half-equations to get the overall balanced equation. Be sure to cancel the electrons and other species that appear on both sides of the equation, and check that the coefficients are reduced to the simplest whole numbers.

$$I^-(aq) + 3H_2O(l) \longrightarrow IO_3^-(aq) + 6H^+(aq) + 6e^-$$

$$Cr_2O_7^{2-}(aq) + 14H^+(aq) + 6e^- \longrightarrow 2Cr^{3+}(aq) + 7H_2O(l)$$

$$I^-(aq) + Cr_2O_7^{2-}(aq) + 8\,H^+(aq) \longrightarrow IO_3^-(aq) + 2Cr^{3+}(aq) + 4H_2O(l)$$

Example 22.10

Balance the following half-reactions in basic solution (add H^+, OH^-, or H_2O as necessary).

1. $Bi^{3+}\,(aq) \longrightarrow BiO_3^-\,(aq)$
2. $Br_2\,(aq) \longrightarrow BrO_3^-\,(aq)$
3. $CrO_4^{2-}\,(aq) \longrightarrow Cr(OH)_4^-\,(aq)$

Solution

1. Balance the equation $Bi^{3+}(aq) \longrightarrow BiO_3^-(aq)$.

   Step 1: Balance the equation as if H^+ were present:

   $$Bi^{3+}(aq) \longrightarrow BiO_3^-(aq)$$

Balance O with H_2O:

$3\,H_2O(l) + Bi^{3+}(aq) \longrightarrow BiO_3^-(aq)$

Balance H using H^+ :

$3\,H_2O(l) + Bi^{3+}(aq) \longrightarrow BiO_3^-(aq) + 6\,H^+(aq) + 2e^-$

Step 2: Since we know that the reaction is taking place in a basic solution, we must add 6 OH^- to both sides of the equation. This will neutralize the 6 H^+ on the right side.

$$3\,H_2O(l) + Bi^{3+}(aq) + 6\,OH^-(aq) \longrightarrow BiO_3^-(aq) + 6\,H^+(aq) + 6\,OH^-(aq) + 2e^-$$

Combine $H^+ + OH^-$ to form H_2O on the right side:

$$3\,H_2O(l) + Bi^{3+}(aq) + 6\,OH^-(aq) \longrightarrow BiO_3^-(aq) + 6\,H_2O(aq) + 2e^-$$

Eliminate 3 H_2O from both sides of the equation:

$Bi^{3+}(aq) + 6\,OH^-(aq) \longrightarrow BiO_3^-(aq) + 3\,H_2O(aq) + 2e^-$

2. Balance $Br_2(aq) \longrightarrow BrO_3^-(aq)$.

Step 1: Balance the equation as if H^+ were present:

Balance Br:

$Br_2(aq) \longrightarrow 2\,BrO_3^-(aq)$

Balance O with H_2O:

$6\,H_2O(l) + Br_2(aq) \longrightarrow 2\,BrO_3^-(aq)$

Balance H using H^+ :

$6\,H_2O(l) + Br_2(aq) \longrightarrow 2\,BrO_3^-(aq) + 12\,H^+(aq) + 10e^-$

Step 2: Since we know that the reaction is taking place in a basic solution, we must add 12 OH^- to both sides of the equation. This will neutralize the 12 H^+ on the right side.

$$6\,H_2O(l) + Br_2(aq) + 12\,OH^-(aq) \longrightarrow 2\,BrO_3^-(aq) + 12\,H^+(aq) + 12\,OH^-(aq) + 10e^-$$

Combine $H^+ + OH^-$ to form H_2O on the right side of the equation :

$$6\,H_2O(l) + Br_2(aq) + 12\,OH^-(aq) \longrightarrow 2\,BrO_3^-(aq) + 12\,H_2O(aq) + 10\,e^-$$

Eliminate 6 H_2O from both sides of the equation :

$Br_2(aq) + 12\,OH^-(aq) \longrightarrow 2\,BrO_3^-(aq) + 6\,H_2O(aq) + 10e^-$

3. Balance $CrO_4^{2-}(aq) \longrightarrow Cr(OH)_4^-(aq)$.

Step 1: Balance the equation as if H^+ were present.

Cr and O are balanced:

$CrO_4^{2-}(aq) \longrightarrow Cr(OH)_4^-(aq)$

Balance H with H^+. Also, balance the charge:

$CrO_4^{2-}(aq) + 4\,H^+(aq) + 3e^- \longrightarrow Cr(OH)_4^-(aq)$

Step 2: We know that the reaction is taking place in a basic solution. Therefore we must add 4 OH^- to both sides of the equation. This will neutralize the 4 H^+ on the left side.

$$CrO_4^{2-}(aq) + 4\,H^+(aq) + 4\,OH^-(aq) + 3e^- \longrightarrow Cr(OH)_4^-(aq) + 4\,OH^-(aq)$$

Combine H^+ and OH^- to form H_2O. This yields the balanced equation.

$$CrO_4^{2-}(aq) + 4\,H_2O(aq) + 3e^- \longrightarrow Cr(OH)_4^-(aq) + 4\,OH^-(aq)$$

Example 22.11

Balance the following net ionic reaction in basic solution using the half-reaction method.

$$MnO_2(s) + Zn(s) \longrightarrow MnO(OH)(s) + Zn(OH)_4^{2-}(aq)$$

Solution

Step 1: Balance the reduction half-reaction as if it occurred in acidic solution.

Mn and O are both balanced.

$$MnO_2(s) \longrightarrow MnO(OH)(s)$$

Balance H with H^+ :

$$MnO_2(s) + H^+(aq) \longrightarrow MnO(OH)(s)$$

Balance charge with electrons:

$$MnO_2(s) + H^+(aq) + e^- \longrightarrow MnO(OH)(s)$$

Step 2: Balance the oxidation half-reaction as if it occurred in acidic solution.

Zn is balanced.

$$Zn(s) \longrightarrow Zn(OH)_4^{2-}(aq)$$

Balance O with H_2O and H with H^+ :

$$Zn(s) + 4\,H_2O(l) \longrightarrow Zn(OH)_4^{2-}(aq) + 4\,H^+(aq)$$

Balance charge with electrons:

$$Zn(s) + 4\,H_2O(l) \longrightarrow Zn(OH)_4^{2-}(aq) + 4\,H^+(aq) + 2e^-$$

Step 3: Multiply the balanced reduction half-equation by 2 to make the number of electrons transferred equal.

$$2\,\left[MnO_2(s) + H^+(aq) + e^- \longrightarrow MnO(OH)(s)\right]$$

$$2\,MnO_2(s) + 2H^+(aq) + 2e^- \longrightarrow 2\,MnO(OH)(s)$$

Step 4: Add the two half-reactions and eliminate any species that appears on both sides of the equation. This yields the balanced equation as if it were in acidic solution.

$$2\,MnO_2(s) + 2\,H^+(aq) + 2e^- \longrightarrow 2\,MnO(OH)(s)$$

$$Zn(s) + 4\,H_2O(l) \longrightarrow Zn(OH)_4^{2-}(aq) + 4\,H^+(aq) + 2e^-$$

$$Zn(s) + 4\,H_2O(l) + 2\,MnO_2(s) \longrightarrow Zn(OH)_4^{2-}(aq) + 2\,H^+(aq) + 2\,MnO(OH)(s)$$

Step 5: Since the reaction is occurring in basic solution, we must add 2 OH^- ions to both sides of the equation to neutralize the 2 H^+.

$$Zn(s) + 4\,H_2O(l) + 2\,MnO_2(s) + 2\,OH^-(aq) \longrightarrow Zn(OH)_4^{2-}(aq) + 2\,H^+(aq) + 2\,OH^-(aq) + 2\,MnO(OH)(s)$$

Combine $2H^+$ and $2OH^-$ to form H_2O:

$$Zn(s) + 4\,H_2O(l) + 2\,MnO_2(s) + 2\,OH^-(aq) \longrightarrow Zn(OH)_4^{2-}(aq) + 2\,H_2O(aq) + 2\,MnO(OH)(s)$$

Eliminate 2 H_2O from both sides. This gives the balanced equation.

$$Zn(s) + 2\,H_2O(l) + 2\,MnO_2(s) + 2\,OH^-(aq) \longrightarrow Zn(OH)_4^{2-}(aq) + 2\,MnO(OH)(s)$$

22.9. Oxidation–Reduction Titration

A redox titration is a volumetric analysis method just like an acid–base titration except that it is based on a redox reaction in which electrons are transferred to an oxidizing agent from a reducing agent. The substance being titrated needs to be 100% in a single oxidation state. Where this is not the case, the substance has to be oxidized or reduced by adding excess reducing or oxidizing agent before initiating the titration. (In this case it is important to remove or destroy the excess agent before the titration to avoid interference).

As in acid–base titration, the equivalence point of a redox titration can be determined visually, by the color change of an indicator, or graphically (from titration curves), by plotting the potential measured by an electrode versus the volume of titrant added. In some cases, the redox reaction is self-indicating, and no external indicator is required because the analyte (titrant) changes color intensely at the equivalence point. An example is the permanganate ion, MnO_4^-, whose color changes from purple to pink as it is reduced to Mn^{2+}. If the titrant is not self-indicating, then a redox indicator is added.

Many inorganic and organic substances can be determined via redox titration, such as H_2O_2, I_2, Fe^{3+}, MnO_4^-, Sn^{2+}, phenol, aniline, carboxylic acid, aldehydes, and ketones.

22.9.1. Calculations involving redox titration

Here we will only deal with basic calculations involving standardization experiments and the determination of an unknown.

Example 22.12

Standardized solutions of potassium permanganate ($KMnO_4$) are frequently used in redox titration. In one experiment, a solution of $KMnO_4$ is standardized by titration with oxalic acid ($H_2C_2O_4$) solution prepared by dissolving 0.260 g in 250 mL of 0.5 M H_2SO_4. The balanced equation for the reaction is:

$$5\,H_2C_2O_4(aq) + 2\,MnO_4^-(aq) + 6\,H^+(aq) \longrightarrow 10\,CO_2(g) + 2\,Mn^{2+}(aq) + 8\,H_2O(l)$$

What is the molar concentration of the permanganate solution if it required 22.40 mL to react completely with the oxalic acid solution?

Solution

1. First we need to calculate the number of moles of oxalic acid that reacted with MnO_4^-. (Note that the volume and concentration of sulfuric acid are not relevant in this calculation.)

$$\text{Moles of } H_2C_2O_4 = (0.260\text{ g})\left(\frac{1\text{ mol } H_2C_2O_4}{90\text{ g } H_2C_2O_4}\right) = 0.00289\text{ mol}$$

2. Calculate the number of moles of $KMnO_4$ present in the 22.40 mL of solution used to reach the end point.
From the balanced equation, 5 mol $H_2C_2O_4$ = 2mol $KMnO_4$. Thus the conversion factor is $\dfrac{2\text{ mol } KMnO_4}{5\text{ mol } H_2C_2O_4}$

$$\text{moles } KMnO_4 = (\text{mol } H_2C_2O_4)\left(\frac{2\text{ mol } KMnO_4}{5\text{ mol } H_2C_2O_4}\right)$$

$$= (0.00289\text{ mol } H_2C_2O_4) \times \left(\frac{2\text{ mol } KMnO_4}{5\text{ mol } H_2C_2O_4}\right) = 0.00116\text{ mol}$$

3. Calculate the molarity of the permanganate solution from the end point volume and the number of moles.

$$\text{Molarity} = \frac{\text{Number of moles}}{\text{Liters of solution}} = \frac{0.0116 \text{ mol}}{0.0224 \text{ L}} = 0.52 \text{ M}$$

Example 22.13

$KMnO_4$ acts as its own indicator when used in titration. A standard 0.05 M solution of $KMnO_4$ was titrated with 50 mL of 0.20 M acidified $FeSO_4$ solution. Calculate the volume of $KMnO_4$ solution required to completely oxidize the $FeSO_4$ solution.

Solution

We can solve this problem with either the conversion factor method or the formula method.

The conversion factor method

1. Write a balanced equation for the reaction:

$$MnO_4^-(aq) + 8\,H^+(aq) + 5\,Fe^{2+}(aq) \longrightarrow 5\,Fe^{3+}(aq) + Mn^{2+}(aq) + 4\,H_2O(l)$$

2. Calculate the number of moles of Fe^{2+} titrated from its molarity and the volume used:

$$\text{moles } Fe^{2+} = (50 \text{ mL})\left(\frac{0.20 \text{ mol}}{1000 \text{ mL}}\right) = 0.010 \text{ mol}$$

3. Using the balanced equation, calculate the moles of MnO_4^- required:

$$1 \text{ mol } MnO_4^- = 5 \text{ mol } Fe^{2+}$$

$$\text{Conversion factor} = \frac{1 \text{ mol } MnO_4^-}{5 \text{ mol } Fe^{2+}}$$

$$\text{moles } MnO_4^- = \left(0.010 \text{ mol } Fe^{2+}\right)\left(\frac{1 \text{ mol } MnO_4^-}{5 \text{ mol } Fe^{2+}}\right) = 0.0020 \text{ mol}$$

4. Now calculate the volume of 0.05 M MnO_4^- which contains 0.0020 mol MnO_4^-:

$$\text{mL of } MnO_4^-(aq) = \left(0.0020 \text{ mol } MnO_4^-\right)\left(\frac{1000 \text{ mL}}{0.05 \text{ mol}}\right) = 40 \text{ mL}$$

The formula method

This problem can also be solved using the formula method.

$$MnO_4^-(aq) + 8\,H^+(aq) + 5\,Fe^{2+}(aq) \longrightarrow 5\,Fe^{3+}(aq) + Mn^{2+}(aq) + 4\,H_2O(l)$$

From the balanced equation:

$$\frac{M_{MnO_4^-} V_{MnO_4^-}}{M_{Fe^{2+}} V_{Fe^{2+}}} = \frac{n_{MnO_4^-}}{n_{Fe^{2+}}} = \frac{1}{5}$$

$$V_{MnO_4^-} = \frac{M_{Fe^{2-}} V_{Fe^{2+}} n_{MnO_4^-}}{M_{MnO_4}^- n_{Fe^{2+}}} = \frac{50\text{ mL} \times 0.20\text{ M} \times 1}{0.05\text{ M} \times 5} = 40\text{ mL}$$

Example 22.14

Exactly 2.05 g of an iron ore was dissolved in an acid solution and all the Fe^{3+} was converted to Fe^{2+}. The resulting solution required 45 mL of 0.05 M $K_2Cr_2O_7$ solution for oxidation.

(a) Write a balanced equation for the redox reaction.
(b) Calculate the mass of Fe in the ore.
(c) What is the percentage of Fe in the original sample?

Solution

(a) The balanced equation is:

$$6\,Fe^{2+}(aq) + Cr_2O_7^{2-}(aq) + 14\,H^+(aq) \longrightarrow 6\,Fe^{3+}(aq) + 2\,Cr^{3+}(aq) + 7\,H_2O(l)$$

(b) To calculate the mass of Fe, we need to calculate the moles of Fe reacted. This can readily be obtained from the moles of $Cr_2O_7^{2-}$ used for the titration.

$$\text{moles } Cr_2O_7^{2-} = (45\text{ mL})\left(\frac{0.05\text{ mol}}{1000\text{ mL}}\right) = 0.00225\text{ mol}$$

From the balanced equation and the moles of $Cr_2O_7^{2-}$, we can calculate moles of Fe.

$$2 \text{ mol } Cr_2O_6^{2-} = 6 \text{ mol } Fe^{2+}$$

$$\text{Conversion factor} = \frac{6 \text{ mol } Fe^{2+}}{2 \text{ mol } Cr_2O_6^{2-}}$$

$$\text{moles } Fe^{2+} = \left(0.00225 \text{ mol } Cr_2O_6^{2-}\right)\left(\frac{6 \text{ mol } Fe^{2-}}{2 \text{ mol } Cr_2O_6^{2-}}\right)$$

$$= 0.00675 \text{ mol}$$

$$\text{Mass of Fe} = \left(55.85 \frac{\text{g}}{\text{mol}}\right)(0.0068 \text{ mol}) = 0.3770 \text{ g}$$

(c) The percentage of Fe in the original sample is obtained as follows:

$$\% \text{ Fe} = \frac{\text{Mass of Fe}}{\text{Mass of sample}} \times 100\%$$

$$= \frac{0.3770}{2.05} \times 100\% = 18.4\%$$

Example 22.15

One thousand milligrams of a new calcium supplement tablet (for patients suffering calcium deficiency) was dissolved in acid, and treated with excess sodium oxalate. The resulting solution was made basic, causing the precipitation of the Ca as the oxalate, CaC_2O_4. The precipitate was filtered, washed, and redissolved in dilute acid. The resulting solution was titrated by 28.50 mL of 0.05 M $KMnO_4$ solution. Determine the calcium content in the tablet.
The equations for the reactions are:

$$CaC_2O_4(s) + 2H^+(aq) \longrightarrow Ca^{2+}(aq) + H_2C_2O_4(aq)$$

$$2\,MnO_4^-(aq) + 5\,H_2C_2O_4(aq) + 6H^+(aq) \longrightarrow 2\,Mn^{2+}(aq) + 10\,CO_2(g) + 8\,H_2O(l)$$

Solution

This is similar to calculations involving back titration. Using the equation showing the reaction between permanganate ion and oxalic acid, we can calculate the moles of oxalic acid that reacted. From the equation showing the production of oxalic acid from the dissolution of calcium oxalate, we can determine the moles of calcium oxalate. Since 1 mol of calcium oxalate contains 1 mol of calcium, the moles and hence the mass of calcium can be calculated.

1. Calculate the moles of oxalic acid:

$$2\,MnO_4^-(aq) + 5\,H_2C_2O_4(aq) + 6\,H^+(aq) \longrightarrow 2\,Mn^{2+}(aq) + 10\,CO_2(g) + 8\,H_2O(l)$$

$$\text{Conversion factor: } \frac{5 \text{ mol } H_2C_2O_4}{2 \text{ mol } MnO_4^-}$$

$$\text{moles } H_2C_2O_4 = (28.50 \text{ mL } MnO_4^-)\left(0.05\frac{\text{mol } MnO_4^-}{1000 \text{ mL } MnO_4^-}\right) \times \left(\frac{5 \text{ mol } H_2C_2O_4}{2 \text{ mol } MnO_4^-}\right) = 0.0036 \text{ mol}$$

2. Calculate the moles of calcium oxalate, and then moles of Ca:

$$CaC_2O_4(s) + 2\,H^+(aq) \longrightarrow Ca^{2+}(aq) + H_2C_2O_4(aq)$$

$$\text{Conversion factor : } \frac{1 \text{ mol } CaC_2O_4}{1 \text{ mol } H_2C_2O_4}$$

$$\text{moles } CaC_2O_4 = (0.0036 \text{ mol } H_2C_2O_4)\left(\frac{1 \text{ mol } CaC_2O_4}{1 \text{ mol } H_2C_2O_4}\right) = 0.0036 \text{ mol}$$

$$\text{moles Ca} = (0.0036 \text{ mol } CaC_2O_4)\left(\frac{1 \text{ mol Ca}}{1 \text{ mol } CaC_2O_4}\right) = 0.0036 \text{ mol}$$

3. Calculate the mass of Ca and the percentage of Ca in the sample:

$$\text{Mass of Ca} = (0.0036 \text{ mol Ca})\left(\frac{40.08 \text{ g Ca}}{1 \text{ mol Ca}}\right) = 0.1428 \text{ g} \quad \text{or} \quad 142.8 \text{ mg Ca}$$

$$\%\ \text{Ca} = \frac{\text{Mass of Ca}}{\text{Mass of sample}} \times 100 = \frac{142.8}{1000 \text{ mg}} \times 100 = 14.28$$

22.10. Problems

1. Calculate the oxidation number for the element indicated in each of the following compounds or ions:

   (a) Si in $CaMg(SiO_3)_2$
   (b) Xe in Na_4XeO_6
   (c) Mn in Mn_2O_7
   (d) Fe in $FePO_4$
   (e) Cl in $Ca(ClO_2)_2$
   (f) O in $SOCl_2$

2. Calculate the oxidation number for the element indicated in each of the following compounds or ions:

   (a) Cr in $Cr_2O_7^{2-}$
   (b) S in $S_4O_6^{4-}$
   (c) V in $HV_6O_{17}^{3-}$
   (d) As in $H_2AsO_4^-$
   (e) Te in $HTeO_3^-$
   (f) Sn in $Sn(OH)_6^{2-}$

3. Consider the following reactions and determine which substance is oxidized and which substance is reduced:

   (a) $4\ Fe + 3\ O_2 \longrightarrow 2\ Fe_2O_3$
   (b) $Cl_2 + 2\ NaI \longrightarrow NaCl + I_2$
   (c) $Mg + H_2SO_4 \longrightarrow MgSO_4 + H_2$
   (d) $CuO + H_2 \longrightarrow Cu + H_2O$

4. For each of the following reactions, determine which substance is oxidized and which is reduced:

   (a) $3\ MnO_2 + 4\ Al \longrightarrow 2\ Al_2O_3 + 3\ Mn$
   (b) $4\ PH_3 + 8\ O_2 \longrightarrow P_4O_{10} + 6\ H_2O$
   (c) $Ca + 2\ H_2O \longrightarrow Ca(OH)_2 + H_2$
   (d) $2\ SO_2 + O_2 \longrightarrow 2\ SO_3$

5. Consider the following reactions. Which substance is the oxidizing agent and which is the reducing agent?

   (a) $2\ FeCl_2(s) + Cl_2(g) \longrightarrow 2\ FeCl_3(g)$
   (b) $Cl_2(g) + H_2S(g) \longrightarrow 2\ HCl(g) + S(s)$
   (c) $C(s) + MgO(s) \longrightarrow CO_2(g) + Mg(s)$
   (d) $2\ Zn(s) + O_2(g) \longrightarrow 2\ ZnO(s)$

6. For each of the following reactions, identify the substance that is oxidized, the substance that is reduced, the oxidizing agent, and the reducing agent.

   (a) $3\ Cu(s) + 8\ HNO_3(aq) \longrightarrow 3\ Cu(NO_3)_2(aq) + 4\ H_2O(l) + 2\ NO(g)$
   (b) $Zn(s) + CuSO_4(aq) \longrightarrow ZnSO_4(aq) + Cu(s)$
   (c) $2\ H_2O(l) + 2\ F_2(g) \longrightarrow 4\ HF(aq) + O_2(g)$
   (d) $Fe_2O_3(s) + 3\ CO(g) \longrightarrow 2\ Fe(s) + 3\ CO_2(g)$

7. Which of the following are oxidation–reduction reactions? Explain your answer in each case.

   (a) $Ca(OH)_2(s) + HNO_3(aq) \longrightarrow Ca(NO_3)_2 + H_2O(l)$
   (b) $Si(s) + 2\ Cl_2(g) \longrightarrow SiCl_4(l)$
   (c) $NiCl_2(aq) + Li_2S(aq) \longrightarrow NiS(s) + 2\ LiCl(aq)$
   (d) $HClO_3(aq) + NH_3(aq) \longrightarrow NH_4ClO_4(aq)$

8. Write a balanced half-reaction for each of the following redox couples in acidic solution:

   (a) $IO_3^- \longrightarrow I_2$
   (b) $O_2 \longrightarrow H_2O_2$
   (c) $S \longrightarrow S_2O_3^{2-}$
   (d) $S_4O_6^{2-} \longrightarrow S_2O_3^{2-}$
   (e) $VO_2^+ \longrightarrow VO^{2+}$
   (f) $CO_2 \longrightarrow H_2C_2O_4$

9. Write a balanced half-reaction for each of the following redox couples in acidic solution:

   (a) $NO_3^- \longrightarrow N_2O_4$
   (b) $Sb_2O_3 \longrightarrow Sb$
   (c) $MnO_2 \longrightarrow Mn^{2+}$
   (d) $HClO \longrightarrow Cl_2$
   (e) $H_2O_2 \longrightarrow H_2O$
   (f) $NO_3^- \longrightarrow HNO_2$

10. Write a balanced half-reaction for each of the following redox couples in alkaline solution:

    (a) $Cu_2O \longrightarrow Cu$
    (b) $MnO_2 \longrightarrow Mn(OH)_2$
    (c) $PbO_2 \longrightarrow PbO$
    (d) $ClO^- \longrightarrow Cl^-$
    (e) $MnO_4^- \longrightarrow MnO_2$
    (f) $Al(OH)_3 \longrightarrow Al$
    (g) $O_3 \longrightarrow O_2$
    (h) $BrO_3^- \longrightarrow Br^-$

11. Balance the following equations in acidic solution using the half-reaction method:

    (a) $Mg + NO_3^- \longrightarrow Mg^{2+} + NO_2$
    (b) $NO_2^- + Cr_2O_7^{2-} \longrightarrow NO_3^- + Cr^{3+}$
    (c) $S^{2-} + NO_3^- \longrightarrow NO + S$

12. Balance the following equations in acidic solution using the half-reaction method:

    (a) $H_2C_2O_4 + MnO_4^- \longrightarrow Mn^{2+} + CO_2$
    (b) $CH_3CH_2OH + Cr_2O_7^{2-} \longrightarrow CH_3COOH + Cr^{3+}$
    (c) $Ca + VO^{2+} \longrightarrow Ca^{2+} + V^{3-}$

13. Balance the following equations in basic solution using the half-reaction method:

    (a) $Zn + MnO_4^- \longrightarrow Zn(OH)_2 + MnO_2$
    (b) $CrO_2^- + IO_3^- \longrightarrow I^- + CrO_4^-$
    (c) $Bi(OH)_3 + Sn(OH)_3^- \longrightarrow Bi + Sn(OH)_6^{2-}$
    (d) $I_2 + Cl_2 \longrightarrow H_3IO_6^{2-} + Cl^-$

14. Balance the following equations in basic solution using the half-reaction method:

(a) $MnO_4^- + C_3H_8O_3 \longrightarrow MnO_4^{2-} + CO_3^{2-}$
(b) $MnO_4^- + SO_3^{2-} \longrightarrow MnO_4^{2-} + SO_4^{2-}$
(c) $MnO_4^- + Br^- \longrightarrow MnO_2 + BrO_3^-$

15. Using the oxidation number method, balance the following equations:

(a) $Cr(OH)_4^- + H_2O_2 \longrightarrow CrO_2^{2-} + H_2O$ (in basic solution)
(b) $H_2S + Na_2Cr_2O_7 + HCl \longrightarrow S + CrCl_3 + H_2O$
(c) $Sb_2S_5 + NO_3^- + H^+ \longrightarrow HSbO_3 + S + NO + H_2O$

16. Balance the following equations using any method:

(a) $Zn + NO_3^- + H^+ \longrightarrow Zn^{2+} + NH_4^+ + H_2O$
(b) $MnO_4^- + Cl^- + H^+ \longrightarrow Mn^{2+} + Cl_2 + H_2O$
(c) $PtCl_6^{2-} + Sb + H_2O \longrightarrow PtCl_4^{2-} + Cl^- + Sb_2O_3 + H^+$
(d) $Hg_2Cl_2 + I^- + OH^- \longrightarrow Hg + Cl^- + IO_3^- + H_2O$
(e) $K_2S_2O_8 + Cr_2(SO_4)_3 + H_2O \longrightarrow K_2SO_4 + K_2Cr_2O_7 + H_2SO_4$
(f) $KMnO_4 + Zn + H_2SO_4 \longrightarrow MnSO_4 + ZnSO_4 + H_2O + K_2SO_4$

17. A solution of acidified $KMnO_4$ was standardized with Fe^{2+} solution as the primary standard. To prepare the Fe^{2+} solution, 0.3554 g of pure iron metal was dissolved in acid and then reduced to Fe^{2+}. The iron solution was titrated with permanganate and required 23.50 cm ^{3} of $KMnO_4$ solution to reach the end point.

(a) Write a balanced ionic equation for the reaction.
(b) What is the concentration of the permanganate solution?

18. 25.00 cm^3 of oxalic acid solution was titrated with acidified $KMnO_4$ solution. If 21.75 cm^3 of the 0.05 M $KMnO_4$ solution was required to reach the end point, what was the molar concentration of the oxalic acid solution?

$$2\ KMnO_4 + 5\ H_2C_2O_4 + 3\ H_2SO_4 \longrightarrow 2\ MnSO_4 + K_2SO_4 + 10\ CO_2 + 8\ H_2O$$

19. A ceric ion (Ce^{4+}) solution was standardized with a solution of H_3AsO_3 as the primary standard. The H_3AsO_3 solution was prepared from 480.5 mg of primary standard As_2O_3 and required 24.95 cm^3 of the cerium solution to reach the end point.

(a) Write a balanced equation for the reaction.
(b) What is the concentration of the cerium solution?

20. 9.55 g of fresh beef was digested in excess concentrated sulfuric acid and the solution made up to 250 cm^3 in a standard flask. 25.00 cm^3 of this solution required 22.50 cm^3 of 0.025 M $KMnO_4$ to reach the end point. Calculate the percentage of iron in the beef sample. (Note that the chemistry here is similar to that in problem 17.)

21. An industrial chemistry student doing his summer internship at the Delta Steel Company was given a sample of iron ore (0.500 g of impure Fe_2O_3) to analyze for the percentage of Fe_2O_3 in the ore. He dissolved the ore sample in boiling dilute nitric acid and evaporated to dryness. He collected the residue and redissolved it in dilute sulfuric acid, filtered off the insoluble residue and passed the solution through a Walden silver reductor. The determination was completed by titrating a sample of the resulting solution with aqueous $KMnO_4$. The solution required 17.5 cm^3 of 0.025 M $KMnO_4$ to reach the end point. Estimate the percentage of Fe_2O_3 in the ore. You may use the following reaction:

$$5\ Fe^{2+} + MnO_4^- + 8\ H^+ \longrightarrow 5\ Fe^{3+} + Mn^{2+} + 4\ H_2O$$

22. Vitamin C, or ascorbic acid ($C_6H_8O_6$), can be found in citrus fruits and fresh vegetables. It was first isolated in its pure form in 1926 by Albert Azent-Gyorgi and Charles King.

    The oxidation of ascorbic acid by iodine or iodate salts is rapid and quantitative. Therefore the exact amount of vitamin C in an unknown sample can be determined by monitoring the amount of iodate used to oxidize a sample containing vitamin C, or by oxidizing the sample with excess iodine followed by titrating the unreacted iodine with sodium thiosulfate.

    In pill form, vitamin C is often compounded with inactive ingredients. A student was asked to determine the ascorbic acid content of a vitamin C sample by using the iodometric titration method. The equations occurring during the oxidation of the acid and titration of the excess iodine are:

    (a) $C_6H_8O_6 + I_2 \longrightarrow C_6H_6O_6 + 2\ H^- + 2I^-$
    (b) $I_2 + 2\ S_2O_3^{2-} \longrightarrow 2I^- + S_4O_6^{2-}$

    25.00 cm^3 of 0.125 M I_2 was used to oxidize a tablet of vitamin C weighing 0.2728 g. At the end of the reaction, the excess iodine required 23.55 cm^3 of 0.200 M $Na_2S_2O_3$ for the titration to reach the blue starch-iodine equivalence point. Calculate the percent of ascorbic acid in the tablet sample.

23

Fundamentals of Electrochemistry

. . . .

Electrochemistry is the branch of chemistry that deals with the interconversion of chemical and electrical energy.

23.1. Galvanic Cells

A *galvanic* (or *voltaic*) *cell* is a chemical system that uses an oxidation–reduction reaction to convert chemical energy into electrical energy (hence it is also known as an electrochemical cell). This process is the opposite of electrolysis (explained in section 23.10), wherein electrical energy is used to bring about chemical changes. The two systems are similar in that both are redox processes; in both, the oxidation takes place at one electrode, the *anode*, while reduction occurs at the *cathode*. Figure 23-1 shows a galvanic cell, indicating the half-reactions at the two electrodes. Electrons flow through the external circuit from the anode (Zn) to the cathode (Cu).

The overall reaction, which is obtained by adding the anodic and cathodic half-cell reactions, is:

$$Zn(s) + Cu^{2+}(aq) \longrightarrow Zn^{2+}(aq) + Cu(s)$$

This cell has a potential of 1.10 V (see next section).

23.2. The Cell Potential

The potential energy of electrons at the anode is higher than at the cathode. This difference in potential is the driving force that propels electrons through the external circuit. The cell potential (E_{cell}) is a measure of the potential difference between the two half-cells. It is also known as the *electromotive force* (emf) of the cell, or, since it is measured in volts, the cell voltage.

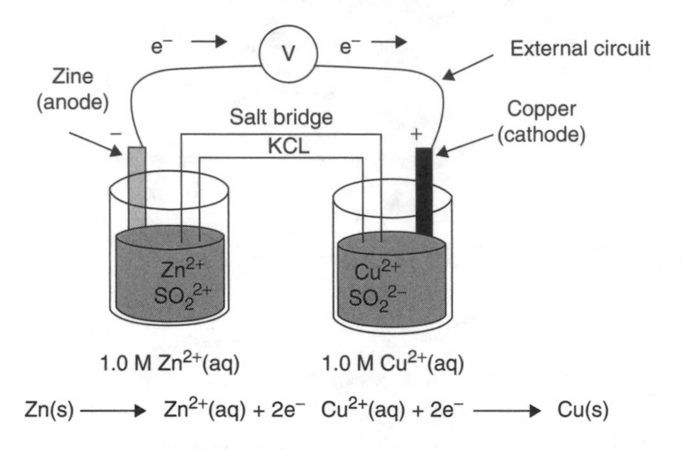

Figure 23-1 A galvanic cell showing zinc and copper half-reactions at the electrodes.

By definition, one volt (V) is the potential difference required to give one joule (J) of energy to a charge of one coulomb (C):

$$1\ \text{V} = \frac{1\ \text{J}}{1\ \text{C}}, \quad \text{or} \quad 1\ \text{C} = \frac{1\ \text{J}}{1\ \text{V}}$$

The following relationships may be useful in solving electrochemistry problems:

$$1\ \text{faraday} = 96{,}500\ \text{C} = \text{charge on 1 mol of electrons}$$

$$1\ \text{faraday} = 96.5\ \frac{\text{kJ}}{\text{V mol e}^-}$$

Hence:

$$96{,}500\frac{\text{C}}{\text{mol e}^-} = 96.5\ \frac{\text{kJ}}{\text{V mol e}^-}$$

23.3. Standard Electrode Potential

An electrochemical cell consists of two half-reactions at different potentials, which are known as *electrode potentials*. The electrode potential for the oxidation half-reaction is called the oxidation potential. Similarly, for the reduction half-reaction, we have the reduction potential. The potential of a galvanic cell is determined by the concentrations of the species in solution, the partial pressures of any gaseous reactants or products, and the reaction temperature. When the electrochemical measurement is carried out under standard-state conditions, the cell potential is called the *standard electrode potential* and is given the symbol E^0. The standard conditions include a concentration of 1 M, gaseous partial pressure of 1 atm, and a temperature of 25°C.

It is impossible to measure the absolute potential value of a single electrode, since every oxidation is accompanied by a reduction. Therefore any measurement is carried out against a reference electrode. The reference electrode that has been selected is the standard hydrogen electrode (SHE). It consists of platinum metal immersed in a 1.0 M H^+ solution. Hydrogen gas at a pressure of 1 atm is bubbled over the platinum electrode. By convention, the SHE is assigned a potential of 0.00 V. The cell reaction is shown below:

$$H_2 \longrightarrow 2H^+ + 2e^- \qquad E^0 = 0.0000 \text{ V (SHE anode)}$$

$$2\,H^+ + 2e^- \longrightarrow H_2 \qquad E^0 = 0.0000 \text{ V (SHE cathode)}$$

$$2\,H^+ + H_2 \longrightarrow 2\,H^+ + H_2 \qquad E^0 = 0.0000 \text{ V}$$

Note that when a cell is made up of an SHE and some other standard electrode, the standard cell potential, E^0_{cell}, is simply the standard potential of the other electrode. For example, consider the copper-SHE cell. The SHE half-cell forms the anode with a potential of 0.000 V. The copper half-cell, with a potential of 0.337 V, serves as the cathode. The reaction that occurs in the cell is:

$$H_2 \longrightarrow 2\,H^+ + 2e^- \text{(anode, oxidation)} \qquad E^0 = 0.000 \text{ V}$$

$$Cu^{2+} + 2e^- \longrightarrow Cu \text{ (cathode, reduction)} \qquad E^0 = 0.337 \text{ V}$$

$$H_2 + Cu^{2+} \longrightarrow 2H^+ + Cu \text{ (net cell reaction)} \qquad E^0 = 0.337 \text{ V}$$

23.3.1. Standard reduction potential

The *standard reduction potential* of any standard electrode is defined as the electromotive force of a cell at 25°C wherein the SHE is the anode and the given electrode is the cathode. The symbol E^0_{red} represents the standard reduction potential, while E^0_{ox} represents the standard oxidation potential. For any given half-cell,

$$E^0_{red} = -E^0_{ox}$$

For example, E^0_{red} for the half-cell $Fe^{2+}(aq) + 2e^- \longrightarrow Fe(s)$ is −0.44 V. Therefore the E^0_{ox} value will be +0.44 V. Values of many standard reduction potentials have been measured. These results, sometimes called the electrochemical series, can be found in many general chemistry textbooks. Representative examples are given in table 23-1.

23.4. The Electrochemical Series

The *electrochemical series* is the arrangement of the standard electrode potentials (for half-cell reactions) in order from the most negative to the most positive

Table 23-1 Selected Standard Electrode Potentials at 25°C

Reduction half-reaction	E^0, *V*
Acidic solution	
$Li(aq)^{+} e^{-} \longrightarrow Li(s)$	−3.04
$K(aq)^{+} e^{-} \longrightarrow K(s)$	−2.924
$Ca^{2+}(aq) + 2e^{-} \longrightarrow Ca(s)$	−2.84
$Na+(aq)\ e^{-} \longrightarrow Na(s)$	−2.713
$Mg^{2+}(aq)\ 2e^{-} \longrightarrow Mg(s)$	−2.356
$Al^{3+}(aq) + 3e^{-} \longrightarrow Al(s)$	−1.676
$V^{2+}(aq) + 2e^{-} \longrightarrow V(s)$	−1.18
$Zn^{2+}(aq) + 2e^{-} \longrightarrow Zn(s)$	−0.763
$Fe^{2+}(aq) + 2e^{-} \longrightarrow Fe(s)$	−0.44
$Cd^{2+}(aq) + 2e^{-} \longrightarrow Cd(s)$	−0.4
$Ni^{2+}(aq) + 2e^{-} \longrightarrow Ni(s)$	−0.25
$Co^{2+}(aq) + 2e^{-} \longrightarrow Co(s)$	−0.28
$Sn^{2+}(aq) + 2e^{-} \longrightarrow Sn(s)$	−0.137
$Pb^{2+}(aq) + 2e^{-} \longrightarrow Pb(s)$	−0.125
$2H^{+}(aq) + 2e^{-} \longrightarrow H_2(g)$	0
$Sn^{4+}(aq) + 2e^{-} \longrightarrow Sn^{2+}(aq)$	0.13
$S(s) + 2H(aq) + 2e^{-} \longrightarrow H_2S(g)$	0.14
$Cu^{2+}(aq) + 2e^{-} \longrightarrow Cu(s)$	0.337
$I_2(s) + 2e^{-} \longrightarrow 2\ I(aq)$	0.535
$O_2(g) + 2H^{-}(aq) + 2e^{-} \longrightarrow H_2O_2(aq)$	0.695
$Fe^{3+}(aq) + e^{-} \longrightarrow Fe^{2+}(aq)$	0.771
$Ag^{+}(aq) + e^{-} \longrightarrow Ag(s)$	0.8
$Br_2(l) + 2e^{-} \longrightarrow 2Br^{-}(aq)$	1.065
$O_2(g) + 4\ H^{+}(aq) + 4e^{-} \longrightarrow 2\ H_2O(aq)$	1.229
$Cl_2(g) + 2e^{-} \longrightarrow 2\ Cl^{-}(aq)$	1.358
$MnO_4^{-}(aq) + 8\ H+(aq) + 5e^{-} \longrightarrow Mn^{2+}(aq) + 4\ H_2O(l)$	1.52
$F_2(g) + 2e^{-} \longrightarrow 2\ F^{-}(aq)$	2.866

or vice-versa. It may also be defined as the arrangement of elements in order of increasing oxidizing or reducing power. Table 23-1 shows the standard electrode potentials of some elements at 25°C. This table may also be called a standard reduction potential table because it is arranged in order of decreasing reducing properties. The higher up an element is on this table, the more readily it loses electrons, and the stronger a reducing agent it is.

23.5. Applications of Electrode Potential

There are several uses of the electrochemical series. The most important ones are:

- *Predicting the feasibility of reactions*: Two half-equations can be combined to give a desired net reaction. The general rule is to write the half-reaction lower in the ECS as given in the standard reduction potential table. The one higher up in the series will only go in the reverse direction. Therefore, write the reverse form (oxidation) of that half-reaction and then add the two equations to obtain the overall reaction and its potential.
- *Predicting displacement reactions*: The series can be used to determine which elements will displace each other from solution. Metals higher up in the series will displace those lower down, since they can readily lose or donate electrons. For example, all metals above hydrogen in the series will displace it from dilute acid.

$$Zn(s) + 2\,H^+(aq) \longrightarrow Zn^{2+}(aq) + H_2(g) \qquad E^0_{cell} = 0.763\ V$$

Generally, the reduced form of an element will reduce the oxidized form of any element below it in the series. Thus, calcium can reduce copper ion to copper metal:

$$Ca(s) + Cu^{2+}(aq) \longrightarrow Ca^{2+}(aq) + Cu(s) \qquad E^0_{cell} = 3.177\ V$$

- *Predicting oxidizing and reducing strength*: The elements with the largest negative E^0_{red} value are stronger reducing agents and most readily lose electrons. On the other hand, elements with largest positive E^0_{red} values, such as many nonmetals (at the bottom of the series), are stronger oxidizing agents.
- *Predicting the deposition of ions in electrolysis*: During electrolysis of a given electrolyte, more than one ion can migrate to the anode or cathode. The position of an ion in the ECS often helps in determining which ion will be preferentially deposited. Cations lower in the series are generally deposited in preference to those higher up in the series.

23.6. Cell Diagrams

A galvanic cell can be represented schematically as:

$$\text{anode}\,|\,\text{anode electrolyte}\,||\,\text{cathode electrolyte}\,|\,\text{cathode}$$

1. The anode (oxidation) is placed on the left and the cathode (reduction) is placed on the right side of the diagram.
2. The single vertical line (|) indicates a boundary between the different phases in a half-cell.

3. The double vertical line(‖) represents a salt bridge separating the two half-cells. It permits the flow of ions between the two cells while preventing the solutions from mixing.
4. Within a half-cell the reactants are listed before the products.
5. The concentrations of ions in solution and the pressure of gases are written in parentheses following the symbol of the ions or molecules. For example, the cell notation for the Zn–Cu cell would be:

$$\mathrm{Zn}\,\big|\,\mathrm{Zn^{2+}(1\ M)}\,\big\|\,\mathrm{Cu^{2+}(1\ M)}\,\big|\,\mathrm{Cu}$$

6. An inert electrode such as platinum or graphite (carbon) is commonly used for a half-cell involving a gas like hydrogen or chlorine. The cell notation uses an additional vertical line to denote the extra phase (the inert electrode) present, as illustrated in the cell notation:

$$\mathrm{Zn(s)}\,\big|\,\mathrm{Zn^{2+}(aq)}\,\big\|\,\mathrm{H^+(aq)}\,\big|\,\mathrm{H_2(g)}\,\big|\,\mathrm{Pt}$$

7. When two ionic species are in the same phase in a cell, commas, rather than vertical lines, separate them. Neither one can be used as an electrode; hence an inert electrode is employed. This is illustrated by the reaction $\mathrm{Fe(s)} + 2\,\mathrm{Fe^{3+}(aq)} \longrightarrow 3\,\mathrm{Fe^{2+}(aq)}$. The cell notation is $\mathrm{Fe(s)}\,\big|\,\mathrm{Fe^{2+}(aq)}\,\big\|\,\mathrm{Fe^{3+}(aq)},\ \mathrm{Fe^{2+}(aq)}\,\big|\,\mathrm{Pt}$.

Example 23.1

Write the cell diagram and the cell reaction for the redox couples Ag^+/Ag and Cu^{2+}/Cu.

$$\mathrm{Cu^{2+}(aq)} + 2\mathrm{e^-} \longrightarrow \mathrm{Cu(s)}, \quad E^0_{\text{cell}} = 0.337\ \mathrm{V}$$

$$\mathrm{Ag^{+}(aq)} + \mathrm{e^-} \longrightarrow \mathrm{Ag\ (s)}, \quad E^0_{\text{cell}} = 0.80\ \mathrm{V}$$

Solution

1. First write the half-cell reactions and then determine the net (cell) reaction.

$$\begin{aligned} &\text{Anode:} && \mathrm{Cu(s)} \longrightarrow \mathrm{Cu^{2+}(aq)} + 2\mathrm{e^-} \\ &\text{Cathode:} && 2\,\mathrm{Ag^+(aq)} + 2\mathrm{e^-} \longrightarrow 2\,\mathrm{Ag(s)} \\ \hline &\text{Cell reaction:} && 2\,\mathrm{Ag^+(aq)} + \mathrm{Cu(s)} \longrightarrow 2\,\mathrm{Ag(s)} + \mathrm{Cu^{2+}(aq)} \end{aligned}$$

2. Next write the cell diagram following the standard notation.

$$\mathrm{Cu(s)}\,\big|\,\mathrm{Cu^{2+}(aq)\ (1\ M)}\,\big\|\,\mathrm{Ag^+(aq)\ (1\ M)}\,\big|\,\mathrm{Ag(s)}$$

Example 23.2

A galvanic cell was constructed by placing a tin electrode into a tin (II) chloride solution in one chamber, and a platinum electrode into an iron (II) chloride solution in another chamber. The two half-cells were separated by a salt bridge. Write the cell equation and then draw a cell diagram for the cell.

$$Sn^{2+}(aq) + 2e^- \longrightarrow Sn\ (s), \qquad E^0_{cell} = -0.137\ V$$

$$Fe^{3+}(aq) + e^- \longrightarrow Fe^{2+}(aq), \qquad E^0_{cell} = 0.771\ V$$

Solution

1. First write the half-cell reactions and then determine the net (cell) reaction:

$$\text{Anode:} \quad Sn(s) \longrightarrow Sn^{2+}(aq) + 2e^-$$

$$\text{Cathode:} \quad 2\ Fe^{3+}(aq) + 2e^- \longrightarrow 2\ Fe^{2+}(aq)$$

$$\text{Cell reaction:} \quad 2\ Fe^{3+}(aq) + Sn(s) \longrightarrow 2\ Fe^{2+}(aq) + Sn^{2+}(aq)$$

2. Next write the cell diagram following the standard notation.

$$Sn(s)\,\big|\,Sn^{2+}(aq)\,\big\|\,Fe^{2+}(aq),\ Fe^{3+}(aq)\,|\,Pt$$

23.7. Calculating E^0_{cell} from Electrode Potential

The value of E^0_{cell} can be obtained by adding the voltages of the two half reactions. Generally:

$$E^0_{cell} = E^0(\text{anode}) + E^0(\text{cathode}) \quad \text{or} \quad E^0_{cell} = E^0_{red} + E^0_{ox}$$

The following rules are helpful:

1. Write the oxidation and reduction half-cell reactions.
2. Obtain the standard electrode potential for each half reaction (i.e., E^0_{red} and E^0_{ox}) from the table of potentials. Recall that $E^0_{ox} = -E^0_{red}$.
3. Add the two half-reactions to generate a net cell reaction, and also add the electrode potentials using the equation:

$$E^0_{cell} = E^0_{red} + E^0_{ox}$$

4. Note that changing the coefficients of the half-equations does not affect the values of E^0 and E^0_{cell}.

Example 23.3

Calculate the E^0_{cell} for a proposed cell that uses the following reaction:

$$Mg(s) + Cu^{2+}(aq) \longrightarrow Mg^{2+}(aq) + Cu(s)$$

Solution

1. First write the half-cell equations for the oxidation and reduction and their standard electrode potentials, recalling that $E^0_{ox} = -E^0_{red}$:

$$\text{Anode:} \quad Mg(s) \longrightarrow Mg^{2+}(aq) + 2e^- \qquad E^0_{ox} = +2.363\ V$$

$$\text{Cathode:} \quad Cu^{2+}(aq) + 2e^- \longrightarrow Cu(s) \qquad E^0_{red} = +0.337\ V$$

2. Obtain the cell potential using the equation:

$$E^0_{cell} = E^0_{red} + E^0_{ox}$$

$$E^0_{cell} = +2.363 + 0.337$$

$$= +2.700\ V$$

Example 23.4

A galvanic cell has an aluminum electrode in aluminum nitrate ($Al(NO_3)_3$) solution and a lead electrode in lead (II) nitrate ($Pb(NO_3)_2$) solution. Calculate E^0_{cell} for the cell.

Solution

1. First write the half-cell equations and their standard electrode potentials, recalling that $E^0_{ox} = -E^0_{red}$.

$$\text{Anode:} \quad 2\ Al(s) \longrightarrow 2\ Al^{3+}(aq) + 6e^- \qquad E^0_{ox} = +1.676\ V$$

$$\text{Cathode:} \quad 3\ Pb^{2+}(aq) + 6e^- \longrightarrow 3\ Pb(s) \qquad E^0_{red} = -0.125\ V$$

2. Obtain the cell potential using the equation

$$E^0_{cell} = E^0_{red} + E^0_{ox}$$

$$E^0_{cell} = +1.676 - 0.125$$

$$= +1.551\ V$$

23.8. The Standard Electrode Potential, the Gibbs Free Energy, and the Equilibrium Constant

The standard Gibbs free energy change, ΔG^0, is related to the thermodynamic equilibrium constant, K, and the standard electrode potential, E^0_{cell}, by the following equations:

$$\Delta G^0 = -2.303RT \log K \quad \text{or} \quad RT \ln K$$

and

$$\Delta G^0 = -nFE^0_{cell}$$

R = the universal gas constant, T = the Kelvin temperature, n = the moles of electrons involved in the half-reaction, and F = the faraday constant, which is equal to 96,500 C per mole, i.e.,

$$1F = 96,500 \frac{\text{C}}{\text{mol}} \quad \text{or} \quad 96.5 \frac{\text{kJ}}{\text{V-mol e}^-}$$

We can relate E^0_{cell} to the equilibrium constant by combining the two equations for ΔG^0 as follows:

$$-nFE^0_{cell} = -RT \ln K \quad \text{or} \quad 2.303 \log K$$

$$\ln K = \frac{nFE^0_{cell}}{RT} \quad \text{or} \quad \log K = \frac{nFE^0_{cell}}{2.303RT}$$

$$E^0_{cell} = \frac{RT}{nF} \ln K \quad \text{or} \quad \frac{2.303\ RT}{nF} \log K$$

If we substitute $R = 8.314$ J/mol-K, $T = 298$ K, and $F = 96{,}500$ C (or J/V-mol e$^-$) into the expression:

$$E^0_{cell} = \frac{2.303RT}{nF} \log K$$

we will obtain a simplified equation for solving problems involving E^0_{cell} and K:

$$E^0_{cell} = \frac{0.0592}{n} \log K$$

23.8.1. Conditions for spontaneous change in redox reactions

The criteria in table 23-2 describe whether a reaction is spontaneous.

Example 23.5

Calculate ΔG^0 and the equilibrium constant, K, at 25°C for the reaction:

$$\text{Ca(s)} + \text{Cu}^{2+}\text{(aq)} \longrightarrow \text{Ca}^{2+}\text{(aq)} + \text{Cu(s)}$$

Table 23-2 Conditions for Spontaneous Change in Redox Reactions

E^0_{cell}	ΔG^0	K	*Reaction direction*
Positive (+)	Negative (−)	>1	Forward (spontaneous as written)
Zero (0)	Zero (0)	~1	At equilibrium
Negative (−)	Positive (+)	<1	Backward (nonspontaneous as written)

Solution

1. First write the half-cell equations and their standard electrode potentials, recalling that $E^0_{\text{ox}} = -E^0_{\text{red}}$.

$$\text{Anode:} \quad \text{Ca(s)} \longrightarrow \text{Ca}^{2+}\text{(aq)} + 2\text{e}^- \qquad E^0_{\text{ox}} = +2.87 \text{ V}$$

$$\text{Cathode:} \quad \text{Cu}^{2+}\text{(aq)} + 2\text{e}^- \longrightarrow \text{Cu(s)} \qquad E^0_{\text{red}} = +0.337 \text{ V}$$

2. Obtain the cell potential using the equation

$$E^0_{\text{cell}} = E^0_{\text{red}} + E^0_{\text{ox}}$$

$$E^0_{\text{cell}} = +2.87 + 0.337$$

$$= +3.207 \text{ V}$$

This reaction is spontaneous since E^0_{cell} is positive.

3. Calculate ΔG^0 using the equation $\Delta G^0 = -nFE^0_{\text{cell}}$ and the E^0_{cell}calculated above:

$$\Delta G^0 = -nFE^0_{\text{cell}}$$

From the half-reaction, $n = 2$.

$$\Delta G^0 = -nFE^0_{\text{cell}} = -2 \times 96{,}500 \frac{\text{J}}{\text{V-mol}} \times 3.207 \text{ V}$$

$$= -618{,}951 \text{ J/mol} \quad \text{or} \quad -618 \text{ kJ/mol}$$

4. Now calculate K from the expression $E^0_{\text{cell}} = \dfrac{0.0592}{n} \log K$:

$$\text{E}^0_{\text{cell}} = \frac{0.0592}{n} \log K,$$

or:

$$\log K = \frac{nE^0_{\text{cell}}}{0.0592} = \frac{2 \times 3.207}{0.0592} = 108.34$$

$$K = \text{antilog } 108.34 = 2.21 \times 10^{108}$$

Example 23.6

In acidic solution, iron (II) ions are easily oxidized by dissolved oxygen gas according to the following equation:

$$O_2(g) + 4\ H^+(aq) + 4\ Fe^{2+} \longrightarrow 2H_2O\ (l)$$

Calculate E^0_{cell}, ΔG^0, and K for the reaction.

Solution

1. First write the half-cell equations and their standard electrode potentials, recalling that $E^0_{ox} = -E^0_{red}$:

$$\text{Anode:}\quad 4Fe^{2+}(aq) \longrightarrow 4Fe^{3+}(aq) + 4e^- \qquad E^0_{ox} = -0.77\ V$$

$$\text{Cathode:}\quad O_2(g) + 4H^+(aq) + 4e^- \longrightarrow 2H_2O\ (l) \qquad E^0_{red} = +1.23\ V$$

2. Obtain the cell potential using the equation:

$$E^0_{cell} = E^0(\text{anode}) + E^0(\text{cathode})$$

$$E^0_{cell} = -0.77 + 1.23$$

$$= +0.46\ V$$

This reaction is spontaneous since E^0_{cell} is positive.

3. Calculate ΔG^0 using the equation $\Delta G^0 = -nFE^0_{cell}$ and the E^0_{cell}calculated above:

$$\Delta G^0 = -nFE^0_{cell}$$

From the half-reaction, $n = 4$.

$$\Delta G^0 = -nFE^0_{cell} = -4 \times 96{,}500\ \frac{J}{V\text{-mol}} \times 0.46\ V$$

$$= -177{,}560\ J/mol \quad \text{or} \quad -177.56\ kJ/mol$$

4. Now calculate K from $E^0_{cell} = \dfrac{0.0592}{n} \log K$:

$$E^0_{cell} = \frac{0.0592}{n} \log K$$

or:

$$\log K = \frac{nE^0_{cell}}{0.0592} = \frac{4 \times 0.46}{0.0592} = 31.1$$

$$K = \text{antilog } 31.1 = 1.2 \times 10^{31}$$

23.9. Dependence of Cell Potential on Concentration (the Nernst Equation)

Potentials for cells not under standard conditions (e.g., concentration other than 1 M) can be derived by considering the dependence of the standard free energy change on concentration, given by the following equation:

$$\Delta G = \Delta G^0 + RT \ln Q$$

Here Q is the reaction quotient, which has the form of an equilibrium constant, except that the concentrations and gas pressures are those that exist in the reaction mixture at any given moment.

If we substitute $\Delta G = -nFE$ and $\Delta G^0 = -nFE^0_{\text{cell}}$ into the above expression, we have $-nFE = -nFE^0_{\text{cell}} + RT \ln Q$.

Solving this equation for E, we obtain the so-called Nernst equation:

$$E = E^0 - \frac{RT}{nF} \ln Q \quad \text{or} \quad E = E^0 - \frac{2.303RT}{nF} \log Q$$

If we express the equation in the common logarithm form and substitute $T = 298$ K as well as the values of R and F, the Nernst equation becomes:

$$E = E^0 - \frac{0.0592}{n} \log Q$$

This equation can be used to calculate the cell potential for nonstandard conditions.

Example 23.7

Calculate the potential of the cell $Zn(s)|Zn^{2+}(10^{-5}M)||Cu^{2+}(0.01\text{ M})|Cu(s)$ at 298 K, if its standard potential (with all concentrations = 1 M) is 1.10 V.

Solution

1. From the cell diagram, write the net cell reaction and determine the number of electrons transferred.

$$\text{Anode:} \quad Zn(s) \longrightarrow Zn^{2+}(aq) + 2e^- \qquad E^0_{\text{ox}} = +0.76\text{ V}$$

$$\text{Cathode:} \quad Cu^{2+}(aq) + 2e^- \longrightarrow Cu(s) \qquad E^0_{\text{red}} = +0.337\text{ V}$$

$$Zn(s) + Cu^{2+}(aq) \longrightarrow Zn^{2+}(aq) + Cu(s) \qquad E^0_{\text{cell}} = 1.10\text{ V}$$

For this reaction, $n = 2$

2. Calculate the reaction quotient, Q.

$$Q = \frac{[Zn^{2+}]}{[Cu^{2+}]} = \frac{10^{-5}\text{ M}}{10^{-2}\text{ M}} = 10^{-3}$$

3. Calculate E_{cell} using the Nernst equation.

$$E_{cell} = E^0_{cell} - \frac{0.0592}{n} \log Q$$

$$E_{cell} = 1.10 - \frac{0.0592}{2} \log 10^{-3}$$

$$E_{cell} = 1.19 \text{ V}$$

So the potential of this cell at 298 K is 1.19 V.

Example 23.8

Using the Nernst equation for the cell diagram

$$Mg(s)\,|\,Mg^{2+}(0.010\text{ M})\,\|\,Sn^{2+}(0.10\text{ M})\,|\,Sn(s)$$

Calculate (a) the cell potential, (b) the ratio of concentrations, $\frac{Mg^{2+}}{Sn^{2+}}$, which would cause the cell potential to be 2.15 V.

Solution

(a) The following steps will be helpful in calculating the cell potential:

1. From the cell diagram, write the half-cells and the net cell reaction; calculate the standard cell potential, and determine the number of electrons transferred.

$$\begin{array}{lll} \text{Anode: } Mg(s) \longrightarrow Mg^{2+}(aq) + 2e^- & & E^0_{ox} = +2.37 \text{ V} \\ \text{Cathode: } Sn^{2+}(aq) + 2e^- \longrightarrow Sn(s) & & E^0_{red} = -0.14 \text{ V} \\ \hline Mg(s) + Sn^{2+}(aq) \longrightarrow Mg^{2+}(aq) + Sn(s) & & E^0_{cell} = +2.23 \text{ V} \end{array}$$

For this reaction $n = 2$.

2. To obtain the nonstandard cell potential we need to calculate the reaction quotient, Q, using the concentrations given:

$$Q = \frac{[Mg^{2+}]}{[Sn^{2+}]} = \frac{10^{-2}\text{ M}}{10^{-1}\text{ M}} = 0.10$$

3. Calculate E_{cell} using this value of Q in the Nernst equation:

$$E_{cell} = E^0_{cell} - \frac{0.0592}{n} \log Q$$

$$E_{cell} = 2.23 - \frac{0.0592}{2} \log 0.10$$

$$E_{cell} = 2.23 - (-0.03 \text{ V}) = 2.26 \text{ V}$$

(b) Make $E_{cell} = 0$, and solve for Q, which is essentially the desired ratio.

1. From part (a) we know that $E^0_{cell} = +2.23$ V and $n = 2$.
2. Calculate E_{cell} using the Nernst equation:

$$E_{cell} = 2.15 \text{ V} = E^0_{cell} - \frac{0.0592}{n} \log\left(\frac{Mg^{2+}}{Sn^{2+}}\right)$$

$$2.15 \text{ V} = 2.23 - \frac{0.0592}{2} \log\left(\frac{Mg^{2+}}{Sn^{2+}}\right)$$

$$0.08 = \frac{0.0592}{2} \log\left(\frac{Mg^{2+}}{Sn^{2+}}\right)$$

$$\log\left(\frac{Mg^{2+}}{Sn^{2+}}\right) = 2.71$$

$$\left(\frac{Mg^{2+}}{Sn^{2+}}\right) = 515$$

Example 23.9

The cell reaction shown below has a standard state potential of +0.58 V:

$$\text{Pt}\left|Fe^{2+}(0.25 \text{ M}), Fe^{3+}(0.85 \text{ M})\right\|Cr_2O_7^{2-}(2.50 \text{ M}),$$

$$Cr^{3+}(3.50 \text{ M}), H^+(1.0 \text{ M})\Big|\text{Pt}$$

Is the reaction spontaneous under these concentration conditions?

Solution

All we need to answer this is the cell potential under the actual conditions. So we can follow the same procedure as in previous examples.

1. From the cell diagram, write the half-cells, and the net cell reaction; calculate the standard electrode potential; and determine the number of electrons

transferred.

$$
\begin{array}{ll}
6\,Fe^{2+}(aq) \longrightarrow 6\,Fe^{3+}(aq) + 6e^- & E^0_{ox} = -0.77\ V \\
14\,H^+(aq) + 6e^- + Cr_2O_7^{2-}(aq) \longrightarrow 2\,Cr^{3+}(aq) + 7\,H_2O(l) & E^0_{red} = +1.33\ V \\
\hline
14\,H^+(aq) + 6\,Fe^{2+}(aq) + Cr_2O_7^{2-}(aq) \longrightarrow 6\,Fe^{3+}(aq) + 2\,Cr^{3+}(aq) & \\
\qquad + 7H_2O(l) & E^0_{cell} = +0.58\ V
\end{array}
$$

For this reaction $n = 6$.

2. To obtain the nonstandard cell potential we need to calculate the reaction quotient, Q, using the concentrations given:

$$Q = \frac{[Fe^{3+}]^6\,[Cr^{3+}]^2}{[Fe^{2+}]^6\left[Cr_2O_7^{2-}\right][H^+]^{14}} = \frac{(0.85)^6\,(3.50)^2}{(0.25)^6\,(2.5)\,(1.0)^{14}} = 7{,}700$$

3. Calculate E_{cell}, inserting the value of Q into the Nernst equation.

$$E_{cell} = E^0_{cell} - \frac{0.0592}{n}\log Q$$

$$E_{cell} = 0.58 - \frac{0.0592}{6}\log 7{,}700$$

$$E_{cell} = 0.58 - 0.04 = 0.54\ V$$

Since the cell potential is positive, the reaction will be spontaneous under these conditions.

23.10. Electrolysis

Electrolysis: The chemical decomposition of an electrolyte in the molten state or in solution by passing an electric current through it.
Electrolyte: A substance which, when molten or in aqueous solution, dissociates into ions and conducts an electric current.
Nonelectrolyte: A substance which, when molten or in solution, neither allows the passage of an electric current nor is decomposed by it.
Anode: The negative electrode through which electrons leave an electrolyte and at which oxidation takes place.
Cathode: The positive electrode through which electrons enter an electrolyte and at which reduction takes place.
Electric current: The flow of electrons through a conductor or an electrolyte.
Coulomb: The unit of electric charge. One coulomb is the quantity of charge or electricity that passes through an electrolytic cell when a current of one ampere (A) flows for one second(s).

23.11. Faraday's Laws of Electrolysis

23.11.1. First law of electrolysis

Faraday's first law of electrolysis states that *the mass (m) of a substance produced at the electrodes during electrolysis is directly proportional to the quantity of charge or electricity (q) that passed through the electrolytic cell.* Mathematically:

$$m \propto q$$

This law implies that the moles of product formed during electrolysis depend on the number of moles of electrons that pass through the electrolyte. The unit of the quantity of electricity is the coulomb (C).

$$q = I \times t$$

$$(\mathrm{C}) = (\mathrm{A}) \times (\mathrm{s})$$

(Note: 1 coulomb = 1 ampere-second.)

Therefore Faraday's first law states that:

$$m \propto I \times t$$

$$m = E \times I \times t$$

where E is a constant called the electrochemical equivalent of the substance reduced or oxidized.

23.11.2. The Faraday constant

The charge on a single electron is 1.602×10^{-19} C. One mole of electrons contains Avogadro's number of electrons (i.e., 6.023×10^{-23} electron). This corresponds to a total charge of

$$1.602 \times 10^{-19}\ \mathrm{C} \times 6.023 \times 10^{-23} = 96500\ \mathrm{C}$$

This quantity of electricity (96,500 C), the charge carried by 1 mol of electrons, is known as the Faraday constant (F).

During electrolysis, 1 faraday of electricity (96,500 C) is needed to reduce and oxidize one equivalent weight (relative atomic mass divided by charge) of the oxidizing and reducing agents.

For example, consider the generation of 1 mol of sodium atoms (i.e., 23 g) as represented by the following half-reaction:

$$\mathrm{Na^+ + e^- \longrightarrow Na}$$

$$1\ \text{faraday} \Rightarrow 1\ \text{mol of e}^- \Rightarrow 1\ \text{atom of Na}$$

$$1\ \text{faraday} \Rightarrow 23\ \text{g Na}$$

Similarly, 3 faradays will be required to liberate 1 mol of a trivalent element such as Al or Cr, because 1 atom requires 3 electrons:

$$Al^{3+} + 3e^- \longrightarrow Al$$

$$3 \text{ faradays} \Rightarrow 3 \text{ mol } e^- \Rightarrow 1 \text{ mol Al}$$

$$3 \text{ faradays} \Rightarrow 27 \text{ g of Al}$$

Example 23.10

Calculate the mass of aluminum that would be deposited on the cathode when a current of 3.5 A is passed through a molten aluminum salt for 120 minutes. (Al = 27 g/mol; 1 faraday = 96,500 C.)

Solution

We can use the following conversions:

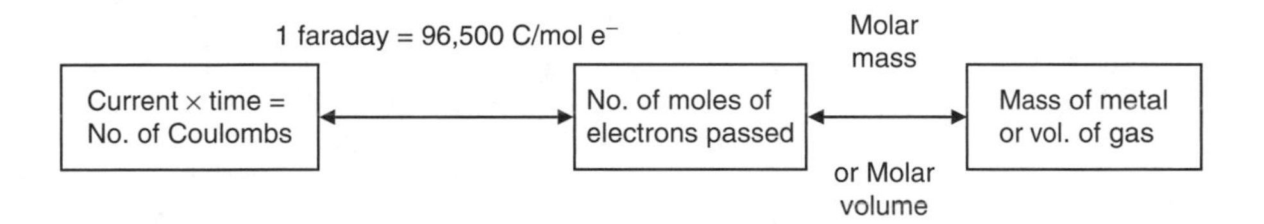

Quantity of electricity, $Q = It = 3.5A \times 120 \text{ min} \times 60 \text{ s/min} = 25{,}200 \text{ C}$.

Cathodic half-reaction:

$$Al^{3+}(l) + 3e^- \longrightarrow Al(s)$$

$$3 \text{ mol of } e^- = 3 \text{ faradays} = 3 \times 96{,}500 \text{ C} = 1 \text{ mol of Al}$$

$$\text{Moles of electrons passed} = \frac{\text{Quantity of electricity (C)}}{1 \text{ faraday}}$$

$$= \frac{25{,}200 \text{ C}}{96{,}500 \text{ C/mol}} = 0.2611 \text{ mol of } e^-$$

3 mol of e^- deposit 1 mol of Al

$$0.2611 \text{ mol of } e^- \text{ deposit} \frac{1 \times 0.2611}{3} \text{ mol of Al} = 0.087 \text{ mol of Al}$$

$$\text{Mass of Al deposited} = \text{mol of Al} \times \text{Molar mass of Al}$$

$$\text{Mass of Al deposited} = 0.087 \text{ mol} \times 27 \frac{\text{g}}{\text{mol}} = 2.35 \text{ g Al}$$

Example 23.11

A current of 3.0 A is passed through a solution of copper (II) sulfate for 1 h. (a) Calculate the mass of copper deposited (cathode reaction: $Cu^{2+}(aq) + 2e^- \longrightarrow Cu(s)$). (b) What is the volume of oxygen gas produced at STP by the oxidation of hydroxide ions? (Anode reaction: $4\ OH^-(aq) \longrightarrow O_2(g) + 2\ H_2O(l) + 4\ e^-$.)

Solution

$$q = It = 3.0\ \text{A} \times 1\ \text{h} \times 60\ \frac{\text{min}}{\text{h}} \times 60 \frac{\text{s}}{\text{min}} = 10{,}800\ \text{C}$$

$$\text{Moles of electrons passed} = \frac{Q}{1\ \text{F}} = \frac{10{,}800}{1 \times 96{,}500} = 0.112\ \text{mol}$$

(a) Cathode half-cell reaction:

$$Cu^{2+}(aq) + 2e^- \longrightarrow Cu(s)$$

$$2\ \text{mol of electrons} = 1\ \text{mol of Cu}$$

$$0.112\ \text{mol of electrons} = 0.112\ \text{mol of e}^- \times \frac{1\ \text{mol Cu}}{2\ \text{mol e}^-} = 0.056\ \text{mol of Cu}$$

$$\text{Mass of Cu deposited} = \text{moles of Cu} \times \text{molar mass of Cu}$$

$$= 0.056\ \text{mol} \times 63.5\ \frac{\text{g}}{\text{mol}} = 3.556\ \text{g}$$

(b) Anode half-cell reaction:

$$4\ OH^-(aq) \longrightarrow O_2(g) + 2\ H_2O(l) + 4e^-$$

$$4\ \text{mol of electrons} = 1\ \text{mol of}\ O_2 = 22.4\ \text{L}\ O_2$$

$$0.112\ \text{mol of electron} = 0.112\ \text{mol of e}^- \times \frac{22.4\ \text{L}\ O_2}{4\ \text{mol e}^-} = 5.6\ \text{L of}\ O_2$$

23.11.3. Second law of electrolysis

Faraday's second law states that *when the same quantity of electricity is passed through different electrolytes, the number of moles or masses of the different ions produced is inversely proportional to their ionic charges.*

If the same quantity of electricity is passed through two electrolytic cells, say, one containing Cu^{2+} ions and the other Ag^+ ions, for the same period of time, then by Faraday's second law of electrolysis:

$$\frac{\text{Number of moles of}\ Cu^{2+}\ \text{liberated}}{\text{Number of moles of}\ Ag^+\ \text{liberated}} = \frac{\text{Charge on}\ Ag^+}{\text{Charge on}\ Cu^{2+}}$$

In other words, the moles of copper liberated will be $\frac{1}{2}$ the moles of silver.

Example 23.12

A current is passed through two electrolytic cells containing solutions of copper (II) sulfate and silver nitrate, both connected in series. If 1.08 g of silver is deposited on the cathode of the cell containing silver nitrate solution, calculate the mass of copper deposited on the cathode of the second cell containing copper (II) sulfate solution.

Solution

Cathode reaction in cell 1, containing $AgNO_3$ solution:

$$Ag^+(aq) + 1e^- \longrightarrow Ag(s)$$

$$1 \text{ mol of } e^- = 1 \text{ mol of } Ag(s)$$

Cathode reaction in cell 2, containing $CuSO_4$ solution:

$$Cu^{2+}(aq) + 2e^- \longrightarrow Cu(s)$$

$$2 \text{ mol of } e^- = 1 \text{ mol of } Cu(s)$$

Since the same quantity of electricity flows through both cells, $q_{cell1} = q_{cell2}$.

$$\text{Number of moles of Ag deposited} = \frac{\text{Mass of Ag in grams}}{\text{Molar mass of Ag}} = \frac{5.4 \text{ g}}{108.0 \text{ g}} = 0.050 \text{ mol}$$

From the above half-reaction:

$$\text{Number of moles of } Ag^+ \text{ liberated} = 0.050 \text{ mol}$$

From Faraday's second law:

$$\frac{\text{Number of moles of } Cu^{2+} \text{ liberated}}{\text{Number of moles of } Ag^+ \text{ liberated}} = \frac{\text{Charge on } Ag^+}{\text{Charge on } Cu^{2+}} = \frac{1}{2}$$

$$\frac{\text{Number of moles of } Cu^{2+} \text{ liberated}}{0.050 \text{ mol } Ag^+} = \frac{1}{2}$$

$$\text{Number of moles of } Cu^{2+} \text{ liberated} = 0.025 \text{ mol}$$

$$\text{Number of moles of Cu deposited} = 0.025 \text{ mol}$$

$$\text{Mass of Cu deposited} = 0.025 \text{ mol} \times 63.5 \frac{\text{g}}{\text{mol}} = 1.588 \text{ g}$$

Example 23.13

A steady current is passed through a series of solutions of $AgNO_3$, $CrCl_3$, $ZnSO_4$, and $NiSO_4$ for 3 h. If 1.5 g of Ag is deposited from the first solution, calculate:

(a) The masses of metals (Cr, Zn, and Ni) deposited simultaneously at the cathodes in the remaining solutions.
(b) The current flowing through the solutions.

Solution

(a) The half-cell reactions at the cathode in the three cells are:

$$Ag^+(aq) + e^- \longrightarrow Ag(s);\ 1F \ \Rightarrow\ 1 \text{ mol of } e^- \ \Rightarrow\ 1 \text{ mol Ag (s)}$$

$$Cr^{3+}(aq) + 3e^- \longrightarrow Cr(s);\ 3F \ \Rightarrow\ 3 \text{ mol of } e^- \ \Rightarrow\ 1 \text{ mol Cr(s)}$$

$$Zn^{2+}(aq) + 2e^- \longrightarrow Zn(s);\ 2F \ \Rightarrow\ 2 \text{ mol of } e^- \ \Rightarrow\ 1 \text{ mol Zn(s)}$$

$$Ni^{2+}(aq) + 2e^- \longrightarrow Ni(s);\ 2F \ \Rightarrow\ 2 \text{ mol of } e^- \ \Rightarrow\ 1 \text{ mol Ni(s)}$$

The same quantity of electricity is passed through all four cells. Therefore:

$$q_{\text{cell Ag}^+} = q_{\text{cell Cr}^{3+}} = q_{\text{cell Zn}^{2+}} = q_{\text{cell Ni}^{2+}}$$

$$\text{Number of moles of Ag deposited} = \frac{\text{Mass of Ag in grams}}{\text{Molar mass of Ag}} = \frac{1.5 \text{ g}}{108.0 \text{ g}} = 0.0139$$

$$\text{Number of moles of Ag}^+ \text{ liberated} = 0.0139$$

From Faraday's second law:

$$\frac{\text{Number of moles of Cr}^{3+} \text{ liberated}}{\text{Number of moles of Ag}^+ \text{ liberated}} = \frac{\text{Charge on Ag}^+}{\text{Charge on Cr}^{3+}} = \frac{1}{3}$$

$$\frac{\text{Number of moles of Cr}^{3+} \text{ liberated}}{0.0139 \text{ mol Ag}^+} = \frac{1}{3}$$

$$\text{Number of moles of Cr}^{3+} \text{ liberated} = 0.0046 \text{ mol}$$

$$\text{Number of moles of Cr deposited} = 0.0046 \text{ mol}$$

$$\text{Mass of Cr deposited} = 0.0046 \text{ mol} \times 52.0 \frac{\text{g}}{\text{mol}} = 0.241 \text{ g}$$

The mass of Zn deposited is obtained similarly:

$$\frac{\text{Number of moles of Zn}^{2+}\text{ liberated}}{0.0139\text{ mol Ag}^{+}} = \frac{1}{2}$$

$$\text{Number of moles of Zn}^{2+}\text{ liberated} = 0.0070\text{ mol}$$

$$\text{Number of moles of Zn deposited} = 0.0070\text{ mol}$$

$$\text{Mass of Zn deposited} = 0.0070\text{ mol} \times 65.4\frac{\text{g}}{\text{mol}} = 0.4550\text{ g}$$

Mass of Ni deposited:

$$\text{Moles of Ni deposited} = \text{moles of Zn (from half-cell equation)} = 0.0070$$

$$\text{Mass of Ni deposited} = 0.0070\text{ mol} \times 58.7\frac{\text{g}}{\text{mol}} = 0.4110\text{ g}$$

So the masses of Cr, Zn, and Ni deposited at the various cathodes are 0.2410 g, 0.4550 g, and 0.411 g, respectively.

(b) Current flowing through the solution:
Using the Ag^{+} half-cell we have:

$$1\text{ mol of Ag} = 1\text{ mol of e}^{-} = 1 \times 96{,}500\text{ C}$$

$$0.0139\text{ mol of Ag} = 1341.35\text{ C}$$

$$\text{Quantity of electricity } q = It$$

$$q = 1341.35\text{ C}; \quad t = 3\text{ h} = 3\text{ h} \times 60\,\frac{\text{min}}{\text{h}} \times 60\frac{\text{s}}{\text{min}} = 10{,}800\text{ s}$$

$$I = \frac{Q}{t} = \frac{1341.35\text{ C}}{10{,}800\text{ s}} = 0.124\text{ A}$$

23.12. Problems

1. Write the cell diagram and the cell reaction for the following redox couples:

   (a) Cu^{2+}/Cu and Sn^{2+}/Sn
   (b) Ag^{+}/Ag and Zn^{2+}/Zn
   (c) Fe^{2+}/Fe and Mg^{2+}/Mg
   (d) Pb^{2+}/Pb and Sn^{2+}/Sn

2. Write the cell diagrams corresponding to the following reactions:

   (a) $Mn(s) + Ti^{2+}(aq) \longrightarrow Mn^{2+}(aq) + Ti(s)$
   (b) $Mg(s) + Zn^{2+}(aq) \longrightarrow Mg^{2+}(aq) + Zn(s)$
   (c) $Sn^{2+}(aq) + Pb^{4+}(aq) \longrightarrow Sn^{4+}(aq) + Pb^{2+}(aq)$
   (d) $Ca(s) + Cu^{2+}(aq) \longrightarrow Ca^{2+}(aq) + Cu(s)$

3. Write the cell notation for the following overall reactions:

   (a) $Sn(s) + Cu^{2+}(aq) \longrightarrow Sn^{2+}(aq) + Cu(s)$
   (b) $Co^{2+}(s) + Ag^{+}(aq) \longrightarrow Co^{3+}(aq) + Ag(s)$
   (c) $Sn^{2+}(aq) + Fe(s) \longrightarrow Sn(s) + Fe^{2+}(aq)$
   (d) $Zn(s) + 2H^{+}(aq) \longrightarrow Zn^{2+}(aq) + H_2(g)$

4. The action of zinc metal on a solution of copper (II) sulfate is represented by the redox equation:

$$Zn + Cu^{2+}(aq) \longrightarrow Zn^{2+}(aq) + Cu$$

   (a) Write the half-cell reactions for both the oxidation and the reduction.
   (b) Is zinc oxidized or reduced?
   (c) Write the cell notation for the reaction.

5. Write balanced equations for the cells represented by the following short-hand notation.

   (a) $Pb(s)|Pb^{2+}(1\ M)||Br_2(1\ atm)|Br^{-}(aq)|Pt(s)$
   (b) $Zn(s)|Zn^{2+}(1\ M)||Eu^{3+}(aq), Eu^{2+}(aq)|Pt(s)$
   (c) $Co(s)|Co^{2+}(1\ M)||Cu^{2+}(aq)|Cu(s)$
   (d) $Cu(s)|Cu^{2+}(1\ M)||Cl^{-}(aq)|Cl_2(1\ atm)|C(s)$

6. Using the table of standard electrode potentials, calculate the emf (E^0_{cell}) for the following cells:

   (a) $Zn(s)|Zn^{2+}(1\ M)||Sn^{2+}(aq)|Sn(s)$
   (b) $Co(s)|Co^{2+}(1\ M)||Cu^{2+}(aq)|Cu(s)$

7. The standard electrode potential for the galvanic cell $Cd(s)|Cd^{2+}(aq)||Pb^{2+}(aq)|Pb(s)$ is +0.27 V, Given that the standard reduction potential for the Cd^{2+}/Cd half-cell is −0.40 V, calculate the standard reduction potential for the Pb^{2+}/Pb half-cell.

8. Using the table of standard electrode potentials, determine if copper metal will spontaneously react with a solution of tin (II) ions.

9. The standard reduction potential is −1.80 V for the U^{3+}/U couple and −0.61 V for the U^{4+}/U^{3+} couple. What is the potential (emf) for the U^{4+}/U couple?

10. The table of standard electrode potentials can be used to predict chemical reactivity. Will Cu metal reduce Sn(IV) to Sn(II)?

11. Will Fe(II) ion reduce MnO_4^- in acidic solution? *Hint:* Use the table of standard electrode potential and calculate the cell voltage.

12. Calculate the potential of the half-cell reaction:

$$Mg(s) \longrightarrow Mg^{2+}(aq) + 2e^{-} \qquad E^0_{cell} = -2.37\ V$$

when the concentration of Mg^{2+} ion is 0.01 M.

13. Given that $[Fe^{2+}] = 0.05$ M and $[Fe^{3+}] = 1.25$ M, calculate the potential of the half-cell reaction $Fe^{2+}(aq) \longrightarrow Fe^{3+}(aq) + e^-$, $E^0_{cell} = -0.77$ V.
14. Using the table of standard electrode potentials, calculate the electromotive force (emf) for the following cells:

    (a) $Zn(s)|Zn^{2+}(0.200\ M)||Sn^{2+}(0.100\ M)|Sn(s)$
    (b) $Mg(s)|Mg^{2+}(0.010\ M)||Cu^{2+}(0.100)\ M|Cu(s)$
    (c) $Zn(s)|Zn^{2+}(0.002\ M)||H^+(1\ M), H_2(0.1\ atm)|Pt(s)$

15. Will the reactions for the following cells take place as written?

    (a) $Mg(s)|Mg^{2+}(0.020\ M)||Sn^{2+}(0.2500\ M)|Sn(s)$
    (b) $Cd(s)|Cd^{2+}(0.010\ M)||Cu^{2+}(0.100\ M)|Cu(s)$
    (c) $Ni(s)|Ni^{2+}(0.200\ M)||Hg^{2+}(0.200\ M)|Hg(s)$

16. Calculate the standard Gibbs free energy change and the equilibrium constant at 25°C for the following reaction (*Note:* 1 V = 1 J/C):

$$Cu(s) + 2\,Ag^+(aq) \longrightarrow Cu^{2+}(aq) + 2\,Ag(s)$$

17. Calculate the standard Gibbs free energy change and the equilibrium constant at 25°C for the reaction:

$$2Br^-(aq) + Cl_2(g) \longrightarrow Br_2(l) + 2Cl^-(aq)$$

18. Calculate the standard cell potential, E^0, and the equilibrium constant, K, for the reaction shown below at 25°C:

$$Fe(s) + Cd^{2+}(aq) \longrightarrow Fe^{2+}(aq) + Cd(s)$$

19. Design a battery from aluminum metal and chlorine gas using the couples $Al^{3+}(aq)/Al(s)$, and $Cl_2(g)/Cl^-(aq)$.

    (a) Write a balanced equation for the electrode reaction in the battery.
    (b) Calculate the standard electrode potential for the battery.
    (c) What is the voltage of the battery if the pressure of chlorine gas is increased from 1 to 5 atm?

20. The net reaction occurring in a lead-acid battery is

$$Pb(s) + PbO_2(s) + H_2SO_4(aq) \longrightarrow 2\,PbSO_4(s) + 2\,H_2O(l)$$

Given that the standard potential E^0 for the cell is 2.04 V, determine the potential of the battery at 25°C when sulfuric acid with a concentration of 10 M is used in the cell.

21. Calculate the value of E^0 for the following reactions and determine if the reactions are spontaneous in the direction written.

    (a) $Zn(s) + Sn^{4+} \longrightarrow Zn^{2+}(aq) + Sn^{2+}(aq)$
    (b) $2\,I^-(aq) + Zn^{2+}(aq) \longrightarrow I_2(s) + Zn(s)$
    (c) $Cl_2(g) + V(s) \longrightarrow 2\,Cl^-(aq) + V^{2+}(aq)$
    (d) $I_2(s) + 2\,Br^-(aq) \longrightarrow 2\,I^-(aq) + Br_2(l)$

22. How many faradays are required to reduce 1 mol of the following?

    (a) $Ag^+(aq) + e^- \longrightarrow Ag(s)$
    (b) $Ca^{2+}(aq) + 2e^- \longrightarrow Ca(s)$
    (c) $Al^{3+}(aq) + 3e^- \longrightarrow Al(s)$
    (d) $Fe^{3+}(aq) + 3e^- \longrightarrow Fe(s)$

23. How many faradays are required to oxidize or reduce 1 mol of the following?

    (a) $Mg(OH)_2(s) + 2e^- \longrightarrow Mg(s) + OH^-(aq)$
    (b) $PbSO_4(s) + 2e^- \longrightarrow Pb(s) + SO_4^{2-}(aq)$
    (c) $Sn^{4+}(aq) + 2e^- \longrightarrow Sn^{2+}(aq)$
    (d) $2\,Br^-(aq) \longrightarrow Br_2(l) + 2e^-$

24. Balance the following redox reactions in acidic solution and determine the number of coulombs of electricity required to reduce 1 mol of the oxide or anion:

    (a) $Sb_2O_3(s) \longrightarrow 2Sb(s)$
    (b) $SO_4^{2-}(aq) \longrightarrow H_2SO_3(aq)$
    (c) $IO_3^-(aq) \longrightarrow I_2(s)$
    (d) $H_2O_2 \longrightarrow H_2O$

25. Given that 1 faraday is equivalent to 96,500 C per mole of electron, what is the charge on an individual electron? (*Note:* 1 mol of electrons contains 6.022×10^{23} electrons.)

26. A solution of copper (II) sulfate is electrolyzed between copper electrodes for 2.00 h using a steady direct current of 50 A. Calculate the mass of elemental copper deposited.

27. Aluminum is extracted from its ore by the electrolysis of a molten Al salt. Calculate the mass of Al produced by a current of 5 A flowing for 12 h.

28. When a current of 0.25 A was passed through a divalent metal salt solution for 3.83 h, 2.00 g of the metal was deposited. Determine the atomic mass of the metal. Identify the metal *M* using the periodic table.

29. A steady current of 0.5 A flows through aqueous solutions of zinc nitrate and silver nitrate connected in series. In a particular experiment, 10.5 g of zinc was deposited from the $Zn(NO_3)_2$ electrolytic cell. Calculate:

    (a) the mass of Ag metal deposited at the same time.
    (b) the quantity of electricity (in coulombs) that passed through the electrolytes.
    (c) the time it took to run the experiment.

30. A direct current is passed through three electrolytic cells of copper (II) sulfate, gold (III) nitrate and dilute sulfuric acid joined together in series such that the same quantity of electricity flows through each cell. If 6.355 g of copper are deposited in the first cell, calculate:

    (a) the mass of Au deposited in the second cell, and
    (b) the volume of hydrogen gas liberated from the third cell, measured at STP.

24

Radioactivity and Nuclear Reactions

24.1. Definition of Terms

Nuclide: an atom containing a specified number of protons and neutrons in its nucleus—in other words, any particular atom under discussion.
Unstable nuclide: one that will spontaneously disintegrate or emit radiation, thus giving off energy and altering to some new form (often another element). The new form may also be unstable; often it will be *stable*, that is, with no tendency to disintegrate. Unstable nuclides are also referred to as *radioactive*.
Radioactivity: the spontaneous emission of radiation by elements with unstable nuclei.
Radionuclide: a radioactive (that is, unstable) nuclide.
Radioisotope: another more commonly seen term for radionuclide.
Radioactive decay: the process whereby a radionuclide is converted to another form (usually another element) by emitting radiation.
Parent nuclide: a nuclide undergoing radioactive decay.
Daughter nuclide: the nuclide produced when a parent nuclide decays.
Activity: the rate at which a sample of the material decays, usually expressed as the number of disintegrations per unit time.

24.2. Radioactive Decay and Nuclear Equations

Naturally radioactive elements decay spontaneously by emitting *alpha particles, beta particles*, and *gamma radiation*. Other elements can be induced to decay by bombarding them with high-energy particles; this is known as artificial radioactivity.

Like chemical reactions, equations representing nuclear reactions must be balanced. However, the method for balancing nuclear equations differs from that used for chemical equations. To balance a nuclear equation, the sum of the atomic numbers or particle charges (subscripts) and the sum of the mass numbers (superscripts) on both sides of the equation must be equal.

24.2.1. Alpha emission ($^4_2\alpha$)

When a nucleus undergoes alpha decay, it emits a particle that is identical to a helium nucleus, with an atomic number of 2 and a mass number of 4.

Since the emission of an alpha particle from the nucleus results in a loss of 2 protons and 2 neutrons, when writing a nuclear reaction involving an alpha decay, subtract 4 from the mass number and 2 from the atomic number. Alpha-particle decay can be represented by the general equation:

$$^A_ZC \longrightarrow {}^4_2\alpha + {}^{(A-4)}_{(Z-2)}D$$

Parent nuclide — Daughter nuclide

For example, uranium-238 emits an alpha particle and forms thorium-234.

$$^{238}_{92}\text{U} \longrightarrow {}^{234}_{90}\text{Th} + {}^4_2\alpha$$

Remember: To balance a nuclear equation, the sums of the mass numbers and atomic numbers must be the same on both sides of the equation.

Example 24.1

Write balanced equations for the following radioactive nuclides undergoing alpha decay:

1. Bismuth-214
2. Polonium-212
3. Platinum-190

Solution

1. $^{214}_{83}\text{Bi} \longrightarrow {}^{210}_{81}\text{Tl} + {}^4_2\alpha$
2. $^{212}_{84}\text{Po} \longrightarrow {}^{210}_{82}\text{Pb} + {}^4_2\alpha$
3. $^{190}_{78}\text{Pt} \longrightarrow {}^{186}_{76}\text{Os} + {}^4_2\alpha$

24.2.2. Beta emission ($^{\ 0}_{-1}\beta$)

Beta radiation consists of particles with a charge of −1 and mass very close to zero—in other words, electrons. Beta particles are written as $^{\ 0}_{-1}\beta$ or $^{\ 0}_{-1}e$. Nuclear disintegration by beta emission always results in the formation of a new element that has the same mass number as the parent nuclide but whose atomic number is increased by one unit. Beta-particle decay can be represented by the general equation $^A_ZC \longrightarrow {}^{\ 0}_{-1}\beta + {}^{\ \ A}_{(Z+1)}D$

For example, thorium-234 emits a beta particle and forms protactinium-234.

$$^{234}_{90}\mathrm{Th} \longrightarrow {}^{234}_{91}\mathrm{Pa} + {}^{0}_{-1}\beta$$

Example 24.2

What products are formed when (a) radium-226, (b) lead-208, and (c) iodine-131 undergo beta decay?

Solution

(a) $^{226}_{88}\mathrm{Ra} \longrightarrow {}^{0}_{-1}\beta + {}^{226}_{89}\mathrm{Ac}$

(b) $^{207}_{82}\mathrm{Pb} \longrightarrow {}^{0}_{-1}\beta + {}^{207}_{83}\mathrm{Bi}$

(c) $^{131}_{53}\mathrm{I} \longrightarrow {}^{0}_{-1}\beta + {}^{131}_{54}\mathrm{Xe}$

24.2.3. Gamma radiation (γ)

Gamma rays are high-energy electromagnetic radiation of very short wavelength, with no associated mass or charge. Gamma radiation always accompanies other radioactive emissions such as alpha and beta radiations. Radioactive decay by gamma emission leaves both the atomic and mass numbers unchanged. The gamma ray is represented as $^{0}_{0}\gamma$ or simply γ.

24.2.4. Positron emission ($^{0}_{1}e$)

A positron is a particle that has the same mass as an electron or a beta particle but the opposite charge (positive). A positron is represented as $^{0}_{1}e$ or $^{0}_{1}\beta$. It is produced when a proton is converted to a neutron within the nucleus as represented by the equation $^{1}_{1}p \longrightarrow {}^{1}_{0}n + {}^{0}_{1}\beta$.

When a nucleus decays by positron emission, the mass number remains the same but the atomic number decreases by one unit. Positron decay can be represented by the general equation

$$^{A}_{Z}C \longrightarrow {}^{0}_{1}\beta + {}^{\;\;A}_{(Z-1)}D$$

For example, carbon-11 decays by emitting a positron:

$$^{11}_{6}\mathrm{C} \longrightarrow {}^{0}_{1}\beta + {}^{11}_{5}\mathrm{B}$$

24.2.5. Electron capture ($^{0}_{-1}e$)

In electron capture, the nucleus pulls an electron from a low-energy orbital such as the 1s orbital. The electron then combines with a proton to form a neutron:

$$^{0}_{-1}e + ^{1}_{1}p \longrightarrow ^{1}_{0}n$$

When a nucleus decays by electron capture, the mass number remains the same but the atomic number decreases by one unit. Electron capture decay can be represented by the general equation:

$$^{A}_{Z}C + ^{0}_{-1}e \longrightarrow {}^{A}_{(Z-1)}D$$

An example of such process is the reaction:

$$^{18}_{9}\text{F} + ^{0}_{-1}e \longrightarrow ^{18}_{8}\text{O}$$

Example 24.3

Write balanced nuclear equations for the following radioactive decay processes:

1. $^{21}_{11}$Na (positron emission)
2. $^{62}_{29}$Cu (electron capture)
3. $^{82}_{37}$Rb (gamma radiation)

Solution

1. $^{21}_{11}\text{Na} \longrightarrow ^{0}_{1}\beta + ^{21}_{10}\text{Ne}$
2. $^{62}_{29}\text{Cu} + ^{0}_{-1}e \longrightarrow ^{62}_{28}\text{Ni}$
3. $^{82}_{37}\text{Rb} \longrightarrow ^{0}_{0}\gamma + ^{82}_{37}\text{Rb}$ (no change)

24.3. Nuclear Transmutation

Naturally occurring isotopes of some elements like uranium and polonium spontaneously and uncontrollably disintegrate by emitting radiation. These are called radioactive isotopes. On the other hand, stable isotopes can be made artificially radioactive by bombarding them with subatomic particles. The result is the production of other nuclides or isotopes. This type of nuclear reaction is known as *nuclear transmutation*. A nuclear transmutation is brought about by bombarding the nucleus of an atom with high-energy particles, such as a protons (p), neutrons (n), or alpha (α) particles.

An example of a nuclear synthetic reaction is:

$$^{14}_{7}\text{N} + ^{4}_{2}\text{He} \longrightarrow ^{1}_{1}\text{H} + ^{17}_{8}\text{O}$$

The shorthand notation for a nuclear equation such as shown above is written in the following order: the target nucleus, the bombarding particle, the particle ejected, and the nuclide produced. Thus the above example becomes: $^{14}_{7}\text{N}\,(\alpha,\ p)\,^{17}_{8}\text{O}$.

Example 24.4

Write balanced equations for the following nuclear reactions

1. $^{239}_{94}\text{Pu} + ? \longrightarrow ^{242}_{96}\text{Cm} + ^{1}_{0}n$
2. $^{32}_{16}\text{S} + ^{1}_{0}n \longrightarrow ^{1}_{1}\text{H} + ?$
3. $^{235}_{92}\text{U} + ^{1}_{0}n \longrightarrow ? + ^{92}_{36}\text{Kr}, + 3\,^{1}_{0}n$
4. $^{241}_{95}\text{Am} + ^{4}_{2}\text{He} \longrightarrow ? + 2\,^{1}_{0}n$

Solution

1. $^{239}_{94}\text{Pu} + ^{4}_{2}\text{He} \longrightarrow ^{242}_{96}\text{Cm} + ^{1}_{0}n$
2. $^{32}_{16}\text{S} + ^{1}_{0}n \longrightarrow ^{1}_{1}\text{H} + ^{32}_{15}\text{P}$
3. $^{235}_{92}\text{U} + ^{1}_{0}n \longrightarrow ^{141}_{56}\text{Ba} + ^{92}_{36}\text{Kr} + 3\,^{1}_{0}n$
4. $^{241}_{95}\text{Am} + ^{4}_{2}\text{He} \longrightarrow ^{243}_{97}\text{Bk} + 2\,^{1}_{0}n$

Example 24.5

Write balanced reactions for the following nuclear processes, and identify the product nuclides (E):

1. $^{138}_{92}\text{U}\,(^{1}_{0}n,\ ^{0}_{-1}\beta)\,^{A}_{Z}E$
2. $^{24}_{12}\text{Mg}\,(^{4}_{2}\text{He},\ ^{1}_{0}n)\,^{A}_{Z}E$
3. $^{55}_{25}\text{Mn}\,(^{2}_{1}\text{D},\ 2\,^{1}_{0}n)\,^{A}_{Z}E$

Solution

First, determine A and Z by balancing the mass number and the atomic number. Then check the periodic table and use the value of Z to identify the element E.

1. $^{138}_{92}\text{U}\,(^{1}_{0}n,\ ^{0}_{-1}\beta)\,^{A}_{Z}E$

   $^{138}_{92}\text{U} + ^{1}_{0}n \longrightarrow ^{0}_{-1}\beta + ^{A}_{Z}E$

Mass number balance: $138 + 1 = 0 + A$; $A = 139$

Atomic number balance: $92 + 0 = -1 + Z$; $Z = 93$

Check the periodic table for element with $Z = 93$. The element is neptunium. The balanced equation is:

$$^{138}_{92}\text{U} + ^{1}_{0}n \longrightarrow ^{0}_{-1}\beta + ^{139}_{93}\text{Np}$$

2. $^{24}_{12}\text{Mg}\,(^{4}_{2}\text{He},\ ^{1}_{0}n)\,^{A}_{Z}E$

   $^{24}_{12}\text{Mg} + ^{4}_{2}\text{He} \longrightarrow ^{1}_{0}n + ^{A}_{Z}E$

   $A = 27; Z = 14; E = \text{Si}$

   The balanced equation is:

$$^{24}_{12}\text{Mg} + ^{4}_{2}\text{He} \longrightarrow ^{1}_{0}n + ^{27}_{14}\text{Si}$$

3. $^{55}_{25}\text{Mn}\,(^{2}_{1}D,\ 2^{1}_{0}n)\,^{A}_{Z}E$

   $^{55}_{25}\text{Mn} + ^{2}_{1}D \longrightarrow ^{1}_{0}n + ^{A}_{Z}E$

   $A = 56; Z = 26; E = \text{Fe}$. The balanced equation is:

$$^{55}_{25}\text{Mn} + ^{2}_{1}D \longrightarrow ^{1}_{0}n + ^{56}_{26}\text{Fe}$$

24.4. Rates of Radioactive Decay and Half-Life

Radioactive nuclides decay at different rates. Some disintegrate within seconds, while others take million of years. Radioactive decay follows first-order kinetics, and the rate of decay is directly proportional to the number of radioactive nuclei N in the sample:

$$\text{Decay rate} = kN$$

where k is the first-order rate constant, called the decay constant. Like any first-order rate constant, it has the units of 1/time.

By using calculus, the first-order rate law can be transformed into an integrated rate equation in a very convenient form:

$$\ln\left(\frac{N_t}{N_0}\right) = -kt \quad \text{or} \quad \log\left(\frac{N_t}{N_0}\right) = \frac{-kt}{2.303}$$

In this equation, N_0 is the initial number of radioactive nuclei in the sample, N_t is the number remaining after time t, k is the decay constant, and t is the time interval of decay. Note that N can be any measurement of amount, activity, or molar concentration so long as both N and N_0 are in the same units.

24.4.1. Half-life

The half-life of a radioactive nuclide is defined as the time it takes half of any given amount to decay; this number is a constant for any given radionuclide.

Substituting $N_t = \frac{1}{2} N_0$ at time $t = t_{1/2}$ in the integrated rate equation gives the expression:

$$t_{1/2} = \frac{0.693}{k} \quad \text{or} \quad k = \frac{0.693}{t_{1/2}}$$

Example 24.6

The half-life of vanadium-50 is 6×10^{15} years. It disintegrates by β-emission to produce $^{50}_{24}Cr$. What is the rate constant for this disintegration process?

Solution

Use the following equation:

$$k = \frac{0.693}{t_{1/2}}$$

$$t_{1/2} = 6 \times 10^{15}\ \text{yr}$$

$$k = \frac{0.693}{t_{1/2}} = \frac{0.693}{6 \times 10^{15}\ \text{yr}} = 1.16 \times 10^{-16}\ \text{yr}^{-1} \quad \text{or} \quad 3.64 \times 10^{-9}\ \text{s}^{-1}$$

Example 24.7

Radioactive iodine is used in cancer therapy. In particular, large doses of $^{131}_{53}I$ are used in the treatment of thyroid cancer. If the half-life is 8.06 days, how long will it take a sample of $^{131}_{53}I$ to disintegrate to 5% of the original concentration?

Solution

First, calculate the decay constant:

$$k = \frac{0.693}{t_{1/2}} = \frac{0.693}{8.06} = 0.086\ \text{day}^{-1}$$

Use the equation:

$$\log\left(\frac{N_t}{N_0}\right) = \frac{-kt}{2.303}$$

$$\log\left(\frac{5.0}{100}\right) = \frac{-\left(0.086\ \text{day}^{-1}\right)t}{2.303}$$

$$-1.3010 = \frac{-\left(0.086\ \text{day}^{-1}\right)t}{2.303}$$

$$t = 34.8\ \text{days}$$

Example 24.8

The half-life of ${}^{15}_{32}P$, a radioisotope used medically in leukemia therapy, is 14.28 days. It decays by emitting beta particles and gamma rays. How much of a 20 μg ${}^{32}_{15}P$ sample remains in a patient's blood after 60 days?

Solution

First, determine the decay constant:

$$k = \frac{0.693}{t_{1/2}} = \frac{0.693}{14.28\ \ \text{days}} = 0.049\ \text{day}^{-1}$$

Next determine the ratio of N_t to N_0:

$$\log\left(\frac{N_t}{N_0}\right) = \frac{-kt}{2.303}\log\left(\frac{N_t}{N_0}\right) = \frac{-\left(0.049\ \text{day}^{-1}\right)(60\ \text{days})}{2.303}$$

$$= -1.2766\ \frac{N_t}{N_0} = 0.053$$

But:

$$N_o = 20\ \mu\text{g}, \quad \text{so} \quad N_t = N_o \times 0.053 = 20\ \mu\text{g} \times 0.053 = 1.06\ \mu\text{g}.$$

The amount of the radioisotope remaining after 60 days is 1.06 μg ${}^{32}P$.

24.5. Energy of Nuclear Reactions

Like chemical reactions, nuclear reactions involve energy changes. However, the energy changes in nuclear reactions are several orders of magnitude larger than those in chemical reactions. The large amount of energy liberated in a nuclear reaction is due to a loss in mass during the process. The energy equivalent of mass is given by Einstein's equation:

$$E = mc^2$$

Where E is the energy released, m is the mass decrease, and c is the speed of light, 3.0×10^8 m/s.

24.5.1. Mass defect

The nucleus of an atom is made up of protons and neutrons (collectively known as nucleons). When a stable nucleus is formed, the mass of the nucleus is always less than the mass of its constituent protons and neutrons. This difference in mass is known as the *mass defect*, which is due to the conversion of nucleon mass into the energy which binds them together. Mass defect, Δm, is calculated from the following expression:

$$\Delta m = (\text{total masses of all nucleons, } p^+ \text{and } n^0) - (\text{actual mass of atom})$$

or, more simply:

$$\Delta m = \text{mass of reactants} - \text{mass of products}$$

When calculating mass defect, the masses of the nucleons must be expressed in atomic mass units (amu). Recall that:

$$1 \text{ amu} = 1.660 \times 10^{-24}\,\text{g} \quad \text{or} \quad 1.660 \times 10^{-27}\,\text{kg} \quad \text{and} \quad 1\,\text{g} = 6.023 \times 10^{23}\,\text{amu}$$

Example 24.9

Calculate the mass of proton and the mass of the neutron in amu given that the mass of the proton is 1.6724×10^{-24} g, and the mass of the neutron is 1.6747×10^{-24} g.

Solution

Use the conversion factor 6.023×10^{23} amu = 1 g

$$\text{mass of proton} = \left(\frac{6.023 \times 10^{23}\text{ amu}}{1\text{ g}}\right)\left(1.6724 \times 10^{-24}\text{ g}\right) = 1.0073\,\text{amu}$$

$$\text{mass of neutron} = \left(\frac{6.023 \times 10^{23}\text{ amu}}{1\text{ g}}\right)\left(1.6747 \times 10^{-24}\text{ g}\right) = 1.0087\text{ amu}$$

Example 24.10

Calculate the mass defect for the aluminum-27 nucleus if the isotopic mass of $^{27}_{13}Al$ is 26.9815 amu. The mass of a proton is 1.0073 amu and the mass of a neutron is 1.0087 amu.

Solution

Use the expression:

$$\Delta m = (\text{mass of separated neutrons} + \text{protons}) - (\text{mass of nucleus})$$

Aluminum has an atomic number of 13. Therefore, it contains 13 protons. Since the mass number of Al is 27, the number of neutrons will be 14, (27 − 13 = 14). Hence:

$$\Delta m = [(14 \times \text{mass of neutron}) + (13 \times \text{mass of neutron})] - (\text{mass of } ^{27}\text{Al nucleus})$$

$$\Delta m = [(14 \times 1.0087\,\text{amu}) + (13 \times 1.0073\,\text{amu})] - (26.9815\,\text{amu})$$

$$\Delta m = (27.2161\ \ \text{amu}) - (26.9815\,\text{amu}) = 0.2346\,\text{amu}$$

24.5.2. Binding energy

The binding energy of the nucleus is the energy holding the nucleons together. It is defined as the energy released when the neutrons and protons in the nucleus combine to form the nucleus. Binding energy gives a direct measure of the stability of the nucleus. The larger the binding energy, the more stable the nucleus. The binding energy can be calculated by using Einstein's equation:

$$BE = \Delta mc^2$$

The symbol BE is the binding energy, and Δm is the change in mass, or mass defect.

Example 24.11

What is the binding energy (a) in kJ/mol, and (b) kJ/nucleon, of iron-56, which has an atomic mass of 55.9349 amu? The mass of a neutron is 1.0087 amu, and that of a proton is 1.0073 amu.

Solution

Use the expression

$$\Delta m = (\text{mass of separated neutrons} + \text{protons}) - (\text{mass of nucleus})$$

The atomic number of iron is 26, so it has 26 protons. Since the mass number of Fe is 56, the number of neutrons will be 30. Therefore

$$\Delta m = [(30 \times \text{mass of neutron}) + (26 \times \text{mass of proton})]$$
$$- (\text{mass of } ^{56}\text{Fe nucleus})$$
$$\Delta m = [(30 \times 1.0087\ \text{amu}) + (26 \times 1.0073\ \text{amu})] - (55.9349\ \text{amu})$$
$$\Delta m = (56.4508\ \text{amu}) - (55.9349\ \text{amu}) = 0.0.5159\ \text{amu}$$
$$\Delta m \text{ in kg} = (0.5159\ \text{amu})\left(\frac{1.660 \times 10^{-27}\ \text{kg}}{1\ \text{amu}}\right) = 8.56 \times 10^{-28}\ \text{kg}$$

(a) *BE* in kJ/mol:

$$BE = \Delta mc^2 = \left(8.56 \times 10^{-28}\ \text{kg}\right)\left(3.0 \times 10^8\ \frac{\text{m}}{\text{s}}\right)^2 = 7.71 \times 10^{-11}\ \text{J}$$
$$\text{or} \quad 7.71 \times 10^{-14}\ \text{kJ}$$

Now convert this to kJ per mol by using Avogadro's number:

$$BE = \left(7.71 \times 10^{-14} \frac{\text{kJ}}{\text{atom}}\right)\left(6.023 \times 10^{23} \frac{\text{atom}}{\text{mol}}\right) = 4.64 \times 10^{10} \frac{\text{kJ}}{\text{mol}}$$

(b) *BE* in kJ/nucleon:
Iron has 56 nucleons (26 protons + 30 neutrons). Therefore binding energy per nucleon is:

$$BE = \left(7.71 \times 10^{-14}\ \frac{\text{kJ}}{\text{atom}}\right)\left(\frac{1\ \text{atom}}{56\ \text{nucleons}}\right) = 1.38 \times 10^{-15}\ \frac{\text{kJ}}{\text{nucleon}}$$

Example 24.12

When a uranium-235 nuclide is bombarded with a fast-moving neutron in a fission reaction, one possible reaction pathway is the production of zirconium-97 and tellurium-137 according to the equation $^{235}_{92}\text{U} + ^{1}_{0}n \longrightarrow ^{97}_{40}\text{Zr} + ^{137}_{52}\text{Te} + 2\,^{1}_{0}n$.

Calculate the mass defect and the binding energy (in MeV/nucleon) for the daughter nuclides. Which of the two has more binding energy? (*Note*: MeV, millions of electron volts, are an energy measurement often used in atomic physics. $1\ \text{MeV} = 1.602 \times 10^{13}\ \text{J}$.)

(a) ^{97}Zr (atomic mass = 96.9110 amu)
(b) ^{137}Te (atomic mass = 136.9254 amu)

Solution

(a) In ${}^{97}_{40}Zr$, there are 40 protons and 57 neutrons. The mass defect is obtained as

$$\Delta m = [(57 \times \text{mass of neutron}) + (40 \times \text{mass of protons})]$$
$$- (\text{mass of } {}^{97}\text{Zr nucleus})$$

$$\Delta m = [(57 \times 1.0087 \text{ amu}) + (40 \times 1.0073 \text{ amu})] - (96.9110 \text{ amu})$$
$$= 0.8769 \text{ amu}$$

$$\Delta m \text{ in kg} = (0.8769 \text{ amu})\left(\frac{1.660 \times 10^{-27} \text{ kg}}{1 \text{ amu}}\right) = 1.456 \times 10^{-27} \text{ kg}$$

$$BE = \Delta mc^2 = \left(1.456 \times 10^{-27} \text{ kg}\right)\left(3.0 \times 10^8 \frac{\text{m}}{\text{s}}\right)^2 = 1.31 \times 10^{-10} \text{ J}$$

$$\text{or} \quad 1.31 \times 10^{-13} \text{ kJ}$$

BE in MeV:
Use the conversion factor 1 MeV = 1.602×10^{-12} J. Therefore *BE* per nucleon is:

$$BE = \left(\frac{1.31 \times 10^{-10} \text{ J}}{\text{nucleus}}\right)\left(\frac{1 \text{ MeV}}{1.602 \times 10^{-13}}\right)\left(\frac{1 \text{ nucleus}}{97 \text{ nucleon}}\right) = 8.43 \frac{\text{MeV}}{\text{nucleon}}$$

(b) In ${}^{137}_{52}Te$, there are 52 protons and 85 neutrons. The mass defect is

$$\Delta m = [(85 \times \text{mass of neutron}) + (52 \times \text{mass of protons})]$$
$$- (\text{mass of } {}^{137}\text{Te nucleus})$$

$$\Delta m = [(85 \times 1.0087 \text{ amu}) + (52 \times 1.0073 \text{ amu})] - (136.9254 \text{ amu})$$
$$= 1.1937 \text{ amu}$$

$$\Delta m \text{ in kg} = (1.1977 \text{ amu})\left(\frac{1.660 \times 10^{-27} \text{ kg}}{1 \text{ amu}}\right) = 1.982 \times 10^{-27} \text{ kg}$$

$$BE = \Delta mc^2 = (1.982 \times 10^{-27} \text{ kg})\left(3.0 \times 10^8 \frac{\text{m}}{\text{s}}\right)^2 = 1.78 \times 10^{-10} \text{ J}$$

$$\text{or} \quad 1.78 \times 10^{-13} \text{ kJ}$$

BE in MeV:
Use the conversion factor: 1 MeV = 1.602×10^{-12} J. Therefore *BE* per nucleon is:

$$BE = \left(\frac{1.78 \times 10^{-10} \text{ J}}{\text{nucleus}}\right)\left(\frac{1 \text{ MeV}}{1.602 \times 10^{-13}}\right)\left(\frac{1 \text{ nucleus}}{137 \text{ nucleon}}\right) = 8.13 \frac{\text{MeV}}{\text{nucleon}}$$

The Zr nuclide is more stable than the Te nuclide by about 0.304 MeV/nucleon.

Example 24.13

Calculate the amount of energy released when 50 g of deuterium (^{2}H) undergoes fusion to form ^{3}He:

$$^2_1\text{H} + ^2_1\text{H} \longrightarrow ^3_2\text{He} + ^1_0n$$

The atomic masses of the species involved are:

$$^2_1\text{H}\,(2.0140\text{ amu});\quad ^3_2\text{He}\,(3.01605\text{ amu});\quad \text{and}\quad ^1_0n\,(1.0087\text{ amu})$$

Solution

The energy released is obtained using Einstein's equation:

$$\Delta E = \Delta mc^2$$

The mass defect can be calculated using the following expression:

$$\Delta m = (\text{mass of reactants}) - (\text{mass of products})$$

$$\Delta m = \left(2 \times \text{mass of } ^2_1\text{H}\right) - \left(\text{mass of } ^3_2\text{He} + \text{mass of } ^1_0n\right)$$

$$\Delta m = (2 \times 2.0140\text{ amu}) - (3.01603\text{ amu} + 1.0087\text{ amu}) = 0.0033\text{ amu}$$

$$\Delta m \text{ in kg} = (0.0033\text{ amu})\left(\frac{1.660 \times 10^{-27}\text{ kg}}{1\text{ amu}}\right) = 5.48 \times 10^{-30}\text{ kg}$$

$$\Delta E = \Delta mc^2$$

$$\Delta E = \left(5.48 \times 10^{-30}\text{ kg}\right)\left(3.0 \times 10^8\text{ m/s}\right)^2 = 4.93 \times 10^{-13}\text{ J per 2 atoms of } ^2\text{H}$$

or:

$$\Delta E = \left(4.93 \times 10^{-13}\text{ J}\right)\left(\frac{1}{2\text{ atoms } ^2\text{H}}\right) = 2.46 \times 10^{-13}\text{ J/atom of } ^2\text{H}$$

Calculate the number of atoms in 50 g of ^{2}H:

$$\left(50.0\text{ g } ^2\text{H}\right) \times \left(\frac{1\text{mol } ^2\text{H}}{2.01\text{ g}}\right) \times \left(6.023 \times 10^{23}\,\frac{\text{atom}}{\text{mol}}\right) = 1.50 \times 10^{25}\text{ atom}$$

1.5×10^{25} atoms of ^{2}H undergo the fusion reaction. The energy released by this number of atoms is:

$$\Delta E = \left(2.46 \times 10^{-13}\,\frac{\text{J}}{\text{atom}}\right)\left(1.50 \times 10^{25}\text{ atom}\right) = 3.69 \times 10^{12}\text{ J} \quad \text{or} \quad 3.69 \times 10^{9}\text{ kJ}$$

24.6. Problems

1. Write balanced equations for the following radioactive nuclides undergoing alpha decay:

   (a) Uranium-238
   (b) Actinium-228
   (c) Thorium-232
   (d) Radium-226

2. Write balanced equations for the following radioactive nuclides undergoing beta decay:

   (a) Barium-140
   (b) Iodine-131
   (c) Thallium-232
   (d) Sulfur-36

3. Write a balanced nuclear equation for each of the following:

   (a) Alpha emission of bismuth-213
   (b) Electron capture of lead-208
   (c) Beta emission of tungsten-188
   (d) Positron emission of potassium-40

4. Complete the following nuclear reactions:

   (a) ${}^{14}_{7}\text{N} + {}^{4}_{2}\text{He} \longrightarrow ? + {}^{1}_{1}\text{H}$
   (b) ${}^{55}_{25}\text{Mn} + {}^{2}_{1}\text{H} \longrightarrow {}^{56}_{26}\text{Fe} + ?$
   (c) ${}^{235}_{92}\text{U} + {}^{1}_{0}n \longrightarrow {}^{94}_{38}\text{Sr} + ? + 3\,{}^{1}_{0}n$
   (d) ${}^{27}_{13}\text{Al} + {}^{4}_{2}\text{He} \longrightarrow ? + {}^{1}_{0}n$
   (e) $? + {}^{4}_{2}\text{He} \longrightarrow {}^{14}_{7}\text{N} + {}^{1}_{0}n$

5. Complete the following nuclear reactions:

   (a) ${}^{18}_{8}\text{O} + {}^{1}_{1}\text{H} \longrightarrow ? + {}^{1}_{0}n$
   (b) $? + {}^{1}_{0}n \longrightarrow {}^{85}_{36}\text{Kr} + {}^{1}_{1}\text{H}$
   (c) ${}^{14}_{7}\text{N} + {}^{4}_{2}\text{He} \longrightarrow ? + {}^{1}_{1}\text{H}$
   (d) ${}^{11}_{5}\text{B} + ? \longrightarrow {}^{11}_{6}\text{C} + {}^{1}_{0}n$
   (e) ${}^{23}_{12}\text{Mg} \longrightarrow ? + {}^{0}_{+1}e$

6. Write the symbols for the daughter nuclei in the following reactions:

   (a) ${}^{14}_{7}\text{N}\,({}^{4}_{2}\alpha,\ {}^{1}_{1}p)$
   (b) ${}^{68}_{30}\text{Zn}\,({}^{4}_{2}\alpha,\ {}^{1}_{0}n)$
   (c) ${}^{31}_{15}\text{P}\,({}^{1}_{0}n,\ \gamma)$
   (d) ${}^{191}_{77}\text{Ir}\,({}^{4}_{2}\alpha,\ {}^{1}_{0}n)$

7. Complete each of the following and write it as a nuclear equation.

   (a) ${}^{18}_{8}O\,({}^{1}_{0}n,\ {}^{1}_{1}p)?$
   (b) ${}^{85}_{37}Rb\,(?,\ {}^{1}_{0}n)\,{}^{85}_{36}Kr$
   (c) ${}^{58}_{26}Fe\,(2\,{}^{1}_{0}n,\ ?)\,{}^{60}_{27}Co$
   (d) $?\,({}^{4}_{2}\alpha,\ {}^{0}_{-1}\beta)\,{}^{243}_{97}Bk$

8. Using the shorthand notation, write an equation for each of the following nuclear processes:

   (a) ${}^{2}_{1}H + {}^{3}_{1}H \longrightarrow {}^{4}_{2}He + {}^{1}_{0}n$
   (b) ${}^{27}_{13}Al + {}^{4}_{2}He \longrightarrow {}^{30}_{14}Si + {}^{1}_{1}H$
   (c) ${}^{6}_{3}Li + {}^{1}_{0}n \longrightarrow {}^{4}_{2}He + {}^{3}_{1}H$
   (d) ${}^{63}_{29}Cu + {}^{1}_{1}H \longrightarrow {}^{62}_{29}Cu + {}^{2}_{1}H$
   (e) ${}^{40}_{18}Ar + {}^{4}_{2}He \longrightarrow {}^{43}_{19}K + {}^{1}_{1}H$

9. The decay constant for cobalt-60, used in cancer therapy, is 0.131/year. If a 1.0 g sample of cobalt is stored for 25 years, what mass of the isotope will remain?

10. Strontium-90 is one of the nuclides that occur in the fallout from nuclear explosion; its decay constant is 0.0247 per year. What mass of the isotope will remain after a 0.0258 g sample of strontium-90 is kept for 5 years?

11. Phosphorus-32 is a radioisotope used in the treatment of blood cancer. A leukemia patient is subjected to a dose of phosphorus-32 radiation therapy. After 30 days, about 25% of the isotope remains in the patient's body. Calculate the rate of disintegration and the half-life of this isotope.

12. The half-life of ${}^{24}_{11}Na$ is 15.0 h. Calculate:

    (a) the decay constant
    (b) the fraction of the isotope originally present that remains after 6 h
    (c) the mass of the radioactive isotope left by allowing a 10.0 mg sample to decay for 12 h (use the decay constant from (a))

13. At 10:00 am on a given day, a nuclear chemist observed that a certain transuranium element stored in one of the national defense laboratories decayed at a rate of 55,000 counts per minute. At 3:00 pm of the same day the rate of decay had reduced to 15,000 counts per minute. Calculate the half-life of the radioisotope if the rate of decay is proportional to the number of radioatoms in the sample.

14. The production of ammonia through the Haber process is represented by the equation:

$$N_2(g) + 3H_2(g) \longrightarrow 2\,NH_3(g) \qquad \Delta H^0 = -92.2\ kJ$$

Calculate the mass defect (in grams) accompanying the formation of NH_3 from N_2 and H_2 gases.

15. Calculate the mass defect (in g/mol) for the following nuclides

    (a) ^{56}Fe (atomic mass = 55.9207 amu)
    (b) ^{239}Pu (atomic mass = 239.0006 amu)
    (c) ^{210}Po (atomic mass = 209.9368 amu)
    (d) ^{4}He (atomic mass = 4.0015 amu)

16. Determine the mass defect for the following reaction:

$$^{235}_{92}\text{U} + ^{1}_{0}n \longrightarrow ^{142}_{56}\text{Ba} + ^{91}_{36}\text{Kr} + 3^{1}_{0}n$$

from the following masses:

^{235}U (235.0439 amu); ^{142}Ba (141.9164 amu); ^{91}Kr (90.9234 amu); and $^{1}_{0}n$ (1.00866 amu)

17. Calculate the energy released (in MeV) during each of the following fusion reactions:

    (a) $^{1}_{1}\text{H} + ^{2}_{1}\text{H} \longrightarrow ^{4}_{2}\text{He}$
    (b) $^{1}_{1}\text{H} + ^{3}_{2}\text{He} \longrightarrow ^{4}_{2}\text{He} + ^{0}_{1}e$
    (c) $^{3}_{1}\text{H} + ^{2}_{1}\text{H} \longrightarrow ^{4}_{2}\text{He} + ^{1}_{0}n$
    (d) $^{1}_{1}\text{H} + ^{7}_{3}\text{Li} \longrightarrow 2\,^{4}_{2}\text{He}$

    The atomic masses are $^{1}_{1}$H (1.00783 amu); $^{2}_{1}$H (2.01410 amu); $^{3}_{1}$H (3.01605 amu); $^{3}_{2}$He (3.01603 amu); $^{7}_{3}$Li (7.01600 amu); $^{4}_{2}$He (4.00260 amu); and $^{0}_{1}e$ (0.0005486 amu).

18. (a) Write a balanced equation for the nuclear reaction represented by the notation:

$$^{56}_{26}\text{Fe}\,(^{2}_{1}\text{H},\ ^{4}_{2}\alpha)\,^{54}_{25}\text{Mn}$$

    (b) Calculate the energy change in joules per mole of $^{2}_{1}$H.

19. Given that 1 kg = 6.022×10^{26}amu and that 1 J = 1 kg m^{2} s^{-2}, show that 1 amu is equivalent to 931.25 MeV of energy. (Hint: 1 MeV = 1.6×10^{-13} J.)

20. $^{127}_{53}$I is one of the most stable nuclei and has an atomic mass of 126.9004 amu. Calculate the total binding energy and the average binding energy per nucleon for this radioisotope (1 amu = 931.25 MeV).

21. The half-life of $^{201}_{81}$Tl, a radioisotope used for parathyroid imaging, is 3.05 days. Calculate the total binding energy per nucleon for the formation of a $^{201}_{81}$Tl isotope. The atomic mass of Tl is 204.3833.

22. A skeleton suspected to be that of the first Onojie (King) of Opoji was recovered by a local archaeologist from an ancient palace site. Radiocarbon measurements made of the bones showed a decay rate of 18.5 disintegrations of ^{14}C per minute per gram of carbon. Determine the age of the suspected Onojie if carbon from living material taken from the same area decays at the rate of 23.1 disintegrations per minute per gram of carbon. The half-life of ^{14}C is 5715 years.

Solutions

Chapter 1

1. (a) Two (b) Five (c) Three (d) Five (e) Five (f) Five
2. (a) The least certain number here is 5.55. Therefore, round the final answer to 3 significant figures, i.e., $(2.254)(5.55) = 12.5$.
   (b) The least certain number here is 18.322. Round the final answer to 5 significant figures, i.e., $18.322/1.91750 - 9.5552$.
   (c) The least certain number here is 4.3. Round the final answer to 2 significant figures, i.e., $(4.3)(8.51)(20.5360)/6.750 = 11$.
3. (a) 108.8 g (b) 16.7 mL (c) 50.5 (d) 0.00018
4. (a) 5.25×10^5 (b) 9.11×10^{-6} (c) 6.02×10^{23} (d) -1.0×10^{-4}
5. (a) 2.345 (b) 785,000 (c) −0.0000002345 (d) 345 (e) 0.025
6. (a) (3.2×10^4) (b) (2.43×10^{-2}) (c) (6.12×10^3) (d) (7.9×10^5)
7. (a) (4.72×10^{-2}) (b) (1.08×10^{-5}) (c) (9.1×10^3) (d) (1.0×10^{-9}) (e) (1.6×10^7)
8. (a) (4.3×10^2) (b) (2.7×10^{-8}) (c) (1.3×10^3) (d) (4.5×10^1)
9. 979.45
10. 2.851×10^2
11. (a) $x = 1{,}000{,}000$ (b) $x = 4$ (c) $x = 5$ (d) $x = -2$
12. (a) 15 (b) 0.67 (c) 60.00 (d) 12.5 (e) 1.22
13. (a) 4 (b) $\frac{1}{3}$ (c) − 6 (d) − 1
14. 163 kJ
15. 2.94×10^4 L/mol-s
16. (a) $x = -4$ or $x = -3$ (b) $x = 1$ or $x = -7$ (c) $x = \frac{2}{5}$, or $x = \frac{1}{2}$ (d) $x = 0.732$ or $x = -2.732$ (e) $x = -4$ or $x = 3$

Chapter 2

1. (a) 86°F (b) – 320°F (c) 32.1°F (d) 648.9°C (e) – 56°C (f) 78.3°C
2. (a) 573 K (b) 77 K (c) 314 K (d) 927°C (e) 171°F (f) 351 K
3. (a) – 43.6°F, 231 K (b) – 306.4°F, 85 K
4. $T = \frac{5}{9}(T - 32)$ or $T = -40°C = -40°F$
5. 990 L
6. (a) 2.21×10^4 cm^3 (b) 35,400 g or 35.4 kg
7. 9.650×10^3 g
8. 47.8 g/cm^3
9. $V = 406$ mm^3; $r = 4.59$ mm
10. Mass = 5865 g $\approx 6.0 \times 10^3$ g
11. 144 g
12. 5.5×10^6 nm
13. 1.027×10^3 kg/m^3
14. 125 cm^3 olive oil has a greater mass (115 g) than 130 cm^3 ethyl alcohol (103 g)
15. 636 cm^3
16. 7.62×10^4 mm
17. 31,500,000 s
18. 56.1 lb
19. 159 kg
20. (a) 1.92×10^{-18} J (b) 1.92×10^{-21} kJ (c) 4.59×10^{-19} cal
    (d) 1.92×10^{-11} ergs

Chapter 3

1.

	Symbol	*Protons*	*Neutrons*	*Electrons*
(a)	$^{197}_{79}Au$	79	118	79
(b)	$^{10}_{5}B$	5	5	5
(c)	$^{40}_{20}B$	20	20	20
(d)	$^{163}_{66}Dy$	97	42	66

2.

	Symbol	*Protons*	*Neutrons*	*Electrons*
(a)	$^{84}_{26}Kr$	36	4S	36
(b)	$^{24}_{12}Mg$	12	12	12
(c)	$^{69}_{31}Ga$	31	38	31
(d)	$^{75}_{33}As$	33	42	33

3.

	Species	Protons	Neutrons	Electrons
(a)	$^{59}_{27}Co^{2+}$	27	32	25
(b)	$^{24}_{12}Mg^{2+}$	12	12	22
(c)	$^{69}_{31}Ga^{3-}$	31	38	69
(d)	$^{118}_{50}Sn^{2+}$	50	68	48

4. (a) Br^- (b) Cs^+ (c) Mn^{2+}

5.

Symbol	$^{40}_{20}Ca^{2+}$	$^{31}_{15}P^{3-}$	$^{79}_{34}Se^{2-}$	$^{238}_{92}U^{4-}$	$^{15}_{7}N^{3-}$	$^{137}_{56}Ba^{2-}$
Mass number	40	31	79	238	15	137
Protons	20	15	34	92	7	56
Neutrons	20	16	45	146	8	81
Electrons	18	18	36	88	10	54
Net charge	2	−3	−2	4	−3	2

6. (a) $^{56}_{26}Fe$ (b) $^{131}_{53}I$ (c) $^{202}_{80}Hg$ (d) $^{19}_{9}F$

7.

Particle	*Atomic no.*	*Mass no.*	*No. of Protons*	*No. of Neutrons*	*No. of Electrons*	*Net charge*	*Symbol*
A	35	80	35	45	36	−1	Br^-
B	13	27	13	14	10	3	Al^{3+}
C	15	31	15	16	18	−3	P^{3-}
D	24	52	24	28	24	0	Cr
E	17	35	17	18	17	0	Cl
F	54	131	54	77	54	0	Xe
G	81	204	81	123	80	1	Tl^+
H	40	91	40	51	36	4	Zr^{4+}

8.

	Symbol	Protons	Neutrons	Electrons
(a)	$^{32}_{16}S^{2-}$	16	16	18
(b)	$^{128}_{52}Te^{2-}$	35	45	36
(c)	$^{80}_{35}Br^-$	52	76	54
(d)	$^{14}_{7}N^{3-}$	7	7	10

9. 35.45

10. 11.01

11. 16.0
12. 60.2% ^{69}Ga and 39.8% ^{71}Ga

Chapter 4

1. (a) 84 amu (b) 100 amu (c) 120 amu (d) 328 amu (e) 342 amu
2. (a) 342 g/mol (b) 34 g/mol (c) 100.9 g/mol (d) 190.5 g/mol (e) 142 g/mol
3. 339.5 amu
4. 1,005 g/mol
5. 180 amu
6. 162 amu
7. 227 g/mol
8. 154 amu
9. 192.2 amu, $M = \text{Ir}$
10. 31 amu, $X = \text{P}$

Chapter 5

1. (a) 1.0 mol $CaCO_3$ (b) 5.56 mol H_2O (c) 2.50 mol NaOH (d) 1.02 mol H_2SO_4 (e) 3.57 mol N_2
2. (a) 0.092 mol $CaSO_4$ (b) 7.35×10^{-3} mol H_2O_2 (c) 4.76×10^{-4}mol $NaHCO_3$ (d) 0.10 mol H_2SO_4
3. (a) 1.01g MgO (b) 1.55g Na_2O (c) 0.425g NH_3 (d) 0.70g C_2H_4 (e) 4.45g $H_4P_2O_7$
4. (a) 28 g CaO (b) 0.15 g Li_2O (c) 710 g P_4O_{10} (d) 42.0 g C_2H_4 (e) 0.80 g $B_3N_3H_6$
5. (a) 3.32×10^{-2}g Ca (b) 2.33×10^4g Fe (c) 20 g Ca (d) 6.48×10^{-22}g K (e) 1.0×10^{-26}g H
6. (a) 1.20×10^{24} Al atoms and 1.81×10^{24} O atoms
   (b) 1.20×10^{24} H atoms and 6.02×10^{23} S atoms
   (c) 6.02×10^{23} C atoms and 1.20×10^{24} O atoms
   (d) 6.02×10^{23} H atoms and 6.02×10^{23} Cl atoms
7. (a) 8×10^4 mol (b) 415.3 mol (c) 0.19 mol (d) 1.0×10^{-23}mol (e) 5.0×10^{-27}mol
8. 6327 kg
9. (a) 1.77×10^{25} O (b) 9.68×10^{24} O (c) 9.03×10^{24} O (d) 7.76×10^{24} O
10. 2.4×10^{24} Fe^{3+} ions and 3.6×10^{24} SO_4^{2-} ions
11. (a) 0.75 mol Y; 1.50 mol Ba; 2.25 mol Cu (b) 4.52×10^{23} Y atoms; 9.03×10^{23} Ba atoms; 1.36×10^{24} Cu atoms
12. 9.03×10^{23} Si atoms and 1.2×10^{24} N atoms

Chapter 6

1. (a) H_2S: S = 94.1% (b) SO_3: S = 40.0% (c) H_2SO_4: S = 32.7%
   (d) $Na_2S_2O_3$: S = 40.5%
2. (a) K_2CO_3: K = 56.6% , C = 8.7% , O = 34.7%
   (b) $Ca_3(PO_4)_2$: Ca = 38.7% , P = 20.0% , O = 41.3%
   (c) $Al_2(SO_4)_3$: Al = 15.8% , S = 28.1% , O = 56.1%
   (d) C_6H_5OH: C = 76.6% , H = 6.4% , O = 17.0%
3. (a) $C_{20}H_{25}N_3O$: C = 74.3% , H = 7.74% , N = 13.0% , O = 4.95%
   (b) $C_{10}H_{16}N_5P_3O_{13}$: C = 23.7%, H = 3.2%, N = 13.8%, P = 18.3%, O = 41.0%
   (c) $C_6H_4N_2O_4$: C = 42.9%, H = 2.4%, N = 16.7%, O = 38.1%
   (d) $C_6H_5CONH_2$: C = 69.4%, H = 5.8%, N = 11.6%, O = 13.2%
4. Mg = 12.0% , Cl = 34.9% , H = 5.9% , O = 47.2%
5. 6.83 g
6. 38.3 g of C, 22.3 g of N
7. 32.7% Fe
8. 0.6657 g
9. (a) SO_3 (b) $CaWO_3$ (c) CH_2 (d) CCl_2F_2 (e) $Cr_2S_3O_{12}$ or $Cr_2(SO_4)_3$
10. (a) $Mg_2P_2O_7$ (b) $C_2H_3O_5N$ (c) $C_{14}H_9Cl_5$ (d) $Co_3Mo_2Cl_{11}$ (e) $C_3H_2O_3$
11. (a) CH_2 (b) Fe_2S_3 (c) $C_2H_3O_2$ (d) $C_4H_5N_2O$ (e) $Al_2(SO_4)_3$
12. P_4O_{10}
13. Empirical formula = $C_{27}H_{46}O$, molecular formula = $C_{27}H_{46}O$
14. $C_{18}H_{24}O_2$
15. $Na_2CO_3 \cdot 1.225H_2O$
16. $C_3H_8O_2$, $C_9H_{24}O_6$
17. $FeCl_3 \cdot 6H_2O$
18. $C_6H_{10}O_5$; $C_{12}H_{20}O_{10}$
19. $C_2H_4O_3$, $C_3H_{12}O_9$
20. (a) C = 57.14%, H = 6.12%, N = 9.53%, O = 27.23% (b) $C_{14}H_{18}N_2O_5$

Chapter 7

1. (a) +4 for C, −2 for O (b) +6for S, −2 for O (c) +6 for S, −1 for F
   (d) −2 for N, +1 for H (e) +2 for Pb, −2 for O
2. (a) +2 (b) +4 (c) +1 (d) 0 (e) −3
3. (a) −3 (b) −2 (c) P = +3, O = −2 (d) S = +6, O = −2
   (e) Cl = +5, O = −2
4. (a) Mn = +7, O = −2 (b) Cr = +6, O = −2 (c) U = +6, O = −2
   (d) S = +2, O = −2 (e) S = $+\frac{5}{2}$, O = −2

5. (a) $+3$ (b) $+1$ (c) $+2$ (d) $+3$ (e) $+\frac{8}{3}$
6. (a) $+3$ (b) $+4$ (c) $+5$ (d) $+6$ (e) $+5$
7. (a) $+6$ (b) $+3$ (c) $+7$ (d) $+5$ (e) $+3$
8. (a) Mg_3N_2 (b) SnF_2 or SnF_4 (c) H_2S (d) InI_3 (e) $AlBr_3$
9. (a) Li_3N (b) B_2O_3 (c) CaO (d) $RbCl$ (e) Cs_2S
10. (a) $Fe_2(CO_3)_3$ (b) $(NH_4)_3PO_4$ (c) $Ca(NO_3)_2$ (d) $LiClO_4$ (e) $K_2Cr_2O_7$
11. (a) $MnCO_3$ (b) $Sn_3(AsO_4)_2$ (c) $Ca(C_2H_3O_2)_2$ (d) $LiClO_4$ (e) Na_3BO_3
12. (a) SO_2 (b) CO_2 (c) NO_2 (d) N_2O_5 (e) CCl_4 (f) ClO_2
(g) LiI (h) SeO_2 (i) $FeCl_2$ (j) Ba_3P_2
13. (a) V_2O_5 (b) CuS (c) Fe_2S_3 (d) GaN (e) $HgCl_2$
14. (a) Na_2CO_3 (b) NH_4Cl (c) K_3PO_4 (d) $Ca(HSO_4)_2$ (e) $Fe_2(CrO_4)_3$
(f) $Pd_3(PO_4)_2$ (g) $Al(HCO_3)_3$ (h) $K_2Cr_2O_7$ (i) $Fe(OH)_3$ (j) $Mg_3(BO_3)_2$
15. (a) Phosphorus trichloride (b) Phosphorus pentachloride
(c) Carbon monoxide (d) Carbon dioxide (e) Sulfur dioxide
(f) Sulfur trioxide (g) Silicon dioxide (h) Diphosphorus pentasulfide
(i) Dinitrogen pentoxide
16. (a) Nickel (II) nitride (b) Iron (III) chloride (c) Aluminum sulfide
(d) Germanium disulfide (e) Titanium disulfide (f) Calcium hydride
(g) Mercury (II) chloride (h) Mercury (I) phosphide (i) Copper (I) nitride
17. (a) Hydrofluoric acid (b) Hydrobromic acid (c) Hydroselenic acid
(d) Hydrosulfuric acid (e) Hydrocyanic acid
18. (a) Nitric acid (b) Perbromic acid (c) Hypochloric acid (d) Oxalic acid
(e) Phosphorous acid (f) Iodic acid
19. (a) Gallium nitrate, nitric acid (b) Cobalt (II) sulfate, sulfuric acid
(c) Iron (II) acetate, acetic acid (d) Calcium borate, boric acid
(e) Lead (II) oxalate, oxalic acid (f) Chromium (III) phosphate,
phosphoric acid (g) Nickel carbonate, carbonic acid
(h) Iron (III) cyanide, hydrocyanic acid (i) Aluminum iodide, hydroiodic acid
(j) Rubidium bromate, bromic acid
20. (a) Potassium hydroxide (b) Ammonium hydroxide (c) Cobalt
(II) hydroxide (d) Barium hydroxide (e) Chromium (III) hydroxide

Chapter 8

1. (a) $SO_3 + H_2O \longrightarrow H_2SO_4$
(b) $2\,H_2O \longrightarrow 2\,H_2 + O_2$
(c) $PCl_3 + Cl_2 \longrightarrow PCl_5$
(d) $2\,C + O_2 \longrightarrow 2\,CO$
(e) $2\,Mg + O_2 \longrightarrow 2\,MgO$
2. (a) $4\,NH_3 + 3O_2 \longrightarrow 2\,N_2 + 6\,H_2O$
(b) $I_4O_9 \xrightarrow{\Delta} I_2O_5 + I_2 + 2\,O_2$

(c) $K_2S_2O_3 + 4\ Cl_2 + 5\ H_2O \xrightarrow{\Delta} 2\ KHSO_4 + 8\ HCl$
(d) $CS_2 + 3\ O_2 \longrightarrow 2\ N_2 + 2\ SO_2$

3. (a) $CaCO_3 \xrightarrow{\Delta} CaO + CO_2$
(b) $2\ KClO_3 \xrightarrow{\Delta} 2\ KCl + 3\ O_2$
(c) $2\ Li_3N \xrightarrow{\Delta} 6\ Li + N_2$
(d) $2\ H_2O_2 \xrightarrow{\Delta} 2\ H_2O + O_2$

4. (a) $TiCl_4 + 2\ H_2S \longrightarrow TiS_2 + 4\ HCl$
(b) $LaCl_3 \cdot 7H_2O \xrightarrow{\Delta} LaOCl + 2\ HCl + 6\ H_2O$
(c) $3\ CsCl + 2\ ScCl_3 \longrightarrow Cs_3Sc_2Cl_9$
(d) $Li_2CO_3 + 5\ Fe_2O_3 \longrightarrow 2\ LiFe_5O_8 + CO_2$
(e) $2\ La_2O_3 + 12\ B_2O_3 \longrightarrow 4\ LaB_6 + 21\ O_2$
(f) $4\ ScCl_3 + 7\ SiO_2 \longrightarrow 2\ Sc_2Si_2O_7 + 3\ SiCl_4$

5. (a) $2\ NaN_3 \longrightarrow 2\ Na + 3\ N_2$
(b) $(NH_4)_2\ Cr_2O_7 \longrightarrow Cr_2O_3 + N_2 + 4\ H_2O$
(c) $2\ Ag_2CO_3 \longrightarrow 4\ Ag + 2\ CO_2 + O_2$
(d) $2\ NaHCO_3 \longrightarrow Na_2CO_3 + CO_2 + H_2O$
(e) $Al_2(CO_3)_3 \longrightarrow Al_2O_3 + 3\ CO_2$

6. (a) $H_2SO_4 + 2\ KOH \longrightarrow K_2SO_4 + 2\ H_2O$
(b) $2\ H_3PO_4 + 3\ Ba(OH)_2 \longrightarrow Ba_3(PO_4)_2 + 6\ H_2O$
(c) $4\ HBr + Sn(OH)_4 \longrightarrow SnBr_4 + 4\ H_2O$
(d) $HBr + NaOH \longrightarrow NaBr + H_2O$
(e) $H_4P_2O_7 + 4\ NaOH \longrightarrow Na_4P_2O_7 + 4\ H_2O$

7. (a) $Ca(NO_3)_2 + Na_2SO_4 \longrightarrow 2\ NaNO_3 + CaSO_4$
(b) $Mg(NO_3)_2 + Li_2S \longrightarrow MgS + 2\ LiNO_3$
(c) $Pb(NO_3)_2 + Cs_2CrO_4 \longrightarrow PbCrO_4 + 2\ CsNO_3$
(d) $2\ FeCl_3 + 3\ CaS \longrightarrow Fe_2S_3 + 3\ CaCl_2$
(e) $2\ Na_3PO_4 + 3\ Zn(OH)_2 \longrightarrow Zn_3(PO_4)_2 + 6\ NaOH$

8. $2\ Ca_3(PO_4)_2 + 6\ SiO_2 + 5\ C \longrightarrow P_4 + 6\ CaSiO_3 + 5\ CO_2$

9. (a) $C_4H_8 + 6\ O_2 \longrightarrow 4\ CO_2 + 4\ H_2O$
(b) $2\ C_6H_6 + 15\ O_2 \longrightarrow 12\ CO_2 + 6\ H_2O$
(c) $2\ C_6H_{12}O + 17\ O_2 \longrightarrow 12\ CO_2 + 12\ H_2O$
(d) $C_{12}H_{22}O_{11} + 12\ O_2 \longrightarrow 12\ CO_2 + 11\ H_2O$
(e) $2\ C_4H_{10} + 13\ O_2 \longrightarrow 8\ CO_2 + 10\ H_2O$

10. (a) $2\ Fe + 3\ S \xrightarrow{\Delta} Fe_2S_3$
(b) $2\ C_8H_{18} + 25\ O_2 \longrightarrow 16\ CO_2 + 18\ H_2O$
(c) $2\ SO_2 + O_2 \longrightarrow 2\ SO_3$
(d) $3\ Ba(OH)_2 + 2\ H_3PO_4 \longrightarrow Ba_3(PO_4)_2 + 6\ H_2O$
(e) $2\ Al(s) + 6\ HCl(aq) \longrightarrow 2\ AlCl_3(s) + 3\ H_2(g)$

11. (a) $2\ Al + 3\ Cl_2 \longrightarrow 2\ AlCl_3$
(b) $2\ K + 2\ H_2O \longrightarrow 2\ KOH + H_2$
(c) $BCl_3 + 3\ H_2O \longrightarrow B(OH)_3 + 3\ HCl$

(d) $AgNO_3 + NaCl \longrightarrow AgCl + NaNO_3$
(e) $SiF_4 + 8\ NaOH \longrightarrow Na_4SiO_4 + 4\ NaF + 4\ H_2O$
(f) $Ca(OH)_2 + 2\ NH_4Cl \longrightarrow 2\ NH_3 + CaCl_2 + 2\ H_2O$

12. (a) $2\ Si_4H_{10} + 13\ O_2 \longrightarrow 8\ SiO_2 + 10\ H_2O$
(b) $CH_3NO_2 + 3\ Cl_2 \longrightarrow CCl_3NO_2 + 3\ HCl$
(c) $2\ NaF + CaO + H_2O \longrightarrow CaF_2 + 2\ NaOH$
(d) $Al_4C_3 + 12\ H_2O \longrightarrow 4\ Al(OH)_3 + 3\ CH_4$
(e) $2\ TiO_2 + B_4C + 3\ C \longrightarrow 2\ TiB_2 + 4\ CO$
(f) $C_7H_{16}O_4S_2 + 11\ O_2 \longrightarrow 7\ CO_2 + 8\ H_2O + 2\ SO_2$

Chapter 9

1. 127.9 g C_2H_5OH and 122.1 g CO_2
2. 80.00 g of O_2
3. (a) 510.8 g $KClO_3$ (b) 0.0418 mol KCl (c) 0.0627 mol O_2 and 2.01 g O_2
4. 6.10 g MnO_2
5. 3.51 g Si
6. 87.92 g W
7. 7.47 L
8. 4.96 L
9. 44.07 L
10. 4.48 L
11. (a) $2\ C_2H_2 + 5\ O_2 \longrightarrow 2\ H_2O + 4\ CO_2$
(b) 0.505 mol
(c) 9.05 L
12. (a) $2\ Al(OH)_3 + 3\ H_2SO_4 \longrightarrow Al_2(SO_4)_3 + 6\ H_2O$
(b) H_2SO_4 is the limiting reagent
(c) 180 g $Al_2(SO_4)_3$
(d) 17.7 g $Al(OH)_3$
13. (a) $2\ Na_2S_2O_3 + AgBr \longrightarrow Na_3Ag(S_2O_3)_2 + NaBr$
(b) $Na_2S_2O_3$ is the limiting reagent
(c) 0.0484 mol
14. (a) $C_7H_8 + 3\ HNO_3 \longrightarrow C_7H_5N_3O_6 + 3\ H_2O$
(b) HNO_3 is the limiting reagent
(c) Theoretical yield = 60.04 g
(d) 91.6 %
15. (a) $Ca_3P_2 + 6\ H_2O \longrightarrow 3\ Ca(OH)_2 + 2\ PH_3$
(b) 16.5 g of PH_3
16. 14.24 g of aspirin
17. 295.9 g of acetic acid
18. 55.4% Fe, and 44.6% Cu

19. (a) $4\ C_{21}H_{23}O_5N + 101\ O_2 \longrightarrow 84\ CO_2 + 4\ NO_2 + 46\ H_2O$
    (b) O_2 is the limiting reagent, and heroin is the excess
    (c) 8.9 g of heroin unreacted
    (d) 64.4 %
    (e) 40.31 g CO_2 + 2.01 g NO_2 + 9.03 g H_2O = 51 .35 g of products

20. 17.45 kg of Fe

Chapter 10

1. (a) 2.34×10^{-5} nm
   (b) 0.173 nm
   (c) 5.44×10^{12} nm
   (d) 5.75×10^{8} nm

2. (a) 5.34×10^{-2} cm
   (b) 2.65×10^{-5} cm
   (c) 6.64×10^{4} cm
   (d) 8.79×10^{-6} cm

3. $5.09 \times 10^{14} s^{-1}$

4. 3.23 nm

5. $1.70 \times 10^{4}\ cm^{-1}$

6. (a) 4.32×10^{-5} cm (b) $2.31 \times 10^{4}\ cm^{-1}$

7. 2.45×10^{-24} kJ

8. 4.88×10^{-19} J

9. (a) $5.71 \times 10^{14} s^{-1}$ (b) $1.91 \times 10^{4}\ cm^{-1}$ (c) 3.79×10^{-19} J

10.

Frequency (MHz)	Energy (kJ)
60	3.98×10^{-29}
200	1.33×10^{-28}
400	2.65×10^{-28}
600	3.98×10^{-28}
700	4.64×10^{-28}

11. 2.42×10^{-19} J

12. (a) 109,219 cm^{-1}; (b) 2.09×10^{-18}J

13. 2.18×10^{-18} J

14. (a) 2.04×10^{-18}J, 97.3 nm (b) 4.58×10^{-19}J, 434 nm

15. 9.0 nm

16. $2.95 \times 10^{4}\ m\ s^{-1}$

17. 0.0549 Å

18. (a) 2 (b) 8 (c) 18 (d) 32

19. $l =$ 0, 1, 2 corresponding to s, p, and d sublevels

20. (a) $3p$ (b) $3d$ (c) $4f$

21. (a) $n = 2; l = 1; m_l = 1, 0, -1$
    (b) $n = 3; l = 2; m_l = 2, 1, 0, -1, -2$
    (c) $n = 4; l = 1; m_l = 1, 0, -1$
    (d) $n = 4; l = 3; m_l = 3, 2, 1, 0, -1, -2, -3$
22. (a) Na $1s^2 2s^2 2p^6 3s^1$
    (b) P $1s^2 2s^2 2p^6 3s^2 3p^3$
    (c) Ca $1s^2 2s^2 2p^6 3s^2 3p^6 4s^2$
    (d) Rb $1s^2 2s^2 2p^6 3s^2 3p^6 4s^2 3d^{10} 4p^6 5s^1$
    (e) Fe $1s^2 2s^2 2p^6 3s^2 3p^6 4s^2 3d^6$
23. (a) S^{2-} $1s^2 2s^2 2p^6 3s^2 3p^6$
    (b) Ni^{2+} $1s^2 2s^2 2p^6 3s^2 3p^6 4s^0 3d^8$
    (c) Cu^{2+} $1s^2 2s^2 2p^6 3s^2 3p^6 4s^0 3d^9$
    (d) Ti^{4+} $1s^2 2s^2 2p^6 3s^2 3p^6$
    (e) Br^- $1s^2 2s^2 2p^6 3s^2 3p^6 4s^2 3d^{10} 4p^6$
24. (a) O $[He]2s^2 2p^4$
    (b) Zn^{2+} $[Ar]4s^0 3d^{10}$
    (c) Mg $[Ne]3s^2$
    (d) Pb^{2+} $[Xe]6s^2 5d^{10} 4f^{14}$
    (e) V $[Ar]4s^2 3d^3$
    (f) Te^{2-} $[Kr]4d^{10} 5s^2 5p^6$
    (g) Mn^{7+} [Ar]
25. (a) B (b) Ge (c) I (d) Bi (e) Mn (f) P
26. (a) F (b) P (c) Sc (d) Ni (e) C (f) Ti
27. (a) Oxygen

    (b) Sodium

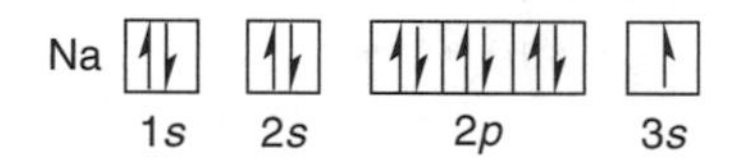

    (c) Titanium

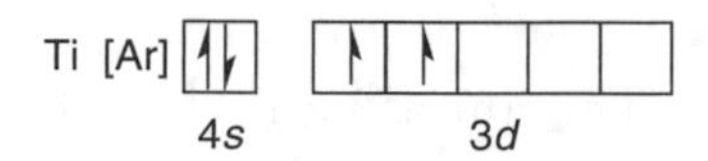

    (d) Zinc

28. (a) *A* = oxygen (O), *B* = magnesium (Mg)
    (b) *A* has 26 subatomic particles; *B* has 38 subatomic particles.

Chapter 11

1. 233.3 mL
2. 0.215 L
3. 3.33×10^4 N m^{-2}
4. 279.8 cm^3
5. 619 cm^3
6. -121.5°C
7. 1.26 L
8. -67.4°C
9. 566.3°C
10. 9.31 atm
11. 240.3 cm^3
12. 9.10 cm^3; $2\ C_4H_{10} + 13\ O_2 \longrightarrow 8\ CO_2 + 10\ H_2O$
13. 13.0 L
14. 6.67 L ClF_3
15. 2.50 L
16. 0.0820 L-atm/mol-K
17. 5.22 L
18. 5.43 L
19. 7.15 mol
20. 44.82 g/mol
21. 22.24 g/mol
22. 2.99 g/L
23. 3.60 g/L
24. (a) 1500 L of O_2
    (b) 1000 L of SO_2
25. 24.37 %
26. C_3H_8
27. (a) C_3H_6O
    (b) $C_3H_6O(g) + 4\ O_2(g) \longrightarrow 3\ CO_2(g) + 3\ H_2O(g)$
28. (a) $2\ KClO_3(s) \longrightarrow 2\ KCl(s) + 3\ O_2(g)$
    (b) 38.41 g $KClO_3$
29. 57.5 L
30. $P_{N_2} = 1483.5$ mmHg; $P_{O_2} = 398.1$ mmHg; $P_{Ar} = 17.7$ mmHg; $P_{CO_2} = 0.7$ mmHg
31. (a) $X_{He} = 0.250$; $X_{Ne} = 0.333$; $X_{Ar} = 0.417$
    (b) $P_{He} = 0.625$ atm; $P_{Ne} = 0.833$ atm; $P_{Ar} = 1.042$ atm

32. (a) $n_{O_2} = 24.3\%$; $n_{N_2} = 42.0\%$; $n_{CO_2} = 33.7\%$
    (b) volume % $N_2 = 40.0\%$
33. (a) N_2 is 1.25 times faster than CO_2
    (b) H_2 is 4 times faster than O_2
    (c) Ne is 1.41 times faster than Ar
34. O_2 is 2.82 times faster than He
35. (a) Rate of diffusion of $SO_2 = 12\ cm^3$, rate of diffusion of Y = 27.43 cm^3
    (b) Molecular mass of Y = 28.0 g/mol
36. (a) 15.3 atm (b) 14.18 atm
37. (a) 66.6 atm (b) 70.0 atm
38. (a) $P_{N_2} = 25.03$ atm; $P_{O_2} = 6.79$ atm; $P_{Ar} = 0.303$ atm; $P_{CO_2} = 0.012$ atm
    (b) $P_{Total} = 32.14$ atm

Chapter 12

1. 0.058 *e*
2. 41.2 %
3. 15.3 D
4. 41.2 kJ/mol
5. 375 mmHg
6. 331.5 K or 58.5°C
7. 31.3 kJ/mol
8. 1001.6 mmHg
9. (a) 35.1 kJ/mol (b) 23 mmHg (c) 353.2 K
10. (a) 29.7 kJ/mol (b) 4381 mmHg (c) 268.3 K
11. 4 Al atoms
12. 2 V atoms per unit cell
13. 22.7 g/cm^3
14. (a) 1 Po atom per unit cell (b) $3.72 \times 10^{-23}\ cm^3$ (c) Density = 9.33 g/cm^3
15. (a) 4 Na^+ and 4 Cl^- (b) 5.65×10^{-8} cm or 5.65 Å
16. 197.3 g/mol
17. 6.05×10^{23} atoms/mol
18. 144 pm
19. 1.99 Å
20. 4 Al atoms per unit cell
21. $r_{Zn^{2-}}/r_{O^{2-}} = 0.529$. ZnO should have a structure in which Zn^{2-} cations fit into the octahedral holes in a closest-packed array of O^{2-} anions. Since the stoichiometry is 1:1, all of the holes should be occupied. ZnO should have a FCC NaCl structure.

22. $r_{In^{3-}}/r_{P^{3-}} = 0.382 \Rightarrow$ InP conforms to the zincblende structure.
23. $d = 3.50$ Å
24. $d = 0.61$ Å
25. For $n = 1$, 2, and 3, the grazing angles are 16°, 33.36°, and 55.59° respectively.
26. $d_{100} = 4.08$ Å, $d_{111} = 2.35$ Å, and $d_{311} = 1.23$ Å
27. (a) 5.42 Å (b) $d_{002} = 2.71$ Å, $d_{211} = 2.21$ Å, and $d_{222} = 1.56$ Å
28. (a) 3.57 Å (b) density = 3.51 g/cm^3

Chapter 13

1. 12%
2. 30.625 g solution
3. 10 g NaCl
4. 1.10 g anhydrous HCl
5. (a) 0.165 M (b) 0.225 M (c) 1.00M (d) 0.25M
6. 0.291 M
7. 11.50 cm^3
8. 1120 cm^3
9. 0.40 N
10. 130.72 cm^3
11. 65.32 g H_3PO_4
12. $\chi_{NaNO_3} = 0.014$; $\chi_{H_2O} = 0.986$
13. $\chi_{CH_3OH} = 0.110$; $\chi_{CH_3COCH_3} = 0.375$; $\chi_{CCl_4} = 0.515$
14. $\chi_{NaCl} = 0.110$, $\chi_{H_2O} = 0.89$
15. 12.12 *m*
16. 2.28 g $FeSO_4$
17. 1.90 *m*
18. 4.89 M
19. 2.40 M; 2.55 *m*
20. 0.52 M
21. 0.24 M
22. 1.33 M
23. Measure 0.965 L (965 mL) of stock concentrated HCl into a 5.00 L volumetric flask. Dilute with distilled water to the 5.0 L mark.
24. (a) 0.0094 M (b) 1.00 g/L
25. (a) $Li^+ = 2.25$ M; $PO_4^{3-} = 0.75$ M
    (b) $Li_3PO_4 = 0.075$ M, $Li^+ = 0.225$ M; $PO_4^{3-} = 0.075$ M
26. (a) 106 g per 100 g H_2O (b) 6.67 mol/L

27. 23.4 g of solute
28. 92.1 g $AgNO_3$
29. 12.07 g of salt
30. 31.03 g of salt
31. (a) 121 g of KNO_3 (b) 57.5 g KNO_3 (c) 560 g $NaNO_3$
    (d) Dissolve 26.25 g of $NaNO_3$ in 25 g of water and mix thoroughly.
    (e) 60 g NH_3 per 100 g of H_2O (f) 40 g of NH_3 liberated.
32. (a) (i) 35°C (ii) 10°C (b) 136 g $NaNO_3$ per 100 g of H_2O
    (c) 51 g of KNO_3 (d) 23.2 g of NH_3
33. 5.3×10^{-4} M
34. 5.3×10^{-6} M
35. (a) 1.3×10^{-5} M (b) 0.144 M

Chapter 14

1. 0.195 M
2. 0.20 M
3. 7.0 g KOH
4. 0.176 g/L
5. (a) 0.165 M (b) 6.609 g/L
6. (a) 0.0836 M (b) atomic mass of $M = 85.47$ (c) $M = \text{Rb}$
7. 87.4%
8. 3.72%
9. (a) 63% (b) 10
10. 133 g/mol
11. 11.52%
12. 16.5 mg natural product/mL
13. 23.5%
14. 81.3%

Chapter 15

1. 23.7 mmHg
2. 23.6 mmHg
3. 264 g sucrose
4. 0.533 *m*
5. 22.03 mmHg
6. 23.4 mmHg

7. 55.5 mmHg
8. 0.085
9. 168.7°C
10. −0.170°C
11. 130 g/mol
12. 238 g/mol
13. 215.0°C
14. (a) 182 g/mol (b) $C_3H_7O_3$, $C_6H_{14}O_6$
15. (a) −0.55°C (b) 41.40°C (c) −0.47°C
16. 443 g/mol
17. $(C_6H_{10}O_5)_{65}$ or $C_{390}H_{650}O_{325}$
18. 1.23 M
19. 3,731 g/mol
20. 0.345 M
21. 29.35 mmHg
22. (a) 46.2 atm (b) 6.90 atm (c) 14 atm

Chapter 16

1. (a) $\text{Rate} = -\frac{1}{2}\frac{\Delta[H_2O_2]}{\Delta t} = \frac{1}{2}\frac{\Delta[H_2O]}{\Delta t} = \frac{\Delta[O_2]}{\Delta t}$

   (b) $\text{Rate} = -\frac{1}{2}\frac{\Delta[N_2O_5]}{\Delta t} = \frac{1}{4}\frac{\Delta[NO_2]}{\Delta t} = \frac{\Delta[O_2]}{\Delta t}$

   (c) $\text{Rate} = -\frac{\Delta[SO_2Cl_2]}{\Delta t} = \frac{\Delta[SO_2]}{\Delta t} = \frac{\Delta[Cl_2]}{\Delta t}$

   (d) $\text{Rate} = -\frac{1}{4}\frac{\Delta[PH_3]}{\Delta t} = \frac{\Delta[P_4]}{\Delta t} = \frac{1}{6}\frac{\Delta[H_2]}{\Delta t}$

2. (a) $\text{Rate} = -\frac{1}{2}\frac{\Delta[NH_3]}{\Delta t} = -\frac{1}{5}\frac{\Delta[O_2]}{\Delta t} = \frac{1}{4}\frac{\Delta[NO]}{\Delta t} = \frac{\Delta[H_2O]}{\Delta t}$

   (b) $\text{Rate} = -\frac{1}{2}\frac{\Delta[NO]}{\Delta t} = -\frac{1}{2}\frac{\Delta[H_2]}{\Delta t} = \frac{\Delta[N_2]}{\Delta t} = \frac{\Delta[H_2O]}{\Delta t}$

   (c) $\text{Rate} = -\frac{\Delta[CH_3CHO]}{\Delta t} = \frac{\Delta[CH_4]}{\Delta t} = \frac{\Delta[CO]}{\Delta t}$

   (d) $\text{Rate} = -\frac{1}{2}\frac{\Delta[SO_2]}{\Delta t} = -\frac{\Delta[O_2]}{\Delta t} = \frac{1}{2}\frac{\Delta[SO_3]}{\Delta t}$

3. 0.0125 M/min
4. 0.024 M/s
5. 0.04 M /h

6. (a) Rate $= k[X]^1[Y]^2$
   (b) $288\ M^{-1}min^{-1}$

7. (a) Rate $= k[NH_4^+]^2[CNO]^1$
   (b) $510.2\ M^{-1}s^{-1}$

8. (a) Rate $= -\frac{k[I^-][OCl^-]}{[OH^-]}$
   (b) First-order

9. (a) Zero-order in $S_2O_3^{2-}$ and first-order in I_2
   (b) Rate $= k[S_2O_3^{2-}]$
   (c) $0.004\ s^{-1}$

10. (a) First-order in H_2O_2, first-order in I^- and second-order in H^+
    (b) Rate = $k[H_2O_2][I^-][H^+]^2$
    (c) $2.13\ M^{-3}s^{-1}$
    (d) $8.3 \times 10^{-7}\ M\ s^{-1}$

11. (a) 0.085 M
    (b) 602 s

12. (a) $4.0 \times 10^{-7}\ M\ s^{-1}$
    (b) 0.243 M

13. (a) $2.67 \times 10^3 s$
    (b) 3.125×10^{-4} M
    $0.053\ h^{-1}$
    (b) 0.014 M
    (c) Fraction remaining $= 0.28$
    (d) 30.4 h

15. $5.8 \times 10^{-4}\ s^{-1}$

16. (a) First-order
    (b) $9.0 \times 10^{-4}\ s^{-1}$

17. (a) 1.01×10^{-4} M
    (b) $t = 327$ s
    (c) $t_{1/2} = 36$ s

18. (a) Reaction is second-order
    (b) $k = 0.08\ M^{-1}\ min^{-1}$
    (c) $t_{1/2} = 84$ min

19. $3.68\ s^{-1}$

20. $T = 731$K or 458°C

21. (a) $E_a = 51.7$ kJ/mol
    (b) $k_{355\ K} = 6.61 \times 10^{-6}s^{-1}$

22. (a) $E_a = 223.9$ kJ/mol
    (b) $k_{25°C} = 3.71 \times 10^{-22}\ s^{-1}$

23. $E_a = 90.4$ kJ/mol; $A = 4.23 \times 10^{12}\ s^{-1}$

24. $E_a = 46.6$ kJ/mol; $A = 2.93 \times 10^7\ s^{-1}$

Chapter 17

1. (a) $K_c = \frac{[SO_3]^2}{[SO_2]^2[O_2]}$ (b) $K_c = \frac{[CO_2][H_2]}{[CO][H_2O]}$ (c) $K_c = \frac{[CH_4][H_2S]^2}{[CS_2][H_2]^4}$

   (d) $K_c = \frac{[NO_2]^2}{[N_2O_4]}$

2. (a) $K_p = \frac{P^2_{SO_3}}{P^2_{SO_2} P_{O_2}}$ (b) $K_p = \frac{P_{CO_2}P_{H_2}}{P_{CO}P_{H_2O}}$ (c) $K_p = \frac{P_{CH_4}P^2_{H_2S}}{P_{CS_2}P^4_{H_2}}$

   (d) $K_p = \frac{P^2_{NO_2}}{P_{N_2O_4}}$

3. (a) $K_c = \frac{[CH_3OH]}{[CO][H_2]^2}$ (b) $K_c = \frac{[CH_3OH][Cl^-]}{[CH_3Cl][OH^-]}$ (c) $K_c = \frac{[Ag(NH_3)_2^-]}{[Ag^+][NH_3]^2}$

   (d) $K_c = \frac{[\left(C_6H_5SO_2\bar{N}CH_3\right)Na^-]}{[C_6H_5SO_2NHCH_3][NaOH]}$

4. (a) $K_p = \frac{P_{PCl_5}}{P_{PCl_3}P_{Cl_2}}$ (b) $K_p = \frac{P^2_{POCl_3}}{P^2_{PCl_3}P_{O_2}}$ (c) $K_p = \frac{P^2_{NO_2}P^8_{H_2O}}{P_{N_2H_4}P^6_{N_2O_2}}$

   (d) $K_p = \frac{P_{NO}P_{SO_3}}{P_{NO_2}P_{SO_2}}$

5. (a) $2\,HF \rightleftharpoons H_2 + F_2$
   (b) $2\,C_2H_2 + 5\,O_2 \rightleftharpoons 4\,CO_2 + 2H_2O$
   (c) $CO + H_2O \rightleftharpoons CO_2 + H_2$
   (d) $2\,Cl_2 + 2\,H_2O \rightleftharpoons 4\,HCl + O_2$

6. (a) $H_2NCH_2CH_2NH_2 + 2H_2O \rightleftharpoons H_2NCH_2CH_2\overset{+}{N}H_3 + OH^-$
   (b) $2\,CH_4 + 2\,H_2S \rightleftharpoons CS_2 + 4\,H_2$
   (c) $CH_2O \rightleftharpoons CO + H_2\ 2\,NO$

7. (a) $K_p = \frac{p_{N_2}p^3_{H_2}}{p^2_{NH_3}}$ (b) $K_p = \frac{p_{CS_2}p^4_{H_2}}{p_{CH_4}p^2_{H_2S}}$ (c) $K_p = \frac{p_{H_2}p_{CO}}{p_{CH_2O}}$ (d) $K_p = \frac{p^2_{NO}}{p_{N_2}p_{O_2}}$

8. (a) $K_p = \frac{p_{O_2}p^4_{HCl}}{p^2_{Cl_2}p^2_{H_2O}}$, $K_p = 3.2 \times 10^{-14}$; $K_p << 1$, hence reactant is favored.

   (b) $K_p = \frac{p^2_{HI}}{p_{H_2}p_{I_2}}$, $K_p = 51$; $K_p >> 1$, hence product is favored.

   (c) $K_c = \frac{[CH_3OH]\left[Cl^-\right]}{[CH_3Cl]\left[OH^-\right]} = 1.0 \times 10^{16}$; $K_c >> 1$, hence product is favored

   (d) $K_p = \frac{p^3_{O_2}}{p^2_{O_3}}$, $K_p = 1.3 \times 10^{57}$; $K_p >> 1$, hence product is favored.

9. (a) $K_c = 3.5 \times 10^6$; $K_c >> 1$, hence SO_3 formation is favored.
   (b) $K_p = 6.33 \times 10^4$
   (c) $K_c = 2.0 \times 10^{-7}$
10. $K_c = 0.23$
11. $K_c = 1.17 \times 10^7$
12. Equilibrium composition: $[CH_3\ CO_2\ H] = 0.33$ mol; $[C_2H_5OH] = 0.33$ mol, $[CH_3CO_2C_2\ H_5\ ] = 0.67$ mol; $[H_2O] = 0.67$ mol
13. Equilibrium composition: $[H_2] = 2.11$ mol; $[I_2] = 0.11$ mol, $[HI] = 3.78$ mol.
14. $[COCl_2] = 0.999$ mol; $[CO] = [Cl_2] = 1.48 \times 10^{-5}$ mol
15. (a) $K_p = P^2_{H_2O}$ (b) $K_p = P^6_{H_2O}$ (c) $K_p = P_{H_2O}$ (d) $K_p = P^{10}_{H_2O}$
16. Vapor pressure $P = 6.1 \times 10^{-8}$ and 0.025 atm
17. $K_c = 8.0 \times 10^{-4}$
18. (a) $K_c = 0.045$ (b) $P = 2.32$ atm (c) $P_{NH_4HS} = 0.58$ atm, $P_{NH_3} = P_{H_2S} = 0.87$ atm (d) $K_p = 1.38$
19. (a) More products formed
   (b) More products formed
   (c) Equilibrium shift from left to right favoring product formation
   (d) No effect on position of equilibrium
20. (a) Less NO formed
   (b) No effect
   (c) More NO formed
   (d) Less NO formed

Chapter 18

1. (a) 5.0 (b) 1.3 (c) 10.7 (d) 6.0
2. (a) 9.0 (b) 12.7 (c) 3.3 (d) 8.0
3. (a) 1.58×10^{-12} M (b) 1.0×10^{-2} M (c) 1.25×10^{-8} M (d) 3.2×10^{-7} M (e) 0.10 M
4. (a) 6.33×10^{-3} M (b) 1.0×10^{-12} M (c) 7.74×10^{-7} M (d) 3.16×10^{-8} M (e) 1.0×10^{-13} M
5. (a) 1.0 (b) 12.3 (c) 0.4 (d) 12.7
6. (a) $[H^+] = 0.025$ M – 0.016 M
   (b) (b) $[H^+] = 1.58 \times 10^{-5}$ M – 1.58×10^{-9} M
   (c) $[H^+] = 5.00 \times 10^{-7}$ M – 2.51×10^{-7} M
   (d) $[H^+] = 1.60 \times 10^{-2}$ M – 7.90×10^{-3} M
   (e) $[H^+] = 4.47 \times 10^{-8}$ M – 3.55×10^{-8} M
   (f) $[H^+] = 1.26 \times 10^{-10}$ M – 7.94×10^{-11} M
   (g) $[H^+] = 1.00 \times 10^{-4}$ M – 3.98×10^{-5} M

7. 8.30×10^{-10} M
8. (a) 18.4 M (b) 1.56
9. At 100°C pH = 6.0 and at 25°C pH = 7.0

   Conclusion:
   (i) pH is higher at 100°C (1 unit more) than at 25°C.
   (ii) Water is neutral at both temperatures because $[H^+] = [OH^-]$.
   (iii) Water at elevated temperatures has more ions in it and so is more reactive.
10. pH = pOH = 6.51
11. (a) 0.1 (b) 2.4 (c) 13.4 (d) 11.5
12. $[C_6H_5CO_2H] = 0.019$ M; $[C_6H_5CO_2^-] = [H_3O^+] = 1.12 \times 10^{-3}$ M; pH = 2.95
13. $k_a = 1.71 \times 10^{-4}$; $[HCO_2H] = 5.8 \times 10^{-4}$ M
14. 2.7
15. pH = 2.2; pOH = 11.8
16. pH = 12
17. pH = 10.8; pOH = 3.2
18. 4.1×10^{-6} M
19. 0.036 %
20. 0.025 M
21. 64.9 %
22. 8.16×10^{-5}
23. 8.87
24. 9.0 and 5.3×10^{-3}%
25. 3.2
26. 3.3×10^{-6}
27. 1.84×10^{-4}M
28. 4.7
29. 8.9 and 6.0×10^{-3}%
30. pH = 10.5, $[CH_3NH_2] = 5.47 \times 10^{-2}$ M; $[CH_3NH_3^+] = 6.53 \times 10^{-2}$ M; $[OH^-] = 3.1 \times 10^{-4}$ M; $[H^+] = 3.3 \times 10^{-11}$ M
31. 4.8
32. The ratio is $\frac{[CH_3CO_2H]}{[CH_3CO_2^-]} = 5.5 \times 10^{-3}$
33. The required ratio is $\frac{[H_2PO_4^-]}{[HPO_4^{2-}]} = 0.51$

    To prepare the phosphate buffer with a pH of 7.5, dissolve 0.51 mol of NaH_2PO_4 and 1.0 mol of Na_2HPO_4 in enough water to make up 1.0 L of solution in a volumetric flask.

34. (a) 2.9×10^{-6}
    (b) The pK_a is fairly close to the pH of interest, so it will be a good buffer.
35. 0.02 mol
36. 8.6
37. $[H_3O^+] = [HCO_3^-] = 3.16 \times 10^{-3}$ M; $[CO_3^{2-}] = 5.6 \times 10^{-11}$ M; $[CO_2] = 0.047$; pH = 2.5.
38. $[H^+] = [HS^-] = 1.41 \times 10^{-4}$ M; $[S^{2-}] = 1.0 \times 10^{-19}$ M; pH = 3.9
39. (a) $[H_3O^+] = [HTeO_3^-] = 0.02$ M
    (b) $[TeO_3^{2-}] = 1.0 \times 10^{-8}$ M
40. (a) Phenolphthalein, litmus, methyl orange
    (b) Phenolphthalein
    (c) Bromophenol blue, methyl orange, methyl red
41. (a) pH = 8.7; pOH = 5.3
    (b) Phenolphthalein or thymol blue
42. 10.2
43. (a) 1.0 (b) 1.2 (c) 1.6 (d) 7.0 (e) 12.2
44. (a) 13.0 (b) 12.0
45. (a) 2.9 (b) 4.6 (c) 8.7 (d) 12.2
46. (a) 11.1 (b) 9.4 (c) 8.7 (d) 1.7

Chapter 19

1. (a) $K_{sp} = [Ag^-]^2[CrO_4^{2-}]$ (b) $K_{sp} = [Ca^{2+}][F^-]^2$ (c) $K_{sp} = [Ca^{2+}][CO_3^{2-}]$
   (d) $K_{sp} = [Ca^{2-}]^3[PO_4^{3-}]^2$ (e) $K_{sp} = [Pb^{2+}]^3[AsO_4^{3-}]^2$
2. (a) $K_{sp} = [Ag^+][I^-]$ (b) $K_{sp} = [Ca^{2+}][C_2O_4^{2-}]$ (c) $K_{sp} = [Hg_2^{2+}][Cl^-]^2$
   (d) $K_{sp} = [Pb^{2+}][I^-]^2$ (e) $K_{sp} = [Fe^{3+}][OH^-]^3$
3. 8.5×10^{-5}
4. 6.63×10^{-11}
5. 2.9×10^{-9} M
6. (a) 2.5×10^{-6} M (b) 7.69×10^{-4} g/L
7. 4.3×10^{-8} M
8. $[NO_3^-] = 0.0833$ M; $[K^+] = 0.1167$ M; $[Ag^+] = 2.7 \times 10^{-6}$ M, and $[CrO_4^{2-}] = 0.0167$ M
9. 3.0×10^{-6} M
10. 4.4×10^{-12} M
11. Hg_2I_2 will precipitate first, since the $[I^-]$ required is less.
12. AgI will precipitate first, since the $[Ag^+]$ required to cause its precipitation is less.

13. 0.006 M
14. Since $Q > K_{sp}$, $PbCO_3$ will precipitate from the solution.
15. CaF_2 will precipitate first, since the $[F^-]$ required is less; $[F^-]$ needed to precipitate BaF_2 is 0.15 M, for CaF_2 is 2.0×10^{-5}M.
16. 1.109 mol
17. 0.3570 M
18. 0.0505 M
19. 6.2×10^{-7}M
20. 1.0×10^{-7}M

Chapter 20

1. 38.85 kJ
2. −8678 kJ/mol
3. 3.36 J/g-°C
4. 29.6°C
5. 23.3 kJ
6. (a) Exothermic
   (b) 690 kJ
   (c) 85.2 g
7. −26,000 kJ
8. (a) The reaction is exothermic.
   (b) $2\,CH_4 + 2\,NH_3 + 3\,O_2 \longrightarrow 2\,HCN + 6\,H_2O + 930\,kJ$
   (c) −465 kJ/mol CH_4
   (d) −1395 kJ
9. (a) Exothermic
   (b) Endothermic
   (c) Exothermic
   (d) Exothermic
   (f) Endothermic
10. (a) −177.4 kJ/mol (b) Exothermic (c) $CaCO_3$ would decompose into CaO and CO_2.
11. (a) −227 kJ/mol (b) Exothermic (c) The reverse reaction, favoring reactants, would occur.
12. 9.65 kJ/mol
13. −2535.5 kJ
14. −390.85 kJ/mol
15. 34 kJ/mol
16. −565.6 kJ

17. −454.8 kJ/mol
18. −1153 kJ
19. −202.3 kJ
20. (a) −713 kJ
    (b) Exothermic
21. (a) −543.1 kJ
    (b) −537.2 kJ/mol
22. (a) 132.6 kJ/mol
    (b) −2694 kJ/mol
    (c) −485.2 kJ/mol
23. 0 kJ/mol
24. 538.6 kJ/mol

Chapter 21

1. − 100 J
2. 172.7 kJ
3. − 7.6 kJ
4. − 1932 kJ
5. − 87.2 kJ
6. −6.81 kJ
7. 176.6 J/K
8. − 145.3 J/K or −72.7 J/K per mole of NO_2
9. (a) Increase
   (b) Increase
   (c) No change
   (d) Decrease
   (e) Increase
   (f) Decrease
10. (a) Negative
    (b) Positive
    (c) Positive
    (d) Positive
    (e) Positive
    (f) Negative
11. (a) 0.7 J/K-mol
    (b) 1.2 J/K-mol
12. (a) −187.1 J/K-mol
    (b) − 306.5 J/K-mol
    (c) 169.8 J/K-mol
13. 112 J/K

14. 129.4 J/K
15. $\Delta S_{Fusion} = 26.3$; $\Delta S_{Vap} = 98.2$ J/K-mol
16. (a) −44.03 J/K-mol
    (b) The reaction is a spontaneous process since ΔG is negative.
    (c) $\Delta G = -102.7$ kJ. Yes, the reaction is spontaneous at this temperature.
17. $\Delta G = -190.6$ kJ. Yes, the reaction is spontaneous at this temperature.
18. $\Delta G = 842.8$ kJ. Therefore the engineer will not make NH_3 under these conditions.
19. (a) $T = 617.5$ K. At all temperatures above 617.5 K, $T\Delta S$ will be larger than ΔG which would make ΔG negative.
    (b) The reaction will be spontaneous at all temperatures, since ΔH^{o}_{rxn} is negative and ΔS^{o}_{rxn} is positive.
20. (a) − 394.9 kJ/mol and − 393.5 kJ/mol
    (b) 15.4 J/mol-K
21. − 53.9 kJ/mol; the reaction will occur as written.
22. 2.8×10^{9}
23. 1.96×10^{-34}
24. − 976 kJ/mol
25. 0.08 atm or 60.8 mmHg
26. 31.9 kJ/mol
27. 8.8 kJ/mol
28. 0.51

Chapter 22

1. (a) +4 (b) +8 (c) +7 (d) +3 (e) +3 (f) − 2
2. (a) +6 (b) +2 (c) +5 (d) +5 (e) +4 (f) +4
3. (a) Fe is oxidized and O_2 is reduced.
   (b) Cl_2 is reduced and NaI is oxidized.
   (c) Mg is oxidized and H_2SO_4 is reduced.
   (d) H_2 is oxidized and CuO is reduced.
4. (a) Al is oxidized and MnO_2 is reduced.
   (b) PH_3 is oxidized and O_2 is reduced.
   (c) Ca is oxidized and H_2O is reduced.
   (d) SO_2 is oxidized and O_2 is reduced.
5. (a) $2\,FeCl_2(aq) + Cl_2(l) \longrightarrow 2\,FeCl_3(aq)$
   RA OA
   (b) $Cl_2(g) + H_2S(g) \longrightarrow 2HCl(g) + S(s)$
   OA RA

(c) $C(s) + MgO(s) \longrightarrow CO_2(g) + Mg(s)$
RA OA

(d) $2\,Zn(s) + O_2(g) \longrightarrow 2\,ZnO(s)$
RA OA

6. (a) $3\,Cu(s) + 8\,HNO_3(aq) \longrightarrow 3\,Cu(NO_3)_2(aq) + 4\,H_2O(l) + 2\,NO(g)$
Oxidized Reduced
RA OA

(b) $Zn(s) + CuSO_4(aq) \longrightarrow ZnSO_4(aq) + Cu(s)$
Ox Red
RA OA

(c) $2\,H_2O(l) + 2\,F_2(g) \longrightarrow 4\,HF(aq) + O_2(g)$
Ox Red
RA OA

(d) $Fe_2O_3(s) + 3\,CO(g) \longrightarrow 2\,Fe(s) + 3\,CO_2(g)$
Red Ox
OA RA

7. (a) Acid–base reaction: no net change in oxidation number.
(b) Redox reaction: there is a net change in oxidation state of reactants and products.
(c) Precipitation reaction: NiS precipitated; no change in oxidation state.
(d) Acid–base reaction: no net change in oxidation state of reactants and products.

8. (a) $2\,IO_3^- + 12\,H^+ + 10e^- \longrightarrow I_2 + 6\,H_2O$
(b) $O_2 + 2\,H^+ + 2e^- \longrightarrow H_2O_2$
(c) $2\,S + 3\,H_2O \longrightarrow S_2O_3^{2-} + 6\,H^+ + 6e^-$
(d) $S_4O_6^{2-} + 2e^- \longrightarrow 2\,S_2O_3^{2-}$
(e) (e) $VO_2^+ + 2\,H^- + e^- \longrightarrow VO^{2+} + H_2O$
(f) $2\,CO_2 + 2\,H^+ + 2e^- \longrightarrow H_2C_2O_4$

9. (a) $2\,NO_3^- + 4\,H^+ + 2e^- \longrightarrow N_2O_4 + 2\,H_2O$
(b) $Sb_2O_3 + 6\,H^+ + 6e^- \longrightarrow 2\,Sb + 3\,H_2O$
(c) $MnO_2 + 4\,H^+ + 2e^- \longrightarrow Mn^{2+} + 2\,H_2O$
(d) $2HClO + 2H^+ + 2e^- \longrightarrow Cl_2 + 2\,H_2O$
(e) $H_2O_2 + 2\,H^+ + 2e^- \longrightarrow 2\,H_2O$
(f) $NO_3^- + 3\,H^+ - 2e^- \longrightarrow HNO_2 + H_2O$

10. (a) $Cu_2O + H_2O + 2e^- \longrightarrow 2\,Cu + 2\,OH^-$
(b) $MnO_2 + 2\,H_2O + 2e^- \longrightarrow Mn(OH)_2 + 2\,OH^-$
(c) $PbO_2 + H_2O + 2e^- \longrightarrow PbO + 2\,OH^-$
(d) $ClO^- + H_2O + 2e^- \longrightarrow Cl^- + 2\,OH^-$
(e) $MnO_4^- + 2\,H_2O + 3e^- \longrightarrow MnO_2 + 4\,OH^-$
(f) $Al(OH)_3 + 3e^- \longrightarrow Al + 3\,OH^-$
(g) $O_3 + H_2O + 2e^- \longrightarrow O_2 + 2\,OH^-$
(h) $BrO_3^- + 3\,H_2O + 6e^- \longrightarrow Br^- + 6OH^-$

11. (a) $Mg + 2\,NO_3^- + 4H^+ \longrightarrow Mg^{2+} + 2\,NO_2 + 2\,H_2O$
    (b) $3\,NO_2^- + Cr_2O_7^{2-} + 8\,H^+ \longrightarrow 3\,NO_3^- + 2\,Cr^{3+} + 4\,H_2O$
    (c) $3\,S^{2-} + 8H^+ + 3\,NO_3^- \longrightarrow 3\,S + 2\,NO + 4\,H_2O$
12. (a) $5\,H_2C_2O_4 + 2\,MnO_4^- + 6H^+ \longrightarrow 2\,Mn^{2+} + 10\,CO_2 + 8\,H_2O$
    (b) $3\,CH_3CH_2OH + 2\,Cr_2O_7^{2-} + 16\,H^+ \longrightarrow 3\,CH_3CO_2H + 4\,Cr^{3+} + 11\,H_2O$
    (c) $Ca + 2\,VO^{2+} + 4H^+ \longrightarrow Ca^{2+} + 2\,V^{3+} + 2\,H_2O$
13. (a) $3\,Zn + 2\,MnO_4^- + 4\,H_2O \longrightarrow 3\,Zn(OH)_2 + 2\,MnO_2 + 2\,OH^-$
    (b) $2\,CrO_2^- + IO_3^- + 2\,OH^- \longrightarrow 2\,CrO_4^{2-} + I^- + H_2O$
    (c) $2\,Bi(OH)_3 + 3\,Sn(OH)_3^- + 3\,OH^- \longrightarrow 2\,Bi + 3\,Sn(OH)_6^{2-}$
    (d) $I_2 + 7\,Cl_2 + 10\,OH^- \longrightarrow 3\,H_3IO_6^{2-} + 14\,Cl^- + 6\,H_2O$
14. (a) $14\,MnO_4^- + 20\,OH^- + C_3H_8O_3 \longrightarrow 14\,MnO_4^{2-} + 3\,CO_3^{2-} + 14\,H_2O$
    (b) $2\,MnO_4^- + SO_3^{2-} + 2\,OH^- \longrightarrow 2\,MnO_4^{2-} + 3\,CO_3^{2-} + 14\,H_2O$
    (c) $2\,MnO_4^- + H_2O + Br^- \longrightarrow 2\,MnO_2 + BrO_3^- + 2OH^-$
15. (a) $2\,Cr(OH)_4^- + 3\,H_2O_2 + 2\,OH^- \longrightarrow 2\,CrO_2^{2-} + 8\,H_2O$
    (b) $3\,H_2S + Na_2Cr_2O_7 + 8\,HCl \longrightarrow 3\,S + 2\,CrCl_3 + 2\,NaCl + 7\,H_2O$
    (c) $3\,Sb_2S_5 + 10\,NO_3^- + 10\,H^+ \longrightarrow 6\,HSbO_3 + 15\,S + 10\,NO + 2\,H_2O$
16. (a) (a) $4\,Zn + NO_3^- + 10\,H^+ \longrightarrow 4\,Zn^{2+} + NH_4^+ + 3\,H_2O$
    (b) $2\,MnO_4^- + 10\,Cl^- + 16\,H^+ \longrightarrow 2\,Mn^{2+} + 5\,Cl_2 + 8\,H_2O$
    (c) $3\,PtCl_6^{2-} + Sb + 3\,H_2O \longrightarrow 3\,PtCl_4^{2-} + 6\,Cl^- + Sb_2O_3 + 6\,H^-$
    (d) $3\,Hg_2Cl_2 + I^- + 6\,OH^- \longrightarrow 6\,Hg + 6\,Cl^- + IO_3^- + 3\,H_2O$
    (e) $3\,K_2S_2O_8 + Cr_2(SO_4)_3 + 7\,H_2O \longrightarrow 2\,K_2SO_4 + K_2Cr_2O_7 + 7\,H_2SO_4$
    (f) $2\,KMnO_4 + 5\,Zn + 8\,H_2SO_4 \longrightarrow 2\,MnSO_4 + 5\,ZnSO_4 + 8\,H_2O + K_2SO_4$
17. (a) (a) $MnO_4^- + 5\,Fe^{2+} + 8\,H^+ \longrightarrow Mn^{2+} + 5\,Fe^{3+} + 4\,H_2O$
    (b) 0.05 M
18. 0.109 M
19. 0.388 M
20. 16.5% Fe
21. 34.9% Fe_2O_3
22. 50%

Chapter 23

1. (a) $Sn(s)|Sn^{2-}(1\,M)\|Cu^{2+}(1\,M)|Cu(s)$
   (b) $Zn(s)|Zn^{2+}(1\,M)\|Ag^-(1\,M)|Ag(s)$
   (c) $Mg(s)|Mg^{2+}(1\,M)\|Fe^{2+}(aq)|Fe(s)$
   (d) $Sn(s)|Sn^{2-}(1\,M)\|Pb^{2-}(1\,M)|Pb(s)$
2. (a) $Mn(s)|Mn^{2+}(1\,M)\|Ti^{2+}(1\,M)|Ti(s)$
   (b) $Mg(s)|Mg^{2+}(1\,M)\|Zn^{2+}(1\,M)|Zn(s)$

(c) $Pt(s)|Sn^{2+}(1\ M),\ Sn^{4+}(1\ M)\|Pb^{4+}(1\ M),\ Pb^{2+}(1\ M)|Pt(s)$
(d) $Ca(s)|Ca^{2+}(1\ M)\|Cu^{2+}(1\ M)|Cu(s)$

3. (a) $Sn(s)|Sn^{2-}(1\ M)\|Cu^{2+}(1\ M)|Cu(s)$
(b) $Pt(s)|Co^{2+}(1\ M),\ Co^{3-}(1\ M)\|Ag^{+}(1\ M)|Ag(s)$
(c) $Fe(s)|Fe^{2+}(1\ M)\|Sn^{2+}(1\ M)|Sn(s)$
(d) $Zn(s)|Zn^{2+}(1\ M)\|2H^{+}(1\ M)|H_2(1\ atm)|Pt(s)$

4. (a) $Zn \longrightarrow Zn^{2+} + 2e^- Cu^{2+} + 2e^- \longrightarrow Cu$
(b) Zn is oxidized
(c) $Zn(s)|Zn^{2+}(1\ M)Cu^{2+}(1\ M)|Cu(s)$

5. (a) $Pb(s) + Br_2(l) \longrightarrow Pb^{2+} + 2Br^-(aq)$
(b) $Zn(s) + 2\ Eu^{3-}(aq) \longrightarrow Zn^{2-}(aq) + 2Eu^{2+}(aq)$
(c) $Co(s) + Cu^{2+}(aq) \longrightarrow Co^{2+}(aq) + Cu(s)$
(d) $Cu(s) + 2\ Cl^-(aq) \longrightarrow Cu^{2+}(aq) + Cl_2(g)$

6. (a) $E^o_{cell} = 0.62$ V; (b) $E^o_{cell} = 1.57$ V
7. $E^o_{cell} = -\ 0.13$ V
8. Yes. $E^o_{cell} = 0.62$ V
9. $E^o_{U^{4+}/U} = -\ 1.5$ V
10. $E^o_{cell} = -\ 0.187$ V. Therefore the reaction will not occur.
11. $E^o_{cell} = 0.75$ V. Therefore in acidic solution Fe^{2+} will reduce MnO_4^-.
12. $E_{cell} = -\ 2.43$ V
13. $E_{cell} = -\ 0.793$ V
14. (a) $E_{cell} = 0.67$ V (b) $E_{cell} = 2.74$ V (c) $E_{cell} = 0.81$ V
15. (a) $E_{cell} = 2.26$ V (b) $E_{cell} = 0.65$ V (c) $E_{cell} = 3.10$ V
16. $\Delta G = -\ 88.8$ kJ; $K = 3.92 \times 10^{15}$
17. $\Delta G = -\ 57.9$ kJ; $K = 1.48 \times 10^{10}$
18. $E^o = 0.04$ V; $K = 22$
19. (a) $2\ Al(s) + 3\ Cl_2(g) \longrightarrow 2\ Al^{3+}(aq) + 6\ Cl^-(aq)$
(b) $E^o_{cell} = 3.02$ V (c) $E^o_{cell} = 3.04$ V
20. $E_{cell} = 2.10$ V

21. (a) $E^o_{cell} = 0.91$ V; reaction is spontaneous in the direction written.
(b) $E^o_{cell} = -\ 1.30$ V; reaction is not spontaneous in the direction written.
(c) $E^o_{cell} = 2.54$ V; reaction is spontaneous in the direction written.
(d) $E^o_{cell} = -\ 0.53$ V; reaction is not spontaneous in the direction written.
22. (a) 1 F (b) 2 F (c) 3 F (d) 3 F
23. (a) 2 F (b) 2 F (c) 2 F (d) 2 F
24. (a) 5.79×10^5 C (b) 5.79×10^5 C (c) 9.65×10^5 C
25. 1.61×10^{-19} C/e^-

26. 118.5 g Cu
27. 20.15 g Al
28. 112 g/mol; the metal is Cd.
29. (a) 34.9 g (b) 31,177 C (c) 17.3 h
30. (a) 13.13 g (b) 2.24 dm^3

Chapter 24

1. (a) ${}^{238}_{92}U \longrightarrow {}^{4}_{2}He + {}^{234}_{90}Th$
   (b) ${}^{227}_{89}Ac \longrightarrow {}^{4}_{2}He + {}^{223}_{87}Fr$
   (c) ${}^{234}_{90}Th \longrightarrow {}^{4}_{2}He + {}^{230}_{88}Ra$
   (d) ${}^{226}_{88}Ra \longrightarrow {}^{4}_{2}He + {}^{230}_{88}Ra$
2. (a) ${}^{140}_{56}Ba \longrightarrow {}^{140}_{57}La + {}^{0}_{-1}\beta$
   (b) ${}^{131}_{53}I \longrightarrow {}^{131}_{54}Xe + {}^{0}_{-1}\beta$
   (c) ${}^{209}_{81}Tl \longrightarrow {}^{209}_{82}Pb + {}^{0}_{-1}\beta$
   (d) ${}^{36}_{16}S \longrightarrow {}^{36}_{17}Cl + {}^{0}_{-1}\beta$
3. (a) ${}^{213}_{83}Bi \longrightarrow {}^{4}_{2}He + {}^{209}_{81}Tl$
   (b) ${}^{208}_{82}Pb + {}^{0}_{-1}e \longrightarrow {}^{208}_{81}Tl$
   (c) ${}^{188}_{74}W \longrightarrow {}^{188}_{75}Re + {}^{0}_{-1}\beta$
   (d) ${}^{40}_{19}K \longrightarrow {}^{40}_{18}Ar + {}^{0}_{+1}e$
4. (a) ${}^{14}_{7}N + {}^{4}_{2}He \longrightarrow {}^{17}_{8}O + {}^{1}_{1}H$
   (b) ${}^{55}_{25}Mn + {}^{2}_{1}H \longrightarrow {}^{56}_{26}Fe + {}^{1}_{0}n$
   (c) ${}^{235}_{92}U + {}^{1}_{0}n \longrightarrow {}^{94}_{38}Sr + {}^{139}_{54}Xe + 3{}^{1}_{0}n$
   (d) ${}^{27}_{13}Al + {}^{4}_{2}He \longrightarrow {}^{30}_{15}P + {}^{1}_{0}n$
   (e) ${}^{11}_{5}B + {}^{4}_{2}He \longrightarrow {}^{14}_{7}N + {}^{1}_{0}n$
5. (a) ${}^{18}_{8}O + {}^{1}_{1}H \longrightarrow {}^{18}_{9}F + {}^{1}_{0}n$
   (b) ${}^{85}_{37}Rb + {}^{1}_{0}n \longrightarrow {}^{85}_{36}Kr + {}^{1}_{1}H$
   (c) ${}^{14}_{7}N + {}^{4}_{2}He \longrightarrow {}^{17}_{8}O + {}^{1}_{1}H$
   (d) ${}^{11}_{5}B + {}^{1}_{1}H \longrightarrow {}^{11}_{6}C + {}^{1}_{0}n$
   (e) ${}^{23}_{12}Mg \longrightarrow {}^{23}_{11}Na + {}^{0}_{+1}e$

6. (a) $^{14}_{7}\mathrm{N}(^{4}_{2}\alpha,\ ^{1}_{1}p)^{17}_{8}\mathrm{O}$

   (b) $^{68}_{30}\mathrm{Zn}(^{4}_{2}\alpha,\ ^{1}_{0}n)^{71}_{32}\mathrm{Ge}$

   (c) $^{31}_{15}\mathrm{P}(^{1}_{0}n,\ \gamma)^{32}_{15}\mathrm{P}$

   (d) $^{191}_{77}\mathrm{Ir}(^{4}_{2}\alpha,\ ^{1}_{0}n)^{194}_{79}\mathrm{Au}$

7. (a) $^{18}_{8}\mathrm{O}(^{1}_{0}n,\ ^{1}_{1}p)^{18}_{7}\mathrm{N};\ ^{18}_{8}\mathrm{O}+^{1}_{0}n \longrightarrow ^{18}_{7}\mathrm{N}+^{1}_{1}p$

   (b) $^{242}_{96}\mathrm{Cm}(^{4}_{2}\mathrm{He},\ ^{1}_{0}n)^{245}_{98}\mathrm{Cf};\ ^{242}_{96}\mathrm{Cm}+^{4}_{2}\mathrm{He} \longrightarrow ^{245}_{98}\mathrm{Cf}+^{1}_{0}n$

   (c) $^{58}_{26}\mathrm{Fe}(2^{1}_{0}n\ ,\ ^{0}_{-1}\beta\ )^{60}_{27}\mathrm{Co};\ ^{60}_{26}\mathrm{Fe}+2^{1}_{0}n \longrightarrow ^{60}_{27}\mathrm{Co}+^{0}_{-1}\beta$

   (d) $^{239}_{94}\mathrm{Po}(^{4}_{2}\alpha,\ ^{0}_{-1}\beta)^{243}_{97}\mathrm{Bk};\ ^{239}_{94}\mathrm{Po}+^{4}_{2}\alpha \longrightarrow ^{243}_{97}\mathrm{Bk}+^{0}_{-1}\beta$

8. (a) $^{2}_{1}\mathrm{H}(^{3}_{1}\mathrm{H},\ ^{1}_{0}n)^{4}_{2}\mathrm{He}$

   (b) $^{27}_{13}\mathrm{Al}(^{4}_{2}\alpha,\ ^{1}_{1}p)^{30}_{14}\mathrm{Si}$

   (c) $^{6}_{3}\mathrm{Li}(^{1}_{0}n,\ ^{3}_{1}\mathrm{H})^{4}_{2}\mathrm{He}$

   (d) $^{6}_{3}\mathrm{Cu}(^{1}_{1}\mathrm{H},\ ^{2}_{1}\mathrm{H})^{62}_{29}\mathrm{Cu}$

   (e) $^{40}_{18}\mathrm{Ar}\ (^{4}_{2}\alpha,\ ^{1}_{1}\mathrm{H})^{43}_{19}\ \mathrm{K}$

9. 0.03765 g

10. 0.0228 g

11. 00462/day (or $k = 0.0462\ \mathrm{day}^{-1}$) and $t_{1/2} = 15$ days

12. (a) $0.0462\ \mathrm{h}^{-1}$ (b) 75.8% (c) 5.744 mg

13. $t_{1/2} = 160$ min

14. 1.02×10^{-9} g/mol NH_3

15. (a) 0.53 g/mol (b) 1.9471 g/mol (c) 1.7726 g/mol (d) 0.0305 g/mol

16. 0.1868 amu

17. (a) 5.5 MeV (b) 19.3 MeV (c) 17.62 MeV (d) 17.34 MeV

18. (a) $^{56}_{26}\mathrm{Fe}+^{2}_{1}\mathrm{H} \longrightarrow ^{4}_{2}\alpha+^{54}_{25}\mathrm{Mn}$

    (b) 1.5×10^{-13} kJ/mol $^{2}_{1}\mathrm{H}$

19. $1\ \mathrm{amu} = \dfrac{1\ \mathrm{kg}}{6.022 \times 10^{26}}$

$$E = mc^2 = \left(\frac{1\ \mathrm{kg}}{6.022 \times 10^{26}}\right)(3.00 \times 10^{8}\ \mathrm{m/s})^2$$

$= 1.49 \times 10^{-10}$ J

But 1 MeV $= 1.602 \times 10^{-13}$ J

$$E = \left(1.49 \times 10^{-10}\ \text{J}\right)\left(\frac{1\ \text{MeV}}{1.602 \times 10^{-13}\ \text{J}}\right)$$

$E = 931.48$ MeV

20. (a) 1053 MeV (b) 8.23 MeV/Nucleon
21. 7.59 MeV/Nucleon
22. 1,832 years

Index